Science Dictionary

Viva

Science Dictionary

JOHN O. E. CLARK
WILLIAM HEMSLEY

Viva Books

New Delhi | Mumbai | Chennai | Kolkata | Bangalore | Hyderabad | Kochi | Guwahati

This edition published
in 2011 by

Viva Books Private Limited
4737/23, Ansari Road, Daryaganj
New Delhi 110002
Tel: 011-4224 2200, Fax: 011- 422 2240
E-mail: vivadelhi@vivagroupindia.net
http://www.vivagroupindia.com

Reprinted 2014

ISBN: 978-81-309-1246-2

Printed and bound in India by Anand Sons, New Delhi.

Preface

This Mini Dictionary is intended for use by two principal groups of readers: students following Combined Science courses to GCSE level; and general readers seeking concise definitions of terms used in the fields of physics, biology and chemistry.

The initial term in an entry appears in **bold**; related terms which have their own entries elsewhere in the Mini Dictionary are also shown in bold.

GCSE and National Curriculum

In order to help readers using this book as a study aid, articles of particular relevance at GCSE level are indicated by ■ at the beginning of the entry. In addition, many articles are coded at the end to key them into National Curriculum Attainment Targets – for example [2/4/c]. The first number of the code gves the Attainment Target, the second the level of attainment, and the final letter indicates the statement within that level.

A

■ **abdomen** *1*. Rear body section of an **arthropod**. *2*. Part of the main body cavity in a **vertebrate** that contains organs other than the **heart** and lungs. In mammals it is separated from the **thorax** by the **diaphragm**.[2/4/a] [2/5/a] [2/7/a]

aberration *1*. In biology, the change in the number or the structure of **chromosomes** in a cell. *See also* **mutation.** [2/6/b] *2*. In physics, the production of an optical image with coloured fringes that occurs because the focal length of a lens is different for different colours of light (chromatic aberration), so that the lens focuses different colours in different planes. *3*. In physics, the production of an image that is not sharp when a beam of light falls on the edges as well as on the centre of a lens (spherical aberration).

abiogenesis Theory of the origin of life which states that living organisms formed from non-living material, either the old-fashioned belief held until 150 years ago (*see* **spontaneous generation**) or the modern one that life-forms gradually evolved from simple cell-like collections of biochemicals.

ABO blood groups Most commonly used classification system for human blood. The blood groups are named after the **antigen** types A and B in red blood cells. The groups are A, B, AB and O. Each blood type has corresponding **antibodies** in the blood **serum**. Group O contains neither antigen. Group A has type A antigens (and type B antibodies); group B has type B antigens (and type A antibodies); group AB has both antigens (and neither antibody); group O has neither antigen (and both antibodies). If blood from a donor is given to a patient, the patient's own blood must not have antibodies to the antigens in the blood that is given or **agglutination** rejection will occur. Group O is therefore a universal donor.

abortion Spontaneous or induced termination of pregnancy, resulting in the expulsion of the foetus. A natural abortion is often termed a miscarriage.

abscisic acid Growth-inhibiting plant **hormone** that causes leaves to be shed **(abscission)**.

abscission The shedding of a leaf, fruit or other part of a plant, caused by the formation of a layer of cork cells (abscission layer) at the base of the part which cuts off its supply of nutrients and water. It is brought about by **abscisic acid**.

absolute alcohol **Ethanol** (ethyl alcohol) that contains no more than 1% water.

absolute configuration Arrangement of groups about an asymmetric atom. *See* **configuration**.

■ **absolute humidity** Amount of water vapour in a given volume of **air**, also called vapour concentration. *See also* **relative humidity**. [3/9/b]

absolute temperature Fundamental temperature scale used in theoretical physics and chemistry. It is expressed in kelvin (K), corresponding to the **Celsius scale**. Zero is taken as **absolute zero**. Alternative name: thermodynamic temperature.

absolute unit Unit defined in terms of units of fundamental quantities such as length, time, mass and electric charge or current.

■ **absolute zero** Lowest temperature theoretically possible, at which a substance has no heat energy whatever. It corresponds to −273.15 °C, or 0 K on the kelvin scale.

absorptance α Ability of a medium to absorb radiation. It is the ratio of the total radiation absorbed by a medium to the total radiation arriving at its surface. Alternative name: absorptivity.

absorption Taking up of matter or energy by other matter; *e.g.* in biology, digested food is absorbed into the bloodstream from the intestines; in physics, cold metal placed in hot water absorbs heat energy, and illuminated objects absorb some of the light that falls on them.

absorption spectrum Characteristic pattern of dark lines that is seen in a **spectrum** produced when light is passed through a selectively absorbing medium.

absorptivity Alternative name for **absorptance**.

abyssal Relating to the ocean floor (*e.g.* abyssal animals, or fauna).

a.c. Abbreviation of **alternating current**.

■ **acceleration** (a) Rate of change of **velocity**; its SI units are ms^{-2} (metres per second per second). If velocity changes at a constant rate from an initial value of u to a final value of v in a time t, acceleration a is given by $a = (v - u)/t$. Alternatively if velocity is represented by v, acceleration is found by differentiating v with respect to time t; *i.e.* acceleration $a = dv/dt$. *See also* **force**. [5/8/a]

acceleration due to gravity *See* **acceleration of free fall**.

acceleration of free fall (g) Acceleration given to an object by the gravitational attraction of the Earth. It has an international standard value of 9.80665 ms^{-2}. Alternative name: acceleration due to gravity.

accelerator *1*. Machine that increases the **kinetic energy** of an object or particle, *e.g.* a **particle accelerator**. *2.* Mechanism that controls the speed of an electric motor or engine (by controlling the amount of current or fuel supplied). Alternative name: throttle. *3.* In biology, a substance that increases the effectiveness of an **enzyme**.

access time Period of time required for reading out of, or writing into, a computer's memory.

acclimatization Way in which an organism adapts to a different or changing environment (*e.g.* the thin air at higher altitudes).

accommodation *1.* Ability of the **eye** to alter its focus by changing the curvature of the lens. *2.* In the eyes of a few animals (*e.g.* **cephalopods**), changing the position of the lens relative to the retina.

acellular Not made up of **cells**; in particular, describing an organism that is not divided into separate cells. *See also* **coenocyte**.

acetal $CH_3CH(OC_2H_5)_2$ Colourless volatile liquid organic compound. Alternative name: 1,1-diethoxyethane.

acetaldehyde CH_3CHO Colourless liquid organic compound with a pungent odour; a simple **aldehyde** made by the oxidation of **ethanol** (ethyl alcohol). Alternative names: acetic aldehyde, ethanal.

acetaldol Alternative name for **aldol**.

acetamide CH_3CONH_2 Colourless deliquescent crystalline organic compound, with a 'mousy' odour. Alternative name: ethanamide.

acetate *1.* Salt of **acetic acid** (ethanoic acid) in which the terminal hydrogen atom is substituted by a metal atom; *e.g.*

copper acetate $Cu(CH_3COO)_2$. *2.* **Ester** of acetic acid in which the terminal hydrogen atom is substituted by a **radical**; *e.g.* ethyl acetate $CH_3COOC_2H_5$. Alternative name: ethanoate.

acetic acid CH_3COOH Colourless liquid **carboxylic acid**, with a pungent odour and acidic properties. It is the acid in vinegar. Alternative name: ethanoic acid.

acetone CH_3COCH_3 Colourless volatile liquid organic compound, with a sweetish odour; it is a simple **ketone**. Alternative name: 2-propanone.

acetonitrile CH_3CN Colourless liquid organic compound, with a pleasant odour. Alternative name: methyl cyanide.

acetophenone $C_6H_5COCH_3$ Colourless liquid organic compound, with a sweet pungent odour. Alternative name: phenyl methyl ketone.

acetyl chloride CH_3COCl Colourless highly refractive liquid organic compound, with a strong odour. Alternative name: ethanoyl chloride.

acetylcholine (ACH) White hygroscopic crystalline organic compound. It is important in the nervous system of many animals, where it transmits impulses between **synapses** of nerves; a **neurotransmitter**. After transmission it is broken down by the enzyme cholinesterase.

acetylene C_2H_2 Colourless organic gas with an ether-like odour (when pure), used as a fuel and in organic synthesis. It is the simplest **alkyne** (olefin). Alternative name: ethyne.

acetyl group The group CH_3CO- (as in acetyl chloride). Alternative name: ethanoyl.

acetylide Type of **carbide** resulting from the interaction of

acetylene and a solution of a heavy metal salt. Most acetylides are unstable and explosive. Alternative name: ethynide.

acetylsalicylic acid Alternative name for the analgesic drug **aspirin**.

ACH Abbreviation of **acetylcholine**.

achene Small dry **indehiscent** fruit that contains a single seed, as in many nuts, dandelion 'seeds' and the 'pips' on strawberries.

achromatic lens Combination of two or more lenses that has a **focal length** that is the same for two or more different **wavelengths** of light. This arrangement largely overcomes chromatic **aberration**.

■ **acid** Member of a class of chemical compounds whose aqueous solutions contain hydrogen ions. Solutions of acids have a **pH** of less than 7. Strong acids dissociate completely (into **ions**) in solution; weak acids only partly dissociate. An acid neutralizes a **base** to form a **salt**, and reacts with most metals to liberate hydrogen gas. [4/5/a]

acid dye Class of chemical colourants that are applied in weak **acid** solutions to wool, silk and polyamides (nylon).

acidic oxide Oxide of a non-metal, usually an **anhydride**, that reacts with water to form an **acid**, and with a **base** to form a resin **salt** and water; *e.g.* sulphur dioxide, SO_2, reacts with water to form sulphurous acid, H_2SO_3. *See also* **basic oxide**.

acidosis Condition in which the **pH** of the blood falls below its normal value of 7.35. It can have various causes, such as a breathing difficulty which prevents the lungs from removing sufficient carbon dioxide from the blood, or a kidney defect that leaves the blood deficient in bicarbonate. *See also* **alkalosis**.

■ **acid rain** Phenomenon caused by the pollution of the atmosphere with sulphur oxides and nitrogen oxides, which are produced largely by burning fossil fuels. The most common of these oxides are **sulphur trioxide** (SO_3), which combines readily with water to form **sulphuric acid** (H_2SO_4), and **nitrogen dioxide** (NO_2), which combines with water to form **nitric acid** (HNO_3). These acids are precipitated with snow and rain. [2/4/c]

acid salt Salt formed when not all the replaceable hydrogen atoms of an **acid** are substituted by a metal or its equivalent; *e.g.* sodium hydrogencarbonate (bicarbonate), $NaHCO_3$, diammonium hydrogenphosphate, $(NH_4)_2HPO_4$.

acid value Measure of the amount of free **fatty acid** present in fat or oil. The value is given as the number of milligrams of potassium hydroxide required to neutralize the fatty acids in 1g of the substance being tested.

acoustics *1.* Science of the production, containment and movement of sound. *2.* Qualities of a building with regard to the behaviour of sound within it.

acoustoelectronics Branch of **electronics** concerned with the use of sound waves at **microwave** frequencies.

acquired characteristics Physical or behavioural characteristics that are acquired during an organism's lifetime (*e.g.* scars following wounds), and which cannot be genetically transferred to any offspring. *See also* **Lamarckism**.

acquired immune deficiency syndrome Virus disease more commonly known by its abbreviation **AIDS**.

acriflavine Orange crystalline organic compound used as an antiseptic. Alternative name: 2,8-diaminoacridine methochloride.

Acrilan Synthetic fibre produced by **copolymerization** of **acrylonitrile** and **vinyl acetate**.

acrolein $CH_2=CHCHO$ Colourless liquid organic compound, with a pungent odour. Alternative name: 2-propenal.

acromegaly Disorder of adults, involving overgrowth of bones, that usually results from continual production of growth hormone after the end of puberty. It may be caused by a tumour affecting the **pituitary**.

acrosome Cap-like structure at the tip of animal **sperm** cells. It contains **enzymes** that dissolve the membrane of the ovum during fertilization.

acrylic acid $CH_2=CH.COOH$ Reactive organic acid with a pungent odour, used to make **acrylic resins**. It is one of the olefin-monocarboxylic acids. Alternative name: propenoic acid.

acrylic resin Transparent **thermoplastic** formed by the polymerization of **ester** or **amide** derivatives of **acrylic acid**. The resins are chiefly used in the manufacture of artificial fibres and for optical purposes, such as making lenses (*e.g.* Acrilan, Perspex).

acrylonitrile $H_2C=CHCN$ Colourless liquid **nitrile** used as a starting material in the manufacture of **acrylic resins**. Alternative names: vinyl cyanide, propenonitrile.

ACTH Abbreviation of **adrenocorticotrophic hormone**.

actin Protein that, together with **myosin**, forms actomyosin in skeletal **muscle**. It exists in two forms: G-actin, which is globular, and its polymer, F-actin, which is fibrous.

actinic radiation Electromagnetic radiation (particularly ultraviolet radiation) that can cause a chemical reaction.

actinide Member of a series of **elements** in Group IIIB of the Periodic Table, of **atomic numbers** 90 to 103 (actinium, at. no. 89, is sometimes also included). All actinides are radioactive. Alternative name: actinoid.

actinium Ac Radioactive element in Group IIIB of the Periodic Table (usually regarded as one of the **actinides**); it has several **isotopes**, with half-lives of up to 21.7 years. It results from the decay of uranium–235. At. no. 89; r.a.m. 227.

Actinomycetes Group of mostly **saprophytic** bacteria that are characterized by branching multicellular filaments (hyphae).

■ **Actinopterygii** Subclass of the Osteichthyes, comprising all the ray-finned fish. [2/3/b] [2/4/b]

action Alternative name for **force**.

action potential Rapid but transient change in the electrical potential across the membrane of an excitable cell (*e.g.* a nerve cell, in which it occurs during the passage of a nerve impulse). *See also* **all-or-none response**.

activated carbon Charcoal treated so as to be a particularly good absorbent of gases.

■ **activation energy** Amount of energy required to initiate the breaking and re-formation of chemical **bonds**, and thus to start a chemical reaction. [4/10/e]

■ **active mass** Concentration of a substance that is involved in a chemical **reaction**. [4/9/b]

active site Part of an **enzyme** molecule to which its **substrate** is bound during **catalysis**.

active transport Energy-dependent movement of a dissolved substance across a cell membrane against a concentration gradient; *e.g.* taking up of nutrients by a plant's roots. This movement is normally in the opposite direction to that in which the substance would move by a passive diffusion process.

actomyosin Substance formed from the association of **actin** and **myosin** in skeletal **muscle**, upon which the contraction mechanism is based.

acupuncture System of alternate medicine that treats disorders or induces anaesthesia by inserting needles into the skin. The needles, which may be rotated or heated, are inserted along lines or meridians that are thought to influence various organs or parts of the body.

acute In medicine, describing a condition or disorder of sudden onset (as opposed to **chronic**).

acyclic Not cyclic; describing a chemical compound that does not contain a ring of atoms in its molecular structure.

acyl group Part of an organic compound that has the formula RCO–, where R is a **hydrocarbon** group (*e.g.* acetyl, CH_3CO-).

adaptation *1.* Characteristic of an organism which improves that organism's chance of survival in its environment. *2.*Change in the behaviour of an organism in response to environmental conditions. *3.* Change in the sensitivity of a sensory mechanism after it has been exposed to a particular continuous stimulus. This allows the mechanism to adjust the sensitivity scale according to the level of the stimulus.

adaptive radiation Mechanism of **evolution** in which a single ancestor gives rise to a number of species that coexist but occupy different ecological niches. Alternative name: divergent evolution.

adder Part of a computer that adds digital signals (addend, augend and a carry digit) to produce the sum and a carry digit.

addiction Physical or psychological dependence on a drug, such as alcohol, tobacco, cocaine or heroin. Treatment of addiction is notoriously difficult, and stopping the drug produces unpleasant withdrawal symptoms. Psychiatric counselling may help.

addition polymer Polymer formed from simple **monomers** by **addition reactions** (*e.g.* many kinds of *plastics*).

addition reaction Chemical reaction in which one substance combines with another to form a third, without any other substance being produced. The term is most commonly used in organic chemistry; *e.g.* for the reaction between hydrogen bromide (HBr) and ethylene (ethyne, $CH_2=CH_2$) to form ethyl bromide (bromoethane, C_2H_5Br). *See also* **substitution reaction**.

additive Substance added to food to modify its colour, flavour or texture, or to preserve it. Some additives are natural substances, but many are synthetic compounds. A few have been implicated as possible causes of medical problems (such as allergy or hyperactivity) in sensitive persons.

address In computing, *1.* identity of a location's position in a memory or store, or *2.* specification of an operand's location.

adduct Product of the chemical combination of two atoms or molecules – *i.e.* of an **addition reaction**.

adenine One of the component bases of **nucleic acids**. It is a **purine** derivative, and pairs with **thymine** in **DNA** and **uracil** in **RNA**. Alternative name: 6-aminopurine.

adenoids Pair of **lymph** glands, more prominent in children than in adults, that are located in and guard the back of the nasal cavity. Repeated infection of the adenoids causes them to become enlarged, and they may then be surgically removed (adenoidectomy). *See also* **tonsil**.

adenosine Compound consisting of the base **adenine** linked to the sugar **ribose**. Its phosphates are important energy carriers in biochemical processes (*see* next three articles).

adenosine diphosphate (ADP) Chemical involved with the biological transfer of chemical energy. Energy is released when it is formed from **adenosine triphosphate** (ATP). ADP is converted to ATP during **respiration** by addition of a phosphate group linked with an energy-rich bond.

adenosine monophosphate (AMP) Chemical involved with the biological transfer of chemical energy. Energy is released when it is formed from **adenosine triphosphate** (ATP). It is converted back to ATP during **respiration**.

adenosine triphosphate (ATP) Chemical that provides the energy for a large number of biological processes, including muscle contraction and the synthesis of many molecules. Chemically, it is an **adenosine** molecule with three attached phosphate groups. It gives off energy when it loses one or two of its phosphate groups (in a process known as **phosphorylation**) to become **adenosine diphosphate** or **adenosine monophosphate**, respectively. *See also* **electron transport**.

adenovirus One of a group of **viruses** that contain DNA and are found in a number of animals, including human beings. They cause infections of the respiratory tract, and may induce cancerous tumours.

ADH Abbreviation of **antidiuretic hormone**. *See also* **vasopressin**.

adhesion Attraction between different substances (at the atomic or molecular level); *e.g.* between water particles and glass, creating a **meniscus**. *See also* **cohesion**.

adiabatic process Process that occurs without interchange of heat with the surroundings.

adipic acid $HOOC(CH_2)_4COOH$ Colourless crystalline organic compound, a major constituent of rosin, used in the manufacture of **nylon**. Alternative names: butanedicarboxylic acid, hexanedioic acid.

adipose tissue Tissue that contains cells specialized for the storage of fat. It gives insulation and acts as an energy reserve.

admittance (*Y*) Property that allows the flow of electric current across a **potential difference**; the reciprocal of **impedance**.

ADP Abbreviation of **adenosine diphosphate**.

adrenal cortex Outer part of the **adrenal gland**.

■ **adrenal gland** Endocrine gland that occurs in vertebrates (in mammals there are two glands, located on the kidneys). Each gland has two sections: the central medulla, which secretes the hormones **adrenaline** and **noradrenaline**; and the outer cortex, which secretes certain **steroid** hormones. Alternative name: suprarenal gland. [2/4/a] [2/5/e] [2/7/a]

adrenalin Hormone secreted by the medulla of the **adrenal gland** and at some nerve endings of the **sympathetic nervous system**. It is produced when the body prepares for violent physical action. Its effects include increased heartbeat, raised levels of sugar (glucose) in the blood and improved muscle action. Alternative names: adrenaline, epinephrine.

adrenergic Releasing or stimulated by **adrenalin** or **noradrenalin**. The term is applied to the **sympathetic nervous system**.

adrenocorticotrophic hormone (ACTH) **Protein** released by the anterior **pituitary**. It controls secretion from the cortex of the **adrenal gland**. Alternative name: corticotrophin.

adsorbent Substance on which **adsorption** takes place.

adsorption Accumulation of **molecules** or **atoms** of a substance (usually a gas) on the surface of another substance (a solid or a liquid).

■ **advanced gas-cooled reactor** (AGR) Type of nuclear **reactor** that uses carbon dioxide as a coolant, as employed in some British power stations. [3/9/c]

advection Transfer of heat or matter (*e.g.* water vapour) associated with the horizontal movement of an air mass through the atmosphere.

adventitious Growing in an abnormal position, *e.g.* a root that develops from a stem.

■ **aerial** Wire, rod or other device that is used to transmit or receive radio waves. The simplest aerials are lengths of wire or ferrite rods employed in domestic radio receivers. Transmitting aerials are often large structures, sometimes more than 100m tall. Microwave aerials usually take the form of a dish. Alternative name: antenna.

aerobe Organism that respires aerobically (using oxygen). *See* **aerobic respiration**.

■ **aerobic** Describing a biochemical process that needs free oxygen in order to take place.

aerobic respiration Process by which cells obtain energy from the **oxidation** of fuel molecules by **molecular oxygen** with the formation of carbon dioxide and water. This process yields more energy than **anaerobic respiration**. *See also* **glycolysis**; **Krebs cycle**.

aerodynamics Study of the motion of solid objects in air, or of moving gases.

aerogenerator Electrical **generator** that is driven by wind power.

■ **aerosol** *1*. Suspension of particles of a liquid or solid in a gas; a type of **colloid**. *2*. Device used to produce such a suspension.

aestivation *1*. In zoology, summer dormancy or inactivity. It is characteristic of desert animals and others that survive very hot or dry periods by sleeping underground or in deep shade; *e.g.* lungfish burrow into mud at the beginning of the dry season and remain there until the rains come.

aetiology Study of the cause of disease. Alternative spelling: etiology. *See also* **epidemiology**.

afferent Leading towards (*e.g.* towards an organ of the body); the term is particularly used of various nerves and blood vessels. *See also* **efferent**.

affinity *1*. In biology, a similarity of structure or form between different species of plants or animals. *2*. In chemistry, the tendency of some substances to combine chemically with others.

afforestation Planting of trees to create new forest, often to combat erosion.

aflatoxin One of the group of **toxins** produced by fungi of the genus *Aspergillus*. Animal feed (*e.g.* cereals, peanuts) contaminated by aflatoxins can cause outbreaks of disease in livestock. Aflatoxins are believed to be **carcinogenic** to human beings.

AFP Abbreviation of **alphafetoprotein**.

afterbirth Alternative name for the **placenta**.

agamospermy Formation of plant seeds and embryos by asexual means (*see* **asexual reproduction**).

agar Complex **polysaccharide** obtained from seaweed, especially that of the genus *Gelidium*. It is commonly employed as a gelling agent in media used for growing **micro-organisms** and in food for human consumption. Alternative name: agar-agar.

agent orange Very poisonous mixture of two weedkillers used as a herbicide in warfare to destroy an enemy's crops and defoliate trees.

agglutination Clumping or sticking together, *e.g.* of red blood cells from incompatible blood groups, or of bacteria. Agglutination is commonly used as an end point in immunological tests. *See* **ABO blood groups**.

agglutinin **Antibody** that causes **agglutination**. The term commonly refers to such an antibody in a person's blood plasma that reacts with the red blood cells of another person's blood.

Agnatha Class of animals that includes the jawless fish (*e.g.* hagfish and lampreys).

agonic line Line drawn on a map that connects places at which the **magnetic declination** is zero.

agonist Substance (or sometimes an organ) that has an action that is complementary to the action of another substance (or organ) in an animal or a plant. The term is particularly used of drugs and muscles. *See also* **antagonist**.

AGR Abbreviation of **advanced gas-cooled reactor**.

agranulocytosis Blood disorder in which there is a deficiency of **granulocytes**, resulting in an impaired immune system.

agriculture Practices of cultivating the land to grow crops and rearing animals for food. *See* **horticulture**.

AIDS Acronym for Acquired Immune Deficiency Syndrome, a disease caused by the HIV (human immunodeficiency) virus, transmitted through the exchange of body fluids such as blood or semen.

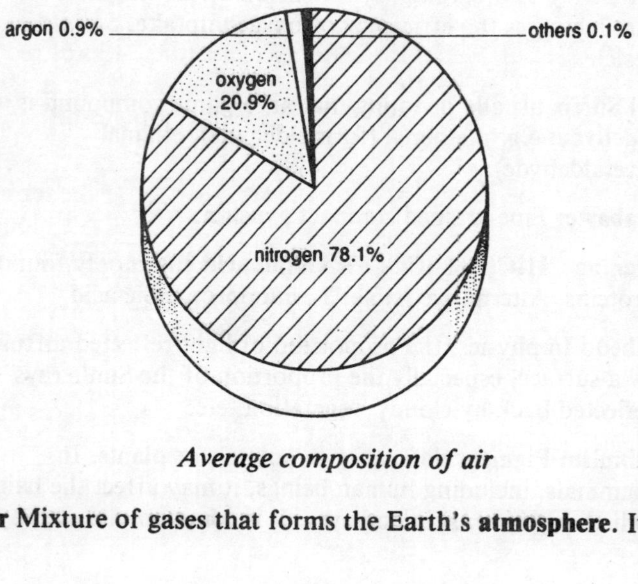

Average composition of air

air Mixture of gases that forms the Earth's **atmosphere**. Its composition varies slightly from place to place – particularly with regard to the amounts of carbon dioxide and water vapour it contains – but the average composition of dry air is (percentages by volume):

nitrogen 78.1
oxygen 20.9
argon 0.9
other gases 0.1%

airbladder Alternative term for **swimbladder**.

■ **air pressure** **Pressure** at a point in the **atmosphere** of the Earth or other planetary body with an atmosphere. *See* **atmospheric pressure**. [5/6/a]

air sac *1.* Cavity in the upper **thorax** of birds that is connected by passages to the lungs, and which increases the efficiency of breathing. *2.* Extension of the **trachea** in insects that increases the efficiency of oxygen uptake. *See also* **vesicle**.

-al Suffix usually denoting that an organic compound is an **aldehyde**; *e.g.* methanal (formaldehyde), ethanal (acetaldehyde).

alabaster Fine-grained compact **gypsum**.

alanine $CH_3C(NH_2)COOH$ **Amino acid** commonly found in proteins. Alternative name: 2-aminopropanoic acid.

albedo In physics, the proportion of light reflected diffusely by a surface, especially the proportion of the Sun's rays reflected back by clouds, vegetation, etc.

albinism Pigment deficiency in animals or plants. In mammals, including human beings, it may affect the hair (which is white), skin and eyes. It is commonly caused by a recessive gene which results in a lack of the enzyme that controls the synthesis of the dark pigment **melanin**.

■ **albumen** White of an egg. [2/7/a]

albumin Soluble **protein** found in many animal fluids, most notably in egg white and **blood serum**.

alchemy Predecessor of chemistry. Its two principal goals were the transmutation of common metals to gold, and the discovery of a universal remedy, the elixir.

■ **alcohol** *1*. Member of a large class of organic compounds that contain **hydroxyl (–OH) groups**. Simple, or primary, alcohols have the general formula ROH, where R is an **alkyl group** (or H in the case of methanol). Secondary alcohols have the formula RR′(CH)OH, and tertiary alcohols are RR′R″COH. An alcohol with two hydroxyl groups is called a **diol**, or glycol. Alcohols react with **acids** to form **esters**. A compound with a hydroxyl group attached to an **aryl group** is a **phenol**. *2*. Alternative name for **ethanol**.

alcoholism Disorder caused by **addiction** to alcoholic drinks. It affects the sufferer's mental functions and leads to a deterioration of physical skills. The heart, liver and nerves may also be affected. If an alcoholic suddenly stops drinking, he or she experiences withdrawal symptoms, including anxiety, tremor and sometimes hallucinations.

aldehyde Member of a large class of organic compounds that have the general formula RCHO, where R is an **alkyl** or **aryl group**. Aldehydes may be made by the controlled oxidation of **alcohols**. Their systematic names end with the suffix *-al* (*e.g.* the systematic name of acetaldehyde, CH_3CHO, is ethanal).

aldimine *See* **Schiff's base**.

aldol $CH_3CHOHCH_2CHO$ Viscous liquid organic compound. It is an **aldehyde**, a condensation product of **acetaldehyde** (ethanal). Alternative names: acetaldol, beta-hydroxy-butyraldehyde.

aldol reaction **Condensation reaction** between two **aldehyde** or two **ketone** molecules that produces a molecule containing an aldehyde (–CHO) group and an alcohol (–OH) group, hence *ald-ol* reaction.

aldose Type of **sugar** whose molecules contain an **aldehyde** (–CHO) group and one or more **alcohol** (–OH) groups.

aldosterone Hormone produced by the cortex of the **adrenal gland**, which affects the rate of carbohydrate **metabolism**. It also helps to control the **electrolyte** balance of the body by allowing the retention of sodium ions and the excretion of potassium ions.

alga Simple plant, which may be **unicellular** or **multicellular** (*e.g.* some seaweeds and the green slime in ponds). Algae contain a variety of photosynthetic pigments (hence brown, green and red seaweeds), and are present in many habitats; most are aquatic.

alginic acid $(C_6H_8O_6)_n$ Yellowish-white organic solid, a polymer of mannuronic acid in the **pyranose** ring form, that occurs in brown seaweeds. Even very dilute solutions of the acid are extremely viscous, and because of this property it has many industrial applications.

ALGOL Acronym for the high-level computer programming language ALGOrithmic Language, used for manipulating mathematical and scientific data.

algorithm Operation or set of operations that are required to effect a particular calculation or to manipulate data in a certain way, usually to solve a specific problem. The term is commonly used in the context of computer programming.

alicyclic Describing a class of organic chemicals that possess properties of both **aliphatic** and **cyclic** compounds (*e.g.* cyclohexane, C_6H_{12}).

alimentary canal Tube in the body of animals along which food passes (moved by **peristalsis**), and in which the food is subjected to physical and chemical **digestion** and is absorbed. Alternative names: digestive tract, gut.

aliphatic Describing a large class of organic chemicals that have straight or branched chain arrangements of their constituent carbon atoms. The class includes **alkanes, alkenes, alkynes** and their derivatives.

alizarin $C_6H_4(CO)_2C_6H_2(OH)_2$ Orange-red crystalline organic compound important in the manufacture of dyes. Alternative name: 1,2-dihydroxy-anthraquinone.

■ **alkali** Substance that is either a soluble **base** or a solution of a base (*e.g.* sodium hydroxide, NaOH). Alkalis have a **pH** of more than 7, and react with **acids** to produce **salts** (and water).

alkali metal One of the elements in Group I of the Periodic Table. They are **lithium, sodium, potassium, rubidium, caesium** and **francium**, which are all soft silvery metals that react vigorously with water.

alkaline Having the properties of an **alkali**.

■ **alkaline earth** One of the elements in Group IIA of the Periodic Table. They are **beryllium, magnesium, calcium, strontium, barium** and **radium**. [4/9/a]

alkaloid One of a group of nitrogen-containing organic compounds that are found in some plants. Many are toxic or medicinal (*e.g.* atropine, digitalis, heroin, morphine, quinine, strychnine).

alkalosis Condition in which the **pH** of the blood and body fluids rises above its normal level of 7.35. It may result from loss of acidic digestive juices through vomiting or from an excess intake of **alkali** (such as bicarbonate). *See also* **acidosis.**

■ **alkane** Member of a group of saturated **aliphatic** hydrocarbons that have the general formula C_nH_{2n+2}; *e.g.* methane (CH_4), ethane (C_2H_6), etc. Alternative name: paraffin.

■ **alkene** Member of a group of unsaturated **aliphatic** hydrocarbons that have carbon-to-carbon double bonds and the general formula C_nH_{2n}; *e.g.* ethene (ethylene, C_2H_4), propene (propylene, C_3H_6), etc. Alternative name: olefin.

alkoxyalkane Alternative name for an **ether**.

alkyl/aryl sulphide *See* **thioether**.

alkyl group Hydrocarbon **radical**, derived from an **alkane** by the removal of one hydrogen atom, that has the generalformula C_nH_{2n+1}; *e.g.* methyl ($-CH_3$), from methane, ethyl ($-C_2H_5$), from ethane, etc.

alkyl halide *See* **halogenoalkane**.

alkyne Member of a group of **unsaturated aliphatic** hydrocarbons that have carbon-to-carbon triple bonds and the general formula C_nH_{2n-2}; *e.g.* ethyne (acetylene, C_2H_2), propyne (methyl acetylene, C_3H_6), etc. Alternative name: acetylene.

allantois Membranous sac that develops in the **embryos** of reptiles, birds and mammals. It is involved in the storage of waste and the provision of food and oxygen. In reptiles and birds, it grows to surround the embryo in the shell; in mammals it becomes incorporated into the **placenta**.

■ **allele** One of the alternative states of a **gene**. In **diploid** cells each gene occurs in two forms; one may be genetically **dominant**, the other **recessive**. Alternative name: allelomorph. [2/7/b]

allergy Abnormal sensitivity of the body to a substance (known as an allergen). Contact with the allergen causes symptoms such as skin rashes, watery eyes, and sneezing. Hay fever is a widespread allergy to the pollen of certain plants.

allogamy Fertilization of the **ovule** of a flowering plant that involves pollen from another flower (whether on the same plant or another plant).

all-or-none response Response of excitable tissue (*e.g.* some nerve cells) that occurs in full only at or above a certain level of stimulus, the threshold, but not at all below that level. *See also* **action potential**.

allotrope One of the forms of an element that exhibits **allotropy**.

■ **allotropy** Existence of different structural forms of an element; *e.g.* graphite and diamond are allotropes of carbon.

■ **alloy** Metallic substance that is made up of two or more elements (usually metals); *e.g.* brass (copper and zinc), bronze (copper and tin), solder (lead and tin) and steel (iron and carbon).

allyl group $CH_2{=}CHCH_2{-}$ **Unsaturated hydrocarbon radical** found in compounds such as allylbromide.

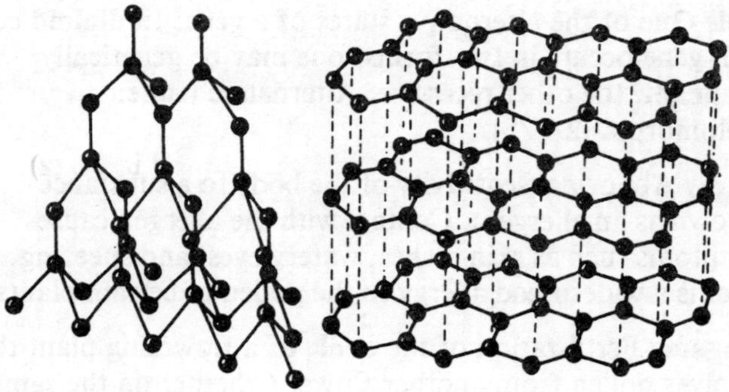

Two main allotropes of carbon

Alnico Trade name for an **alloy** that is composed mainly of aluminium, nickel and cobalt. It is used to make permanent magnets.

■ **alpha decay Radioactive** disintegration of a substance with the emission of **alpha particles**.[4/7/f]

alphafetoprotein (AFP) Protein that occurs in **amniotic fluid**, made in the liver of a foetus. High levels (detected by **amniocentesis**) may indicate a foetal abnormality (*e.g.* spina bifida).

alphanumeric Describing characters or their codes that represent letters of the alphabet or numerals, particularly in computer applications.

■ **alpha particle** Particle that is composed of two **protons** and two **neutrons** (and is thus the equivalent of a helium nucleus). It is produced by **radioactive** decay, and has little penetrative power. It is sometimes represented as α-particle. [4/7/f]

alpha radiation Ionizing radiation consisting of a stream of **alpha particles**.

alpha-iron Iron with a **body-centred cubic** structure.

■ **alternating current** (a.c.) Electric **current** that reverses its direction of flow in periodic cycles (measured in **hertz**). Mains electricity in Britain alternates at a rate of 50 cycles per second; its frequency is 50 Hz. [5/10/b]

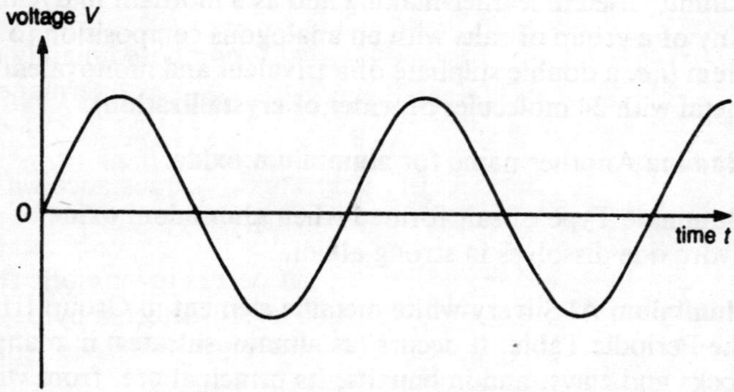

Alternating current varies rapidly with time

■ **alternation of generations** Phenomenon that occurs in the life cycles of certain organisms (*e.g.* mosses, ferns and many coelenterates). A generation that reproduces sexually alternates with a generation that reproduces asexually. Consequently the life cycle is divided into **haploid** and **diploid** phases.

alternator Generator that produces **alternating current** (a.c.) by rotating coils in a magnetic field.

altimeter Instrument for measuring altitude, often a type of **barometer**.

altruism Category of animal behaviour, especially common among social insects, in which older individuals tend to sacrifice themselves, losing their lives if necessary, in order that offspring and other younger individuals may survive or otherwise benefit.

alum *1*. $Al_2(SO_4)_3.K_2SO_4.24H_2O$ Aluminium potassium sulphate, a white crystalline substance (that occurs naturally as kalinite) used in leather-making and as a mordant in dyeing. *2*. Any of a group of salts with an analogous composition to alum (*i.e.* a double sulphate of a trivalent and monovalent metal with 24 molecules of **water of crystallization**).

alumina Another name for **aluminium oxide**.

aluminate Type of salt formed when **aluminium oxide** or hydroxide dissolves in strong alkali.

■ **aluminium** Al Silvery-white metallic element in Group IIIA of the Periodic Table. It occurs (as **aluminosilicates**) in many rocks and clays, and in **bauxite**, its principal ore, from which it is extracted by **electrolysis**. It is a light metal, protected against corrosion by a surface film of oxide. Its alloys are used in the aerospace industry and in light-weight structures. At. no. 13; r.a.m. 26.9815. [4/5/c]

aluminium chloride $AlCl_3$ White or yellowish deliquescent solid, covalently bonded when anhydrous, which fumes in moist air. It is used as a catalyst in the **cracking** of petroleum hydrocarbons.

■ **aluminium oxide** Al_2O_3 White crystalline **amphoteric** compound, the principal source of **aluminium**, which occurs as the mineral **corundum** and in **bauxite**. A thin surface film of aluminium oxide gives aluminium and its alloys their corrosion-resistant properties. Such films can be created by **anodizing**. Alternative name: alumina.

aluminium sulphate $Al_2(SO_4)_3$ White crystalline compound, used as a flocculating agent in water treatment and sewage works, as a **mordant** in dyeing, as a size in paper-making and as a foaming agent in fire extinguishers.

aluminium trimethyl Alternative name for **trimethylaluminium**.

aluminosilicate Common chemical compound in minerals and rocks (*e.g.* clays, mica) consisting of **alumina** and **silica** with water and various bases; such compounds are also formed in glass and various ceramics.

alveolus Minute air sac in the vertebrate **lung**. There are vast numbers of alveoli, and most of the exchange of gases between air and blood takes place within them. *See also* **respiration**.

Alzheimer's disease Condition characterized by progressive degeneration of brain function, often referred to as premature senility. Disorders of speech and memory are the commonest symptoms. Recent research suggests that some cases of the disease may be linked to high levels of dissolved aluminium in drinking water. It was named after the German physician Alois Alzheimer (1864–1915).

AM Abbreviation of **amplitude modulation.**

amalgam Mixture of mercury with one or more other metals, used for dental fillings.

amatol High explosive consisting of a mixture of **nitroglycerine** and **ammonium nitrate**.

ambergris Pale waxy solid with a strong smell which is believed to be produced inside the intestines of diseased sperm whales. Other than whaling, its only source is lumps washed ashore by the tide. It is used in the perfume industry as a fixative.

americium Am Radioactive element in Group IIIB of the Periodic Table (one of the **actinides**). It has several **isotopes,** with half-lives of up to 7,650 years. It is used as a source of alpha-particles. At. no. 95; r.a.m. 243 (most stable isotope).

amide Member of a group of organic chemical compounds in which one or more of the hydrogen atoms of **ammonia** (NH_3) have been replaced by an **acyl group** ($-RCO$); *e.g.* acetamide (ethanamide), CH_3CONH_2. In primary amides one, in secondary amides two and in tertiary amides three of the hydrogens have been so replaced. Alternative name: alkanamides.

aminase One of a group of **enzymes** that can catalyse the **hydrolysis** of **amines**.

amination Transfer of an **amino group** ($-NH_2$) to a compound.

amine Member of a group of organic chemical substances in which one or more of the hydrogen atoms of **ammonia** (NH_3) have been replaced by a **hydrocarbon** group. In primary amines one, in secondary amines two and in tertiary amines three hydrogens have been so replaced.

■ **amino acid** The building blocks of **proteins**, amino acids are organic compounds that contain an acidic **carboxyl group** (–COOH) and a basic **amino group** ($-NH_2$). Twenty amino acids are commonly found in proteins. Those that can be synthesized by a particular organism are known as 'non-essential'; 'essential' amino acids must be obtained from the environment, usually from food.

aminobenzene Alternative name for **aniline**.

aminoethanamide Alternative name for **guanidine**.

amino group Chemical group with the general formula –NRR′, where R and R′ may be **hydrogen** atoms or organic **radicals**; the commonest form is $-NH_2$. Compounds containing amino groups include **amines** and **amino acids**. *See also* **amide**.

aminoisovaleric acid Alternative name for **valine**.

aminoplastic resin Synthetic **resin** derived from the reaction of **urea, thiourea** or **melamine** with an **aldehyde,** particularly **formaldehyde** (methanal).

aminotoluene Alternative name for **toluidine**.

ammeter Instrument for measuring electric current, usually calibrated in **amperes**. The common moving-coil ammeter is a type of **galvanometer**.

■ **ammonia** NH_3 Colourless pungent gas, which is very soluble in water (to form ammonium hydroxide, or ammonia solution, NH_4OH) and alcohol. It is formed naturally by the bacterial decomposition of proteins, purines and urea; made in the laboratory by the action of alkalis on ammonium salts; or synthesized commercially by fixation of nitrogen. Liquid ammonia is used as a refrigerant. The gas is the starting material for making **nitric acid** and nitrates. *See also* **Haber process.**

ammonia-soda process Alternative name for **Solvay process.**

ammonia solution Alternative name for **ammonium hydroxide**.

ammonite Coiled-shelled **cephalopod** mollusc, now extinct, that was common in the seas of the Mesozoic era.

ammonium carbonate $(NH_4)_2CO_3$ Unstable white crystalline compound which decomposes spontaneously to produce **ammonia,** used in smelling salts. The substance known in industry as ammonium carbonate is usually a double salt consisting of ammonium hydrogencarbonate (bicarbonate) and ammonium aminomethanoate (carbamate). Alternative name: sal volatile.

ammonium chloride NH_4Cl Colourless or white crystalline compound, used in dry batteries, as a **flux** in soldering and as a **mordant** in dyeing. Alternative name: sal ammoniac.

■ **ammonium hydroxide** NH_4OH **Alkali** made by dissolving **ammonia** in water, giving a solution that probably contains hydrates of ammonia. It is used for making soaps and fertilizers. 880 ammonia is a saturated aqueous solution of ammonia (density 0.88 g cm^{-3}). Alternative name: ammonia solution.

ammonium ion The ion NH_4^+, which behaves like a metal ion.

ammonium nitrate NH_4NO_3 Colourless crystalline compound, used as a fertilizer and in explosives.

ammonium phosphate $(NH_4)_3PO_4$ Colourless crystalline compound, used as a fertilizer (when it adds both nitrogen and phosphorus to the soil).

ammonium sulphate $(NH_4)_2SO_4$ White crystalline compound, much used as a fertilizer.

amniocentesis Test in which a sample of **amniotic fluid** is taken from the **amnion** surrounding a foetus. The cells that this contains are examined for foetal abnormalities, particularly hereditary disorders.

amnion Innermost membrane that envelops an **embryo** or **foetus** (in mammals, reptiles and birds) and encloses the fluid-filled amniotic cavity.

amniote Any of the higher vertebrates in which the **embryo** is surrounded by an **amnion** containing fluid (*i.e.* reptiles, birds, mammals). The development of the amnion, which permits gas exchange, was the evolutionary step that first enabled eggs to be laid on dry land.

amniotic fluid Liquid that occurs in the **amnion** surrounding a foetus.

■ **amoeba** Single-celled organism, a protozoan, that moves and feeds by the projection of **pseudopodia**.

amoebocyte **Cell** that demonstrates amoeboid movement. Such cells may be normally present in body fluids. *See also* **leucocyte**.

amorphous Without clear shape or structure.

AMP Abbreviation of **adenosine monophosphate**.

■ **ampere** (A) Basic SI unit of electric **current**. [5/5/b]

Ampère's law Relationship that gives the **magnetic induction** at a point due to a given **current** in terms of the current elements and their positions relative to the point. It was named after the French physicist André Marie Ampère. Alternative name: Laplace law.

■ **amphetamine** Drug that stimulates the **central nervous system**. It was once much used for the treatment of depression and to lessen appetite, but is now seldom prescribed because of its addictive properties.

■ **Amphibia** Class of semi-aquatic **vertebrates** that evolved from fish. It includes frogs, toads, newts and salamanders. Amphibians are characterized by the possession of four five-toed legs and an inner ear. Eggs are laid in water, and fertilization is external. The larval stage is a free-swimming tadpole. [2/3/b]

amphibian Member of the class **Amphibia**.

amphipod Member of the **Amphipoda**.

Amphipoda Order of small **crustaceans** that lack a **carapace** (*e.g.* sandhoppers, beach-hoppers and water lice). There are more than 4,000 species throughout the world, from Arctic seashores to the damp soils of tropical rainforests.

amphoteric Describing a chemical compound with both basic and acidic properties. *E.g.* aluminium oxide, Al_2O_3, dissolves in acids to form aluminium salts and in alkalis to form aluminates.

■ **amplifier** Device that can magnify a physical quantity, such as electrical **current** or mechanical force. An electronic amplifier increases the **amplitude** of the input signal, to produce a **gain**.

amplitude Maximum value attained by a quantity that varies in periodic cycles; *i.e.* the maximum displacement from its mean position, usually equal to half its total displacement.

amplitude modulation (AM) Form of radio transmission used on the long and medium wavebands, in which the information content (the sound being broadcast) is conveyed by variations in the **amplitude** of the **carrier wave**.

■ **amu** Abbreviation of **atomic mass unit**. [4/8/d]

amylase Member of a group of **enzymes** that digest **starch** or **glycogen** to **dextrin, maltose** and **glucose**. Amylases are present in digestive juices and micro-organisms. Alternative name: diastase.

amyl group $C_5H_{11}-$ Monovalent **alkyl group**. Alternative name: pentyl group.

amyl nitrite $C_5H_{11}ONO$ Pale brown volatile liquid organic compound, often used in medicine to dilate the blood vessels of patients with some forms of heart disease (*e.g.* angina).

amylose Polysaccharide sugar, a polymer of **glucose** that occurs in **starch**.

amylum Alternative name for **starch**.

■ **anabolic steroid** Compound that is concerned with **anabolism**. Commonly used anabolic steroids are synthetic male **sex hormones** (androgens) which promote protein synthesis (hence their use by some athletes wishing to build up muscle). *See also* **steroid**.

anabolism Phase of **metabolism** that is concerned with the building up (or biosynthesis) of molecules. *See also* **catabolism**.

anaemia General term for any disorder characterized by abnormally low levels of **haemoglobin** in the blood. There are many different forms, which may be inherited or result from an infection or bodily disorder. A frequent cause is lack of iron in the diet.

anaerobe Organism that respires anaerobically. *See* **anaerobic respiration**.

anaerobic Describing a biochemical reaction that takes place in the absence of free oxygen.

■ **anaerobic respiration** Process by which organisms obtain energy from the breakdown of food molecules in the absence of oxygen (*e.g.* in plants **fermentation**, in which sugar is broken down to alcohol). In animals muscle cells respire anaerobically to form lactic acid. Both processes yield less energy than **aerobic respiration.** [2/7/a]

anaesthetic Drug that induces overall insensibility (general anaesthetic) or loss of sensitivity in one area (local anaesthetic).

analgesic Drug that relieves pain without causing loss of consciousness.

analog computer Computer that represents numerical values by continuously variable physical quantities (*e.g.* voltage, current), as opposed to a **digital computer**.

analog/digital converter Device that converts the output of an **analog computer** into digital signals for a **digital computer.**

analogous In biology, describing structural features that have similar functions but which have developed completely independently in different plant or animal groups (*e.g.* the wings of birds and insects). *See also* **homologous**.

analysis *See* **chromatography**; **qualitative analysis**; **quantitative analysis**; **thermal analysis**; **volumetric analysis**.

anaphase Stage in **mitosis** and **meiosis** (cell division) in which **chromosomes** migrate to opposite poles of the cell by means of the **spindle**.

anaphylaxis Sudden severe reaction (an **allergy**) to a drug or venom, requiring urgent medical treatment. Alternative name: anaphylactic shock.

anastomosis *1.* Natural vessel that connects two blood vessels. *2.* Artificially made connection between two body tubes (*e.g.* in the gut).

anatomy Study of the structural forms and minute structures of **plants** and **animals**.

anchor ring Alternative name for **torus**.

■ **AND gate** Computer logic element that combines two binary input signals to produce one output signal according to particular rules. Alternative name: AND element.

androecium The male parts in a flower (the stamens) or a moss.

androgen Type of **hormone** that is associated with the development of male characteristics in vertebrates; a male sex hormone.

-ane Suffix usually denoting that an organic compound is an **alkane** or **cycloalkane**; *e.g.* methane, cyclohexane.

anechoic Without echo. The term usually describes any structure, or room, that has been lined with sound-absorbent material so that sound reflection is reduced to a minimum.

anemometer Instrument for measuring the speed of wind.

aneurin Alternative name for **thiamine** (vitamin B_1).

■ **angiosperm** Member of a major group of (flowering) plants, a section of the **Spermatophyta,** whose characteristics include the possession of flowers and the production of seeds contained in a fruit. [2/4/b]

■ **angle of incidence** Angle that a ray or any other straight line makes with the normal to a surface at which it arrives.

■ **angle of reflection** Angle made by a ray with the normal to a surface from which it is reflected at the point of reflection.

■ **angle of refraction** Angle made by a ray refracted at a surface separating two media with the normal to the surface at the point of **refraction**.

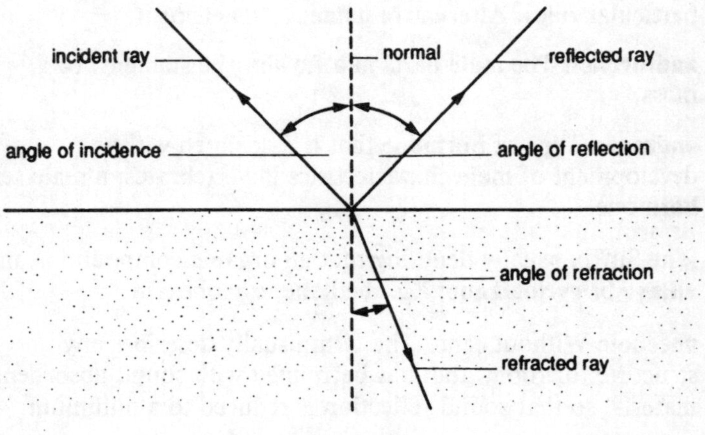

Angles of incidence, reflection and refraction

■ **angstrom** (Å) Unit of **wavelength** for **electromagnetic radiation**, including **light** and **X-rays**, equal to 10^{-10}m. It was named after the Swedish physicist A.J. Ångström (1814–74). [5/9/a]

angular frequency Number of vibrations per unit time, multiplied by 2π, of an oscillating body, usually expressed in radians per second. Alternative name: pulsatance.

angular momentum (*L*) For a rotating object, the cross-product of a **vector** from a specified reference point to a particle and the particle's **linear momentum**.

anhydride Chemical compound formed by removing water from another compound (usually an **acid**); *e.g.* an acid anhydride or **acidic oxide**.

anhydrite Naturally occurring **calcium sulphate**, used to make fertilizers. *See also* **gypsum**.

anhydrous Describing a substance that is devoid of moisture, or lacking **water of crystallization**.

aniline $C_6H_5NH_2$ Colourless oily liquid organic compound, one of the basic chemicals (feedstock) used in the manufacture of dyes, pharmaceuticals and plastics. Alternative names: aminobenzene, phenylamine.

■ **animal** Member of a large kingdom of organisms that feed **heterotrophically** on other organisms or organic matter. Animals are usually capable of locomotion and movement from place to place and react quickly to stimuli (because they have sense organs and a nervous system). Locomotion and sensitivity are in proportion to their need to hunt for food. Animal cells have limited growth, no **chlorophyll**, and are surrounded by a cell membrane. [2/3/b] [2/4/b]

animal starch Alternative name for **glycogen**.

■ **anion** Negatively charged **ion**. [4/10/c]

anisogamy **Fertilization** of **gametes** of slightly different types or sizes.

anisotropic Describing a substance that has different properties with respect to velocity of light transmission, conductivity of heat and electricity, and compressibility in different directions of its matrix.

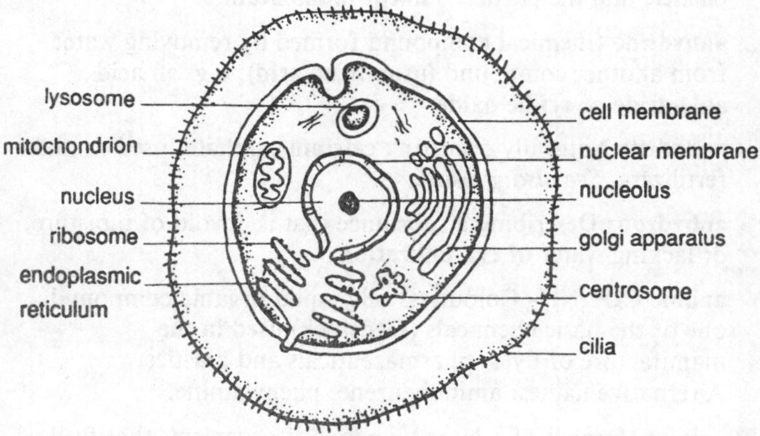

An animal cell

annealing Process of bringing about a desirable change in the properties of a metal or glass (*e.g.* making it tougher), by heating it to a predetermined temperature, thus altering its microscopic structure.

■ **annelid** Worm that is a member of the phylum **Annelida.** [2/3/b] [2/4/b]

Annelida Phylum of invertebrates that consists of segmented worms (the true worms). The segmentation is both internal and external (earthworms and leeches are annelids). Annelid classes include **Oligochaeta** and **Polychaeta**.

annihilation Result of a collision between a particle and its **antiparticle**, accompanied by the evolution of energy. After the collision, both particles cease to exist.

annual Plant that completes its life cycle (from germination, through flowering and fruiting to death) in a single season.

annual ring Yearly addition of **xylem** tissue to the stem of a woody plant, by means of which the age of the plant can be estimated. Alternative name: growth ring.

annulus *1*. In biology, any ring-shaped structure. *2*. Ring of tissue around the stem of the fruiting body of certain **fungi** after expansion of the **pileus**.

anode Positive terminal of an **electrolytic cell** or **thermionic valve**.

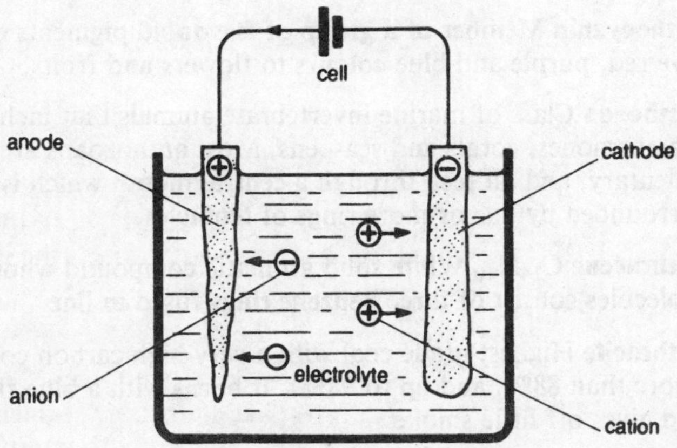

Anode is the positive electrode

anodizing Process of coating a metal (*e.g.* aluminium) with a thin layer of oxide by **electrolysis**.

ANS Abbreviation of **autonomic nervous system**.

antacid Substance used medicinally to combat excess stomach acid (*e.g.* various compounds of **magnesium**).

antagonistic muscles Muscles that work in pairs, one flexing (bending) a limb and the other straightening it.

antenna *1.* In biology, long thin jointed sensory structure in **arthropods** and some **fish**; a feeler. *2.* In radio, an alternative name for an **aerial**.

anterior Near the head of an **animal**. *See also* **posterior**.

■ **anther Pollen**-producing part of a **stamen**. [2/4/a] [2/5/a]

antheridium Male **gamete**-producing organ in lower plants.

antherozoid Male sex cell (gamete) in an **antheridium**.

anthocyanin Member of a group of **flavonoid** pigments that give red, purple and blue colours to flowers and fruits.

Anthozoa Class of marine invertebrate animals that includes sea anemones, corals and sea-pens. Most anthozoans are sedentary, and all feed through a central mouth which is surrounded by one or more rings of tentacles.

anthracene $C_{14}H_{10}$ White solid aromatic compound whose molecules consist of three **benzene rings** fused in line.

anthracite Highest-grade coal with a very high carbon content (more than 88%, and up to 98%). It burns with a blue flame, and gives off little smoke.

anthraquinone Yellow crystalline organic compound important in the manufacture of dyes.

anthrax Bacterial disease that affects many animals, including human beings, and which is frequently fatal. The disease-causing bacterium can form **spores**, and remain dormant in the soil or on stored animal hides for many years.

■ **anthropoid** Man-like. The term strictly refers to the members of the suborder Anthropoidea, which consists of monkeys, apes and man. [2/3/b] [2/4/b]

anthropology Study of the human race. Physical anthropology is concerned with human evolution, social anthropology with behaviour.

antibiotic Member of a group of chemical substances that are by-products of **metabolism** in certain moulds and micro-organisms. Antibiotics cause the destruction of other micro-organisms, and are used as drugs to kill bacteria. *See also* **penicillin**.

antibody Highly specific molecule produced by the **immune system** in response to the presence of an **antigen,** which it neutralizes. *See also* acquired **immunity**.

anticoagulant Chemical substance (*e.g.* coumarin, heparin, warfarin) that prevents blood from clotting. *See also* **coagulant**.

antidiuretic hormone (ADH) **Hormone** secreted from the posterior **pituitary** and synthesized in the **hypothalamus**. In **mammals** it stimulates the reabsorption of water in the kidney, and thus diminishes the volume of **urine** produced. *See also* **vasopressin**.

antigen Foreign substance that stimulates the production of **antibodies**. Most antigens are proteins (*e.g.* bacterial toxins), but they may be other **macromolecules** (*e.g.* bacteria, pollen, transplanted tissue).

■ **antihistamine** Chemical substance that inhibits the action of **histamine** by blocking its site of action. It may be used to treat an **allergy**.

antimatter Matter that is composed of **antiparticles**.

antimony Sb Blue-white semimetallic element in Group VA of the Periodic Table. It is used to impart hardness to lead-tin alloys and as a **donor** impurity in **semiconductors.** Its compounds, used as pigments, are poisonous. At. no. 51; r.a.m. 121.75.

■ **antioxidant** Compound used to delay the **oxidation** of substances such as food by molecular oxygen. Most antioxidants are organic compounds; natural ones are found in vegetable oils and in some fruits (as ascorbic acid, or vitamin C).

■ **antiparticle** Subatomic particle that corresponds to another particle of equal mass but opposite **electric charge** (*e.g.* a positron is the antiparticle of the electron).

antiseptic Substance that prevents sepsis by killing bacteria or preventing their growth. Alternative name: germicide.

antiserum Blood **serum** containing **antibodies,** used in vaccines to treat or prevent a disease or to combat animal venom (*e.g.* snakebite).

antitoxin Type of **antibody** against a toxoid produced in the body by a disease organism or by **vaccination.**

antrum *1.* Cavity in a bone (*e.g.* a sinus). *2.* Part of the stomach next to the pylorus.

■ **anus** Posterior opening of the **alimentary canal,** through which the undigested residue of digestion is passed. [2/4/a] [2/5/a] [2/7/a]

■ **aorta** Principal **artery** that takes oxygenated blood from the heart to all parts of the body other than the **lungs.** [2/4/a] [2/7/a]

apatite Mineral that consists mainly of **calcium phosphate**, used as a source of **phosphorus** and for making fertilizers.

aperture Effective diameter of a lens or lens system. Its reciprocal is the **f-number**.

Apgar score Method of assessing the health of a newborn baby by assigning 0 to 2 points to the quality of heartbeat, breathing, muscle tone, skin colour and reflexes. It was named after the American physician Virginia Apgar (1909–).

apical growth Alternative name for **primary growth**.

apical meristem Zone of cell division in **vascular plants**. It consists of a group of cells at the tip of the stem or root from which all new tissues of the plant are formed.

apoenzyme Inactive **enzyme** that consists of a protein and a non-protein portion. It needs a **coenzyme** in order to function. *See also* **holoenzyme**.

apoplast Parts of a plant that do not consist of living tissue. These are **cell walls, xylem** and spaces between cells.

aposematic coloration Alternative name for **warning coloration**.

apparent depth Perceived depth of a liquid, which is different from its actual depth because of the **refraction** of light.

appendix Vestigial outgrowth of the **caecum** in some mammals. In human beings its full name is vermiform appendix.

Appleton layer Upper level of the **ionosphere** (at about 200km altitude by day, and at 300km by night), which reflects radio waves in the medium waveband back to Earth. It was named after the British physicist Edward Appleton (1892–1965). Alternative name: F-layer.

applications program Computer **program** written by the user for a specific purpose, *e.g.* record-keeping or stock control.

aqua fortis Obsolete term for concentrated **nitric acid**.

aquamarine Blue form of the mineral **beryl**.

aqua regia Mixture of one part concentrated **nitric acid** to three parts concentrated **hydrochloric acid**, so called because it dissolves the 'noble' metals gold and platinum.

aqueous Dissolved in water, or chiefly consisting of water.

aqueous humour Liquid between the **lens** and **cornea** of the eye.

aqueous solution Solution in which the **solvent** is water.

arabinose $C_5H_{10}O_5$ Crystalline **pentose sugar** derived from plant **polysaccharides** (such as gums), sometimes used as a culture medium in bacteriology.

arachnid Member of the class **Arachnida**; *e.g.* a spider.

■ **Arachnida** Class of mainly terrestrial **arthropods**. Arachnids have a combined head and **thorax** (called a prosoma or cephalothorax) and an **abdomen**. The prosoma bears four pairs of legs, but no wings or antennae. Members of the class include mites, ticks, spiders and scorpions. [2/3/b] [2/4/b]

arachnoid membrane One of the three thin membranes that cover the **brain**, the other two being the pia mater and the dura mater.

aragonite Fairly unstable mineral form of **calcium carbonate** ($CaCO_3$).

archegonium Spore-producing tissue from which the spore mother cells are formed in lower plants, such as ferns.

Archimedes' principle When a body is immersed in a fluid it has an apparent loss in weight equal to the weight of the fluid it displaces. It was named after the Greek mathematician Archimedes (287?–212 B.C.).

arc lamp Type of lamp that uses an **electric arc** to generate light.

area Measure of the size of a surface. The areas of some common figures are as follows (where l = length, h = height or altitude, r = radius):

square l^2
rectangle lh
parallelogram lh
triangle $\frac{1}{2}lh$
circle πr^2
sphere $4\pi r^3$
cone πrs (curved surface, s = slant height)
$\pi rs + \pi r^2$ (total surface)
cylinder $2\pi rh$ (curved surface) $2\pi rh + 2\pi r^2$ (total surface)

arenaceous In biology, describing plants that grow best in sandy soils, and animals that live mainly in sand.

■ **areola** *1*. Dark skin on the breast surrounding a nipple. *2*. Part of the iris of the **eye**, bordering the pupil.[2/4/a] [2/5/a] [2/7/a]

areolar tissue Connective tissue made up of cells separated by bundles of fibres embedded in **mucin**.

argentiferous Silver-bearing, usually applied to mineral deposits.

arginine $C_6H_{14}N_4O_2$ Colourless crystalline **essential amino acid** of the alpha-ketoglutaric acid family.

■ **argon** Ar Inert gas element in Group 0 of the Periodic Table (the **rare gases**). It makes up 0.9%of air (by volume), from which it is extracted. It is used to provide an inert atmosphere in electric lamps and discharge tubes, and for welding reactive metals (such as aluminium). At. no. 18; r.a.m. 39.948. [4/8/a] [4/9/a]

arithmetic unit Part of a computer's **central processor** that performs arithmetical operations (addition, subtraction, multiplication, division).

■ **armature** *1*. Moving part (rotor) of a d.c. electric motor or **generator**. *2*. Piece of iron or steel placed across the poles of a permanent magnet in order to preserve its properties. Alternative name: keeper. *3*. Moving metallic part that closes a magnetic circuit in an electric bell or relay. [3/6/d]

aromatic compound Member of a large class of organic chemicals that exhibit **aromaticity**, the simplest of which is **benzene**.

aromaticity Presence in an organic chemical of five or more carbon atoms joined in a ring that exhibits **delocalization** of electrons, as in **benzene** and its compounds. All the carbon-carbon bonds are equivalent.

arrow-worm One of a group of finned worm-like animals that make up the phylum **Chaetognatha**.

arsenate Salt or **ester** of **arsenic acid** (H_3AsO_4). Alternative name: arsenate (V).

arsenic As Silver-grey semimetallic element in Group VA of the Periodic Table which exists as several **allotropes**. It is used in alloys, as a **donor** impurity in **semiconductors** and in insecticides and drugs; its compounds are very poisonous. The substance known as white arsenic is arsenic (III) oxide (arsenious oxide), As_4O_6. At. no. 33; r.a.m. 74.9216.

arsenic acid H_3AsO_4 Tribasic acid from which **arsenates** are derived; an aqueous solution of arsenic (V) oxide, As_2O_5. Alternative names: orthoarsenic acid, arsenic (V) acid.

arsenide Compound formed from **arsenic** and another metal; *e.g.* iron (III) [ferric] arsenide, $FeAs_2$.

arsenious acid H_3AsO_3 Tribasic **acid** from which **arsenites** are derived; an aqueous solution of arsenic (III) oxide (arsenious oxide, white arsenic), As_4O_6. Alternative names: arsenic (III) acid, arsenous acid.

arsenite Salt or **ester** of **arsenious acid** (H_3AsO_3). Alternative name: arsenate (III).

arsine H_3As Colourless highly poisonous gas with an unpleasant odour. Organic derivatives, in which **alkyl groups** replace one or more hydrogen atoms, are also called arsines. Alternative name: arsenic (III) hydride.

arteriole Small **artery**.

■ **artery Blood vessel** that carries oxygenated blood from the heart to other tissues. An exception is the pulmonary artery, which carries deoxygenated blood to the lungs. [2/4/a] [2/5/a] [2/7/a]

arthropod Member of the phylum **Arthropoda**.

Arthropoda Phylum of invertebrates, the largest phylum in the animal kingdom, that consists of animals with jointed appendages, a well-defined head and usually a hard **exoskeleton** made of **chitin**. Arthropods have representatives in every habitat and include arachnids, crustaceans and insects.

■ **artificial insemination** *1*. Artificial implantation of **semen** containing sperm into the female **cervix**. *2*. Transfer of a fertilized **ovum** from the reproductive tract of one female to that of a host mother. [2/10/a]

artificial radioactivity Radioactivity in a substance that is not normally radioactive. It is created by bombarding the substance with ionizing **radiation**. Alternative name: induced radioactivity.

■ **artificial selection** Causing the production of offspring that are most commercially useful by choosing the parents, in both plants and animals. Alternative name: selective breeding. [2/9/c] [2/9/b] [2/10/a]

Artiodactyla Order of mammals that comprises the even-toed (usually two toes) **ungulates**. Artiodactyls include antelopes, bison, buffaloes, camels, cattle, deer, giraffes, goats, pigs and sheep. All are **herbivores** and many are **ruminants**. *See also* **Perissodactyla**.

aryl group **Radical** that is derived from an **aromatic compound** by the removal of one hydrogen atom; *e.g.* phenyl, C_6H_5-, derived from **benzene**.

asbestos Fibrous variety of a number of rock-forming **silicate** minerals that are heat-resistant and chemically inert.

Ascomycetes Important class of **fungi** in which the **spore**-producing body is an **ascus**. It includes morels, truffles, the fungal part of most **lichens**, and many **yeasts**. Alternative name: Ascomycotina.

Ascomycotina Alternative name for **Ascomycetes**.

ascorbic acid $C_6H_8O_6$ White crystalline water-soluble **vitamin** found in many plant materials, particularly fresh fruit and vegetables. It is a natural **antioxidant**. Alternative name: vitamin C.

ascus Cell in fungi in which haploid **spores** are formed. *See* **Ascomycetes.**

aseptic Free from disease-causing micro-organisms (particularly bacteria).

asexual reproduction Reproduction that does not involve **gametes** or **fertilization**. There is a single parent, and all offspring are genetically identical (*see* **clone**). *See also* **vegetative propagation**. [2/10/a]

aspartame Artificial sweetener (a dipeptide) that is 200 times as sweet as ordinary sugar (sucrose), but does not have the bitter aftertaste characteristic of **saccharin**.

aspirator Apparatus that produces suction in order to draw a gas or liquid from a vessel or cavity.

■ **aspirin** $CH_3COO.C_6H_4COOH$ Drug that is commonly used as an **analgesic**, antipyretic (to reduce fever) and anti-inflammatory. Alternative name: acetylsalicylic acid. [2/1/c]

assimilation *1*. General term for the process by which food is made and used by **plants**. *2*. Process of turning food into body substances after it has been digested (*e.g.* in animals excess glucose is turned into glycogen or fat, amino acids are made into proteins).

astatine At Radioactive element in Group VIIA of the Periodic Table (the **halogens**). It has several **isotopes**, of half-lives of 2×10^{-6} sec to 8 hr. Because of their short half-lives, they are available in only minute quantities. At. no. 85; r.a.m. (most stable isotope) 210.

aster phase Alternative name for **metaphase**.

astigmatism *1*. Vision defect caused by irregular curvature of the lens of the **eye**, so that light does not focus properly.

2. Failure of an optical system, such as a **lens** or a **mirror**, to focus the image of a point as a single point.

astronomical unit (AU) Measure used for distances within the Solar System. One astronomical unit is equal to the mean distance between the Earth and the Sun (149,599,000 km). [3/4/e]

astrophysics Branch of astronomy concerned with the structure, physical properties and behaviour of the Universe and heavenly bodies within it; *e.g.* luminosity, size, mass, density, temperature and chemical composition.

atlas In anatomy, the first (cervical) vertebra, which joins the skull to the spine and articulates with the **axis**, allowing nodding movements of the head. *See also* **axis**.

atmosphere *1.* Air or gases surrounding the Earth or other heavenly body. The Earth's atmosphere extends outwards several thousand kilometres, becoming increasingly rarefied until it merges gradually into **space**. For composition, *see* **air**. *2.* (atm.) Unit of **pressure**, equal to 101,325 **pascals (Pa)**, equivalent to 760 mm Hg.

atmospheric pressure Force exerted on the surface of the Earth (and any other planetary body), and on the organisms that live there, by the weight of the **atmosphere**. Its standard value at sea-level is 101,325 pascals, or 760 mm Hg, and it decreases with altitude, but is subject to local and temporary variation. It is measured with a **barometer**.

■ **atom** Fundamental particle that is the basic unit of matter. An atom consists of a positively charged **nucleus** surrounded by negatively charged **electrons** restricted to **orbitals** of a given energy level. Most of the **mass** of an atom is in the nucleus, which is composed principally of **protons** (positively charged) and **neutrons** (electrically neutral); hydrogen is exceptional in having merely one proton in its nucleus. The number of electrons is equal to the number of protons, and this is the **atomic number**. The chemical behaviour of an atom is determined by how many electrons it has, and how they are transferred to, or shared with, other atoms to form chemical bonds. *See also* **atomic mass**; **Bohr theory**; **isotope**; **subatomic particle**. [4/7/e]

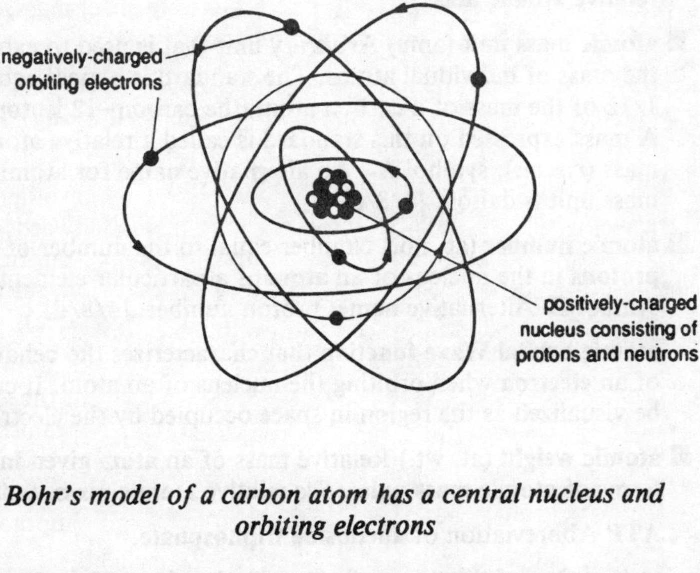

Bohr's model of a carbon atom has a central nucleus and orbiting electrons

atomic bomb Explosive device of great power that derives its energy from **nuclear fission**. Alternative names: atom bomb, nuclear bomb.

atomic clock Time-keeping device in which the operation is controlled by an atomic or molecular process. Clocks based on the frequency with which an atom of caesium changes state in a magnetic field are accurate to within one second per 30,000 years.

atomic energy Energy that is released by **nuclear fission** or **nuclear fusion**. Alternative name: **nuclear energy**.

atomicity Number of **atoms** in one molecule of an element; *e.g.* argon has an atomicity of 1, nitrogen 2, ozone 3.

atomic mass Alternative name for **atomic weight**. *See also* **relative atomic mass**.

■ **atomic mass unit** (amu) Arbitrary unit that is used to express the mass of individual atoms. The standard is a mass equal to 1/12 of the mass of a carbon atom (the carbon−12 **isotope**). A mass expressed on this standard is called a relative atomic mass (r.a.m.), symbol A_r. An alternative name for atomic mass unit is dalton. [4/8/d]

■ **atomic number** (at. no.) Number equal to the number of **protons** in the nucleus of an **atom** of a particular element, symbol Z. Alternative name: proton number. [4/8/d]

atomic orbital Wave function that characterizes the behaviour of an **electron** when orbiting the nucleus of an atom. It can be visualized as the region in space occupied by the electron.

■ **atomic weight** (at. wt.) Relative mass of an **atom** given in terms of **atomic mass units**. *See* **relative atomic mass**. [4/8/d]

ATP Abbreviation of **adenosine triphosphate**.

atrium One of the thin-walled upper chambers of the **heart**. Alternative name: auricle.

atrophy Wasting away of a **cell** or **organ** of the body.

atropine White crystalline poisonous **alkaloid** that occurs in deadly nightshade. Alternative name: belladonna.

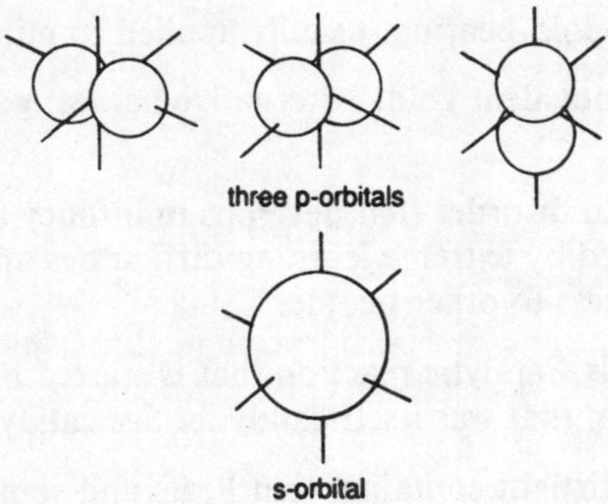

Atomic orbitals

attenuation Loss of strength of a physical quantity (*e*.g. a radio signal or an electric current) caused by **absorption** or **scattering**.

AU Abbreviation of **astronomical unit**.

auditory Relating to the **ear** or hearing.

■ **auditory canal** Tube that leads from the outer **ear** to the ear drum (tympanum). [2/4/a] [2/5/a] [2/7/a]

auditory nerve Nerve that carries impulses concerned with hearing from the inner **ear** to the brain.

Auger effect Radiationless ejection of an **electron** from an **ion**. It was named after the French physicist Pierre Auger (1899–). Alternative name: **autoionization**.

auric Trivalent gold. Alternative names: gold (III), gold (3+).

auricle Alternative name for **atrium**.

auriferous Gold-bearing, usually applied to mineral deposits.

aurous Monovalent gold. Alternative names: gold (I), gold (1+).

autism Brain disorder that develops in infancy and which is characterized by extreme learning difficulties and a lack of responsiveness to other people.

■ **autocatalysis** Catalytic reaction that is started by the products of a reaction that was itself catalytic. *See* **catalyst**.

autoclave Airtight container that heats and sometimes agitates its contents under high-pressure steam. Autoclaves are usually used for sterilizing surgical instruments, and in biotechnology.

autogamy Uniting of **gametes** produced by the same **cell**. Alternative name: **self-fertilization.**

autoionization Spontaneous ionization of excited **atoms**, **molecules** or fragments of an **ion**. Alternative name: pre-ionization. *See also* **Auger effect**.

autolysis Breakdown of the contents of an animal or plant cell by the action of **enzymes** produced within that cell.

autonomic nervous system (ANS) In vertebrates, the part of the **nervous system** that is not under voluntary control and carries nerve impulses to the **smooth muscles**, glands, heart and other organs. It is subdivided into the **parasympathetic** and **sympathetic nervous systems**.

auto-oxidation **Oxidation** caused by the unaided **atmosphere**.

autoradiography Technique for photographing a specimen by injecting it with radioactive material so that it produces its own image on a photographic film or plate.

autosome Any **chromosome** other than one of the **sex chromosomes**.

autotroph Organism that lives using **autotrophism**.

autotrophism In bacteria and green plants, the ability to build up food materials from simple substances, *e.g.* by **photosynthesis**. *See also* **heterotrophism**.

auxin Any substance, including various hormones, that controls the growth of plants.

avalanche diode Alternative name for **Zener diode**.

■ **Aves** Class of vertebrates made up of birds. [2/3/b]

■ **Avogadro constant** (L) Number of particles (atoms or molecules) in a **mole** of a substance. It has a value of 6.02×10^{23}. It was named after the Italian scientist Amedeo Avogadro (1776–1856). Alternative name: Avogadro's number.

■ **Avogadro's law** Under the same conditions of pressure and temperature, equal volumes of all gases contain equal numbers of **molecules**. Alternative name: Avogadro's hypothesis.

Avogadro's number Alternative name for **Avogadro constant**.

avoirdupois System of weights, still used for some purposes in Britain, based on a pound (symbol lb), equivalent to 2.205kg, and subdivided into 16 ounces, or 7,000 grains. In science and medicine it has been almost entirely replaced by **SI units**.

■ **axis** *1.* In biology, central line of symmetry of an organism. *2.* In anatomy, the second (cervical) vertebra, which articulates with the **atlas** and allows the head to turn from side to side. *3.* In constructing a graph, the vertical or horizontal line calibrated in numbers or units. [1/7/c] [1/9/b]

axon Thread-like outgrowth of a **nerve cell** (neurone), the main function of which is to carry impulses towards the cell body. Sensory neurones have long axons and short dendrons; motor neurones have short axons and long dendrons. *See also* **dendron**.

azeotrope Mixture of two liquids that boil at the same temperature.

azide One of the **acyl group** compounds, or salts, derived from hydrazoic acid (N_3H). Most azides are unstable, and heavy metal azides are explosive.

azo dye Member of a class of dyes that are derived from amino compounds and have the $-N=N-$ **chromophore** group. They are intensely coloured (usually red, brown or yellow) and account for a large proportion of the synthetic dyes produced. Azo dyes can be made as acid, basic, direct or mordant dyes. Their use as food colourants has been questioned because of their possible effects on sensitive people, particularly children.

azomethine *See* **Schiff's base**.

B

Babbitt metal Tin alloy (containing also some antimony and copper) used for lining bearings in machinery. It was named after the American engineer Isaac Babbitt (1799–1862).

Babo's law The lowering of the **vapour pressure** of a **solution** is proportional to the amount of **solute**. It was named after the German chemist Lambert Babo (1818–99).

Bacillariophyta Class of **algae** that are characterized by possession of cell walls impregnated with **silica**. Composed of two halves, they are microscopic, unicellular marine or freshwater organisms. Alternative name: diatoms.

bacillus *1.* Descriptive term for any rod-shaped **bacterium**. *2.* Specifically, *Bacillus* is a genus of spore-producing bacteria.

backbone Alternative name for **vertebral column**.

■ **back e.m.f. Electromotive force** produced in a circuit that opposes the main flow of current; *e.g.* in an electrolytic cell because of **polarization** or in an electric motor because of **electromagnetic induction**. [5/7/c]

■ **background radiation Radiation** from natural sources, including outer space (cosmic radiation) and radioactive substances on Earth (*e.g.* in igneous rocks such as granite). [4/7/f]

backing store Computer store that is larger than the main (immediate access) memory, but with a longer access time.

bacteria Plural of **bacterium**.

bactericide Substance that can kill bacteria. *See also* **bacteriostatic**.

bacteriology Study of **bacteria**, their effects on organisms and their uses in agriculture and industry (*e.g.* in **biotechnology**).

bacteriophage Virus that infects **bacteria**. When inside a cell it replicates using its host's **enzymes**; the release of new viruses may disintegrate the cell. Bacteriophages have been used extensively in research on **genes**. Alternative name: phage.

bacteriostatic Substance that inhibits the growth of bacteria without killing them. *See also* **bactericide**.

■ **bacterium** Member of a group of microscopic organisms distinct from the plant and animal kingdoms. The shape of bacteria varies according to species, *e.g.* rod-shaped (bacillus), spherical (coccus), spiral (spirillum or spirochaete). Multiplication is fast, usually by fission, every 20 minutes in favourable conditions. Between them, the various species can utilize almost any type of organic molecule as food and inhabit almost any environment; there are those that can even use inorganic elements such as sulphur. The activities of some bacteria are of great significance to man, *e.g.* **fixation of nitrogen** in certain plants, as agents of decay, and their medical importance as disease-causing agents (pathogens).

■ **Bakelite** Trade name for a **thermosetting** synthetic plastic made by the **condensation** of **phenol** with **formaldehyde** (methanal). It is used in the manufacture of electrical fittings and other plastic products. [4/8/b]

baking powder Mixture of a **hydrogencarbonate** (bicarbonate) and a weak **acid** which on heating or the addition of water produces bubbles of carbon dioxide; the bubbles cause a cake mixture or dough to rise. A common composition is a mixture of sodium hydrogencarbonate (baking soda) and tartaric acid.

baking soda Alternative name for **sodium hydrogencarbonate** (sodium bicarbonate).

balance *1.* Apparatus for weighing things accurately; types include a beam balance, spring balance and substitution balance. *2.* Sense supplied by organs within the semicircular canals of the **inner ear**.

balanced diet Food that contains nutrients – fat, carbohydrate, protein, vitamins, mineral salts, water and roughage (fibre) – in the correct proportions for good health.

baleen Fibrous material that hangs in plates inside the mouths of certain species of whales, where it acts to filter plankton (their food) from sea-water. Alternative name: whalebone.

ballistic missile Ground-to-ground missile that moves under gravity after the initial powered and guided stage.

ballistics Study of projectiles moving under the force of gravity only.

Balmer series Visible atomic **spectrum** of hydrogen, consisting of a unique series of energy emission levels which appears as lines of red, blue and blue-violet light. It is the key to the discrete energy levels of electrons (*see* **Bohr theory**).

■ **bar** Unit of pressure defined as 10^5 newtons per square metre, or pascals (Pa); equal to approximately one atmosphere. [5/6/a]

barbiturate Sedative and hypnotic drug derived from barbituric acid, formerly used in sleeping pills but now generally employed only as a fast-acting anaesthetic and to treat epilepsy.

bar chart Graph in which data is presented as bars or blocks ranged along an axis. *See also* **histogram**.

barium Ba Silver-white metallic element in Group IIA of the Periodic Table (the **alkaline earths**), obtained mainly from the mineral barytes (**barium sulphate**). Its soluble compounds are poisonous, and used in fireworks; its insoluble compounds are used in pigments and medicine. At. no. 56; r.a.m. 137.34.

barium sulphate $BaSO_4$ White crystalline insoluble powder, used as a pigment and as the basis of 'barium meal' to show up structures in X-ray diagnosis (because it is opaque to X-rays).

■ **bark** Outer surface of plant **stems** and **roots** which have undergone secondary growth or thickening. It results from the activity of cork **cambium** and gives the plant a protective outer surface of dead **cells**. [2/3/d] [2/5/a]

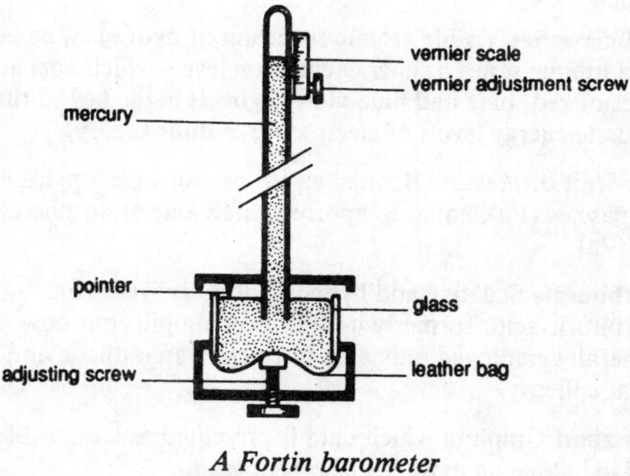

A Fortin barometer

■ **barometer** Instrument for measuring atmospheric pressure, much used in meteorology. [5/6/a]

Barr body Condensed **X-chromosome**, seen in cells of female mammals, due to one or other of the two X-chromosomes in each cell being inactivated. It was named after the Canadian biologist M. Barr (1908–).

barrel Unit of volume. In the oil industry, 1 barrel = about 159 litres (35 gallons); in brewing, 1 barrel = 32 gallons.

baryon One of a group of subatomic particles that include **protons**, **neutrons** and **hyperons**, and which are involved in strong interactions with other particles.

basal ganglion Region of grey matter within the white matter that forms the inner part of the cerebral hemispheres of the **brain**. Alternative name: basal nucleus.

■ **basal metabolic rate** (BMR) Minimum amount of energy on which the body can survive, measured by oxygen consumption and expressed in kilojoules per unit body surface.

basalt Dark-coloured, usually black, fine-grained igneous rock. Deposits that occur on the Earth's surface are generally restricted to solidified lava.

■ **base** *1*. In biology, one of the **nucleotides** of **DNA** or **RNA** (*i.e.* adenine, cytosine, guanine, thymine or uracil). *2*. In chemistry, a member of a class of chemical compounds whose aqueous solutions contain OH^- **ions**. A base neutralizes an **acid** to form a **salt** (*see also* **alkali**).

■ **base code** Sequence of bases on a strand of **DNA** that determines the type of information carried by a **gene**. [2/10/b]

base metals Metals that corrode, oxidize or tarnish on exposure to air, moisture or heat; *e.g.* copper, iron, lead.

base pairing Specific pairing between complementary **nucleotides** in double-stranded **DNA** or **RNA** by **hydrogen bonding**; *e.g.* in DNA, guanine pairs with cytosine and adenine pairs with thymine. *See also* **purine**; **pyrimidine**.

basic In chemistry, having a tendency to release **hydroxide** (OH^-) ions. *See* **base**.

basic oxide Metallic oxide that reacts with water to form a **base**, and with an **acid** to form a **salt** and water; *e.g.* calcium oxide, CaO, reacts with water to form calcium hydroxide, $Ca(OH)_2$. *See also* **acidic oxide**.

basic salt Type of **salt** that contains **hydroxide** (OH^-) ions; *e.g.* basic lead carbonate, $2PbCO_3.Pb(OH)_2$.

Basidiomycetes Major class of **fungi** that includes rusts and smuts, in which the **spores** form in a specialized club-shaped cell known as the basidium which develops inside the fruiting body. Alternative name: Basidiomycotina.

basidium *See* **Basidiomycetes**.

■ **battery** Device for producing electricity (direct current) by chemical action; alternative name: cell. *See* **Daniell cell; dry cell; Leclanché cell; primary cell; secondary cell**. *See also* **fuel cell; solar cell**.

Baumé scale Scale of **relative density (specific gravity)** of liquids, commonly used in continental Europe. It was named after the French chemist Antoine Baumé (1728–1804).

bauxite Earthy mineral form of **alumina** (aluminium oxide, Al_2O_3) and the chief ore of **aluminium**.

BCG Abbreviation of bacillus Calmette-Guérin, a **vaccine**

used against tuberculosis. It was named after two French bacteriologists, Albert Calmette (1863–1933) and Camille Guérin.

bearing In surveying and telecommunications, horizontal angle between a line and a reference direction, measured clockwise from north.

beats Variation in volume that occurs when two notes of nearly equal frequency sound simultaneously.

Beckmann thermometer Mercury thermometer used for accurately measuring very small changes or differences in temperature. The scale usually covers only 6 or 7 degrees. It was named after the German chemist Ernst Beckmann (1853–1923).

■ **becquerel** (Bq) SI unit of radioactivity, equal to the number of **atoms** of a **radioactive** substance that disintegrate in one second. It was named after the French physicist Henri Becquerel (1852–1908). 1 Bq = 2.7×10^{-11} curies (the former unit of radioactivity). [4/7/f]

Beer's law Concerned with the absorption of light by substances, it states that the fraction of incident light absorbed by a solution at a given **wavelength** is related to the thickness of the absorbing layer and the concentration of the absorbing substance. Alternative name: Beer-Lambert law.

■ **beetle** Member of the **Coleoptera** order of insects, which typically have hard wing cases over membranous flight wings and biting mouthparts in the adult. The larvae may be active predators or sluggish grubs. [2/3/b] [2/4/b]

beet sugar Alternative name for **sucrose**.

behaviourism One branch of psychology, which concentrates exclusively on observable actions and takes no account of mental events (thoughts, emotions, etc.) that cannot be experienced by the observer (the psychologist).

bel (B) Unit representing the ratiŏ of two amounts of power, *e.g.* of sound or an electronic signal, equal to 10 **decibels**. It was named after the Scottish inventor Alexander Graham Bell (1847–1922).

■ **Benedict's test** Food test used to detect the presence of a **reducing sugar** by the addition of a solution containing sodium carbonate, sodium citrate, potassium thiocyanate, copper sulphate and potassium ferrocyanide. A change in colour from blue to red or yellow on boiling indicates a positive result. It was named after the American chemist S. Benedict (1884–1936). *See also* **Fehling's test**. [2/9/a]

benign In medicine, describing a **tumour** that does not spread or destroy the tissue in which it is located; not life-threatening. *See also* **malignant**.

benthos Organisms that live on the bottom of a sea or lake.

bentonite Kind of clay, used as an **adsorbent** and in paper-making.

benzene C_6H_6 Colourless inflammable liquid **hydrocarbon**, the simplest **aromatic compound**. It is used as a solvent and in the manufacture of plastics.

1,3-benzenediol Alternative name for **resorcinol**.

benzene ring Cyclic (closed-chain) arrangement of six carbon atoms, as in a molecule of **benzene**. Molecules containing one or more benzene rings display **aromaticity**.

benzodiazepine Member of a group of drugs used as antidepressants and anticonvulsants.

benzoic acid C_6H_5COOH White crystalline organic

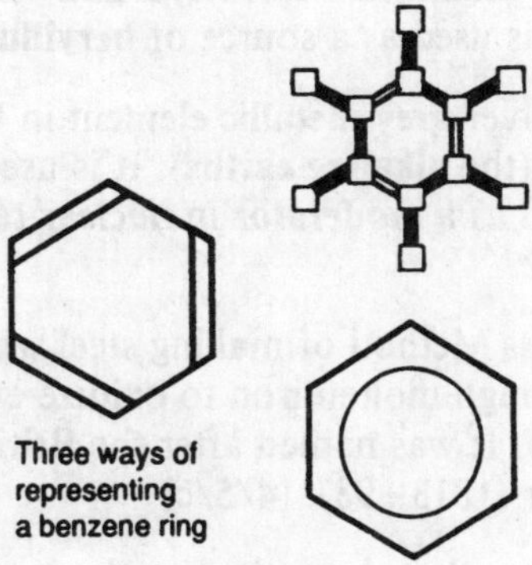

Various representations of the benzene ring

compound, used as a food preservative because it inhibits the growth of yeasts and moulds.

benzole Alternative name for **benzene**.

benzpyrene Cyclic organic compound, found in coal-tar and tobacco smoke, which has strong carcinogenic properties.

benzpyrrole Alternative name **indole**.

berkelium Bk Radioactive element in Group IIIB of the Periodic Table (one of the **actinides**). It has several **isotopes**, made by alpha-particle bombardment of americium-241. At. no. 97; r.a.m. (most stable isotope) 247.

berry Fruit consisting of a fleshy **pericarp** containing seeds;

e.g. cucumber, grape, lemon, melon, tomato (the date is an unusual berry in having only one seed). Botanically, blackberries, raspberries etc. are not true berries.

beryl Beryllium aluminium silicate, a mineral which when not of gem quality is used as a source of **beryllium**.

beryllium Be Silver-grey metallic element in Group IIA of the Periodic Table (the **alkaline earths**). It is used for windows in X-ray tubes and as a **moderator** in nuclear reactors. At. no. 4; r.a.m. 9.0122.

■ **Bessemer process** Method of making steel which involves blowing air through molten iron to oxidize excess carbon and other impurities. It was named after the British engineer Henry Bessemer (1813–98). [4/5/c]

beta-blocker Drug that slows the heartbeat, used to treat hypertension (high blood pressure).

■ **beta decay Disintegration** of an unstable **radioactive** nucleus that involves the emission of a **beta particle**. It occurs when a neutron emits an electron and is itself converted to a proton, resulting in an increase of one proton in the nucleus concerned and a corresponding decrease of one neutron. This leads to the formation of a different element (*e.g.* beta decay of the radio-isotope carbon-14 produces nitrogen and an electron). [4/8/d]

beta particle High-velocity electron emitted by a **radioactive** nucleus undergoing **beta decay**. It is sometimes represented as β-particle.

■ **beta radiation** Radiation, consisting of beta particles (electrons), due to **beta decay**. [4/8/d]

bicarbonate Alternative name for **hydrogencarbonate**.

biceps Flexor muscle in the upper arm that has two points of origin and one insertion. It is one of a pair of **antagonistic muscles**, the triceps being the other (extensor) muscle.

bichromate Alternative name for **dichromate**.

biconcave Describing a lens that is concave on both surfaces.

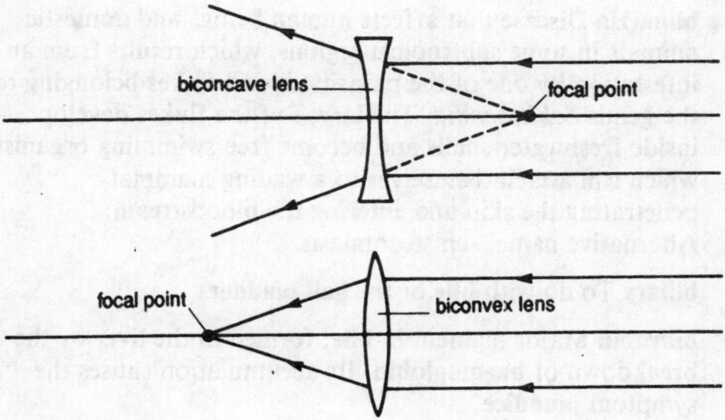

Biconvex and biconcave lenses

biconvex Describing a lens that is convex on both surfaces.

biennial Plant whose life-cycle lasts for two years. Flowering and seed production occur in the second year of life.

bilateral symmetry Type of symmetry in which a shape is symmetrical about a single plane (each half being a mirror image of the other). *E.g.* most vertebrates are bilaterally symmetrical. *See also* **radial symmetry**.

■ **bile** Alkaline mixture of substances produced by the liver (from waste products when blood cells are broken down) and stored in the **gall bladder,** which passes it to the **duodenum,** where it emulsifies fats (preparing them for digestion) and neutralizes acid. Its (yellowish) colour is due to bile pigments (*e.g.* **bilirubin**). [2/4/a] [2/5/a] [2/7/a]

bile duct Tube that carries **bile** from the gall bladder to the **duodenum.**

bilharzia Disease that affects human beings and domestic animals in some subtropical regions, which results from an infestation by one of the parasitic blood flukes belonging to the genus *Schistosoma*. The larvae of the flukes develop inside freshwater snails and become free-swimming organisms which can attach themselves to a wading mammal, penetrating the skin and entering the bloodstream. Alternative name: schistosomiasis.

biliary To do with **bile** or the **gall bladder**.

bilirubin Major pigment in **bile,** formed in the liver by the breakdown of **haemoglobin**. Its accumulation causes the symptom **jaundice**.

billion Number now generally accepted as being equivalent to 1,000 million (10^9). Formerly in Britain a billion was regarded as a million million (10^{12}).

■ **bimetallic strip** Strip consisting of two metals with different thermal expansions joined together, which as a result bends when heated; used in **thermostats**. [3/7/c]

binary compound Chemical compound whose molecules consist of two different atoms.

binary fission Method of reproduction employed by many single-celled organisms, in which the so-called mother cell divides in half (by **mitosis**), forming two identical, but independent, daughter cells. It is a type of **asexual reproduction**.

binding energy Energy required to cause a **nucleus** to decompose into its constituent **neutrons** and **protons**.

binomial nomenclature System by which organisms are identified by two Latin or pseudo-Latin names. The first is the name of the **genus** (generic name), the second is the name of the **species** (specific name), *e.g. Homo sapiens*. Alternative term: Linnaean system (after the Swedish botanist Carolus Linnaeus, 1707–78). *See also* **classification**.

bioassay Method for quantitatively determining the concentration of a substance by its effect on living organisms, *e.g.* its effect on the growth of **bacteria**.

biochemical oxygen demand (BOD) Oxygen-consuming property of natural water because of the organisms that live in it.

biochemistry Study of the **chemistry** of living organisms.

■ **biodegradable** Describing a substance that breaks down or decays by the action of living organisms, especially **saprophytic** bacteria and fungi. Through biodegradation, organic matter is recycled. [2/2/d] [2/6/d]

biodegradation Breakdown or decay of substances by the action of living organisms.

bioenergetics Study of the transfer and utilization of energy in living systems. *See also* **adenosine triphosphate**.

■ **bioengineering** Application of engineering science and technology to living systems. [2/10/c]

biofeedback Method of controlling a bodily process (*e.g.*

heartbeat) that is not normally subject to voluntary control by making the person concerned aware of measurements from instruments monitoring that process.

biogenesis Theory that living organisms may originate only from other living organisms, as opposed to the theory of **spontaneous generation**. *See also* **abiogenesis**.

biological clock Hypothetical mechanism in plants and animals that controls periodic changes in internal functions and behaviour independently of environment, *e.g.* diurnal rhythms and hibernation patterns.

biology Study of living organisms and their relation to the non-living environment. Its two major divisions are botany (the study of plants) and zoology (the study of animals).

bioluminescence Emission of visible light by living organisms, *e.g.* certain bacteria, fungi, fish and insects. It is produced by **enzyme** reactions in which chemical energy is converted to light.

■ **biomass** Total mass of living matter in a given environment or **food chain** level. [2/7/d]

biome Subdivision of the living planet (the ecosphere), the largest area which ecology can deal with conveniently; an ecological region broadly equivalent to one of the climatic regions, *e.g.* rainforest or desert.

biometry Application of mathematical and statistical methods to the study of living organisms.

bionic Describing an artificial device or system that has the properties of a living one.

biophysics Use of ideas and methods of **physics** in the study of living organisms and processes.

biopsy Small sample of cells or tissue removed from a living subject for laboratory examination (usually as an aid to diagnosis).

biorhythm One of the cyclic pattern of 'highs' and 'lows' which some people believe govern each person's emotional, physical and intellectual behaviour. Most scientists are sceptical of the concept because it is based on the individual's subjective perceptions of these qualities. *See also* **circadian rhythm**.

biosphere Region of the Earth and its **atmosphere** which may be inhabited by living organisms.

biosynthesis Formation of the major molecular components of cells, *e.g.* **proteins**, from simple components.

biotechnology Utilization of living organisms for the production of useful substances or processes, *e.g.* in **fermentation** and milk production.

biotin **Coenzyme** that is involved in the transfer of carbonyl groups in biochemical reactions, such as the metabolism of fats; one of the B vitamins.

■ **bird** Vertebrate animal of the class **Aves**, characterized by having feathers, laying hard-shelled eggs and (usually) the power of flight. [2/3/b] [2/4/b]

Birkeland-Eyde process Method for fixation of **nitrogen** in which air is passed over an electric arc to form **nitrogen monoxide** (nitric oxide). It was named after the Norwegian chemists Kristian Birkeland (1867–1913) and Samuel Eyde (1866–1940).

birth Process by which young (*e.g.* of a mammal) separates from its mother. Alternative name: parturition.

bisexual Having both male and female characteristics.

bismuth Bi Silvery-white metallic element in Group VA of the Periodic Table. It is used as a liquid metal coolant in nuclear reactors and as a component of low-melting point lead alloys. Its soluble compounds are poisonous; its insoluble ones are used in medicine. At. no. 83; r.a.m. 208.98.

bistable circuit Alternative name for a **flip-flop**.

bisulphate Alternative name for **hydrogensulphate**.

bisulphite Alternative name for **hydrogensulphite**.

bit Amount of information that is required to express choice between two possibilities. The term is commonly applied to a single digit of binary notation in a computer. The word is an abbreviation of *binary digit*, either 1 or 0, the only two digits in binary notation.

■ **bitumen** Solid or tarry mixture of **hydrocarbons** obtained from coal, oil, etc., commonly used for surfacing paths and roads. [4/8/b]

bituminous coal Second-quality coal, with a carbon content above 65% but also containing large quantities of gas, water and coal tar. It is the most widely used domestic and industrial coal.

biuret $NH_2CONHCONH_2$ Colourless crystalline organic compound made by heating **urea**.

biuret test Test used to detect **peptides** and **proteins** in solution by treatment of **biuret** with **copper sulphate** and **alkali** to give a purple colour.

bivalent Having a **valence** of two. Alternative name: divalent.

bivalve Animal of the class **Bivalvia**.

Bivalvia Class of **molluscs** comprising shellfish with a pair of hinged shells; *e.g.* clams, cockles, mussels, oysters and scallops. Alternative names: Lamellibranchiata, Pelecypoda.

black body Hypothetical full or complete absorber and radiator of **radiation**.

bladder *1*. In lower plants, *e.g.* bladderwort, a modified leaf that catches small aquatic animals. *2*. In animals, a membranous sac containing gas or fluid, *e.g.* the air-filled swim bladder of bony fish, and the gall bladder and urinary bladder of mammals. *See also* **vesicle**.

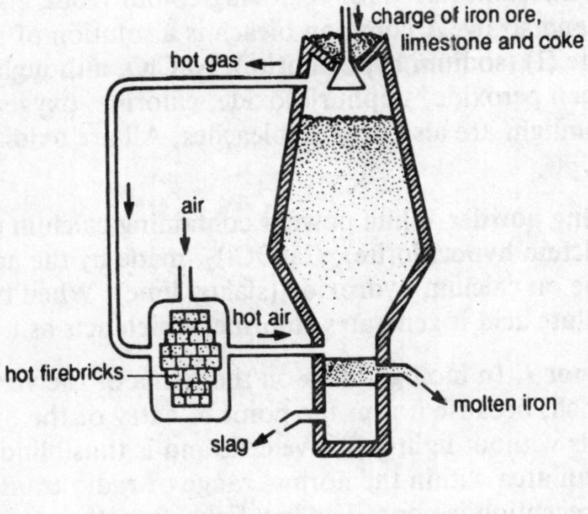

Blast furnace and its air supply

■ **blast furnace** Furnace that can reach a high operating temperature, necessary for extracting iron from iron ore. The furnace is loaded with iron ore, coke and limestone. Hot air is blown into the mixture. Molten **pig iron** is run off from the bottom of the furnace. [4/5/c]

blasting glycerin Alternative name for **gelignite**.

blastocyst Early stage in the development of a mammalian **embryo**, formed by successive cleavages of the fertilized ovum. It consists of a solid ball of cells surrounded by a hollow sphere of cells, and goes on to form the **gastrula**. Alternative name: blastula.

blastula Hollow sphere composed of a single layer of **cells** produced by cleavage of a fertilized **ovum** in **animals**. In mammals, it is also called a blastocyst.

bleach Substance used for removing colour from, *e.g.*, cloth, paper and straw. A common bleach is a solution of sodium chlorate (I) (sodium hypochlorite), NaClO, although hydrogen peroxide, sulphur dioxide, chlorine, oxygen and even sunlight are also used as bleaches. All are **oxidizing agents**.

bleaching powder White powder containing calcium chlorate (I) (calcium hypochlorite), $Ca(OCl)_2$, made by the action of chlorine on calcium hydroxide (slaked lime). When treated with dilute acid it generates chlorine, which acts as a **bleach**.

blind spot *1.* In biology, area on the **retina** of the vertebrate **eye** which, because it is at the point of entry of the optic nerve, is without light-sensitive cells and is thus blind. *2.* In radio, an area within the normal range of radio transmission where reception is poor. The low field strength is usually caused by an interference pattern produced by man-made obstructions or geographical features.

blink microscope Instrument for comparing two very similar photographs, *e.g.* of stars or bacteria. The photographs are viewed side by side, one with each eye, and are rapidly concealed and uncovered. The brain's attempts to superimpose the two images reveals any slight differences between them.

■ **blood** Fluid in the bodies of animals that circulates and transports oxygen and nutrients to cells, and carries waste products from them to the organs of excretion. It also transports hormones and the products of digestion. Essential for maintaining uniform temperature in warm-blooded animals, blood is made up mainly of **erythrocytes, leucocytes, platelets,** water and **proteins**. *See also* **haemocyanin; haemoglobin**. [2/5/a]

■ **blood cells** *See* **erythrocyte; leucocyte; lymphocyte.** [2/4/a] [2/5/a] [2/7/a]

blood clot Coagulated blood which may help wound healing or be the cause of blocking an artery (possibly leading to an infarction or stroke). Alternative name: thrombus.

blood group *See* **ABO blood groups.**

blood plasma Straw-coloured fluid part of **blood** in which blood cells are suspended. Its importance is for transporting blood cells, plasma **proteins, urea**, sugars, mineral salts, hormones and carbon dioxide as bicarbonates in solution.

blood poisoning Alternative name for **septicaemia.**

blood pressure Pressure of blood flowing in the arteries. It varies between the higher value of the systolic pressure (when the heart's ventricles are contracting and forcing blood out of the heart) and the lower value of the diastolic pressure (when the heart is filling with blood). It is affected by exercise, emotion, certain illnesses and various drugs. *See also* **heartbeat**.

blood serum Fluid part of **blood** from which all blood cells and **fibrin** have been removed. It may contain **antibodies**, and such serums are used as **vaccines**.

blood sugar The energy-generating sugar **glucose**, whose level in the blood is controlled by the hormones **insulin** and **adrenalin**.

blood vascular system System consisting of the **heart, arteries, veins** and **capillaries**. The heart acts as a central muscular pump which propels oxygenated **blood** from the lungs along arteries to the tissues; deoxygenated blood is carried along veins back to the heart and lungs.

■ **blood vessel** An **artery**, **vein** or **capillary**. Many of the larger blood vessels have muscular walls, whose contraction and relaxation aids blood flow. [2/4/a] [2/5/a] [2/7/a]

bloom Alternative name for **flower**.

blue-green alga Alternative name for a member of the **Cyanophyta**.

blue vitriol Alternative name for **copper (II) sulphate**.

BMR Abbreviation of **basal metabolic rate**.

body cavity In **triploblastic** animals, an internal cavity bounded by the body wall. It contains the **viscera**.

body-centred cube Crystal structure that is cubic, with an **atom** at the centre of each cube. Each atom is surrounded by eight others.

■ **Bohr theory** Atomic theory that assumes all **atoms** are made up of a central positively charged **nucleus** surrounded by orbiting planetary **electrons**, and which incorporates a **quantum theory** to limit the number of allowed **orbitals** in which the electrons can move. Each orbital has a characteristic energy level, and emission of electromagnetic radiation (*e.g.* light) occurs when an electron jumps to an orbital at a lower energy level (*see* **Balmer series**). It was named after the Danish physicist Niels Bohr (1885–1962). [4/8/d]

boiling point (b.p.) Temperature at which a liquid freely turns into a vapour; the **vapour pressure** of the liquid then equals the external pressure on it.

Boltzmann constant (k) Equal to R/L = 1.3806×10^{-23} J K^{-1} (joule per kelvin), where R = **gas constant** and L = **Avogadro constant**. It was named after the Austrian physicist Ludwig Boltzmann (1844–1906).

bomb calorimeter Apparatus for measuring the heat energy released during the combustion of a substance (such as a fuel).

■ **bond** Link between two **atoms** in a **molecule**. *See* **coordinate bond; covalent bond; ionic bond**. [4/10/e]

bond energy Energy involved in **bond** formation.

bond length Distance between the **nuclei** of two **atoms** that are chemically bonded.

■ **bone** Skeletal substance of **vertebrates**. It consists of **cells** (osteocytes) distributed in a matrix of **collagen** fibres impregnated with a complex salt (bone salt), mainly **calcium phosphate**, for hardness. Cells are connected by fine channels that permeate the matrix. Larger channels contain **blood vessels** and **nerves**. Some bones are hollow and contain **bone marrow**. There are about 206 bones in an adult human skeleton. [2/3/a]

bone marrow Soft tissue that fills the centre of some **bones**. It is responsible for the manufacture of the majority of **blood**

components, including **erythrocytes** (red blood cells) and **lymphocytes** (cells involved in the immune response).

bony fish Alternative name for **Osteichthyes.**

boracic acid Alternative name for **boric acid**.

borane Any of the boron hydrides, general formula B_nH_{n+2}, which have unusual chemical bonding (with too few electrons for normal covalent bonds).

borax $Na_2B_4O_7.10H_2O$ White amorphous compound, soluble in water, which occurs naturally as **tincal**. It is used in the manufacture of enamels and heat-resistant glass. Alternative names: disodium tetraborate, sodium borate.

boric acid H_3BO_3 White crystalline compound, soluble in water, which occurs naturally in volcanic regions of Italy. It has **antiseptic** properties. Alternative name: boracic acid.

boron B Amorphous, non-metallic element in Group IIIA of the Periodic Table. Because of its high neutron absorption it is used for **control rods** in nuclear reactors. Important compounds include **borax** and **boric acid**. At. no. 5; r.a.m. 10.81.

borosilicate glass Heat-resistant glass of low thermal expansion, made by adding boron oxide (B_2O_3) to **glass** during manufacture.

■ **Bosch process** Industrial process for manufacturing **hydrogen** by the catalytic **reduction** of steam with **carbon monoxide**. It is used to produce hydrogen for the **Haber process**. It was named after the German chemist Carl Bosch (1874–1940).

Bose-Einstein statistics Statistical mechanics of systems of identical particles that have their wave functions unaltered if any two particles are interchanged.

boson Subatomic particle, *e.g.* **alpha particle**, **photon**, that obeys **Bose-Einstein statistics** but does not obey the **Pauli exclusion principle**. Atomic **nuclei** of even mass numbers are also bosons.

■ **botany** Study of **plants** and plant life. [2/1/b]

botulism Severe food poisoning, an often fatal disorder caused by eating food contaminated with a **toxin** produced by the bacterium *Clostridium botulinum* (which is destroyed by adequate cooking). Botulinus toxin is one of the most potent poisons known.

boundary layer Disturbed region which forms around any surface moving through a fluid; the transitional zone throughout which the surface affects the motion of the fluid.

Bourdon gauge Instrument for measuring fluid **pressure** which consists of a coiled flattened tube that tends to straighten when pressure inside it increases. The movement of the end of the tube is made to work a pointer that moves around a scale. It was named after the French engineer Eugène Bourdon (1808–84).

■ **bowel** Alternative name for **intestine**. [2/4/a] [2/5/a] [2/7/a]

■ **Bowman's capsule** Dense ball of capillary blood vessels that cover the closed end of every **nephron** in the **kidney.** From it leads the uriniferous tubule. It was named after the British physician William Bowman (1816–92). [2/7/a]

Boyle's law At constant temperature the volume of a gas V is inversely proportional to its pressure p; *i.e.* pV = a constant. It was named after the Irish chemist Robert Boyle (1627–91).

Brachiopoda Phylum of marine shellfish that evolved before the **molluscs**. Many species are now extinct, but other brachiopods remain as some of the oldest 'living fossils'.

bract Leaf-like structure in whose axil a **flower** grows.

Bragg's law When **X-rays** are refracted by the layers of atoms in a **crystal**, the maximum intensity of the refracted ray occurs when $\sin\theta = n\lambda/2d$, where d = distance separating the layers, θ = complement of the **angle of incidence**, n = an integer and λ = **wavelength** of the X-rays. It was named after the British physicist William Henry Bragg (1862–1942).

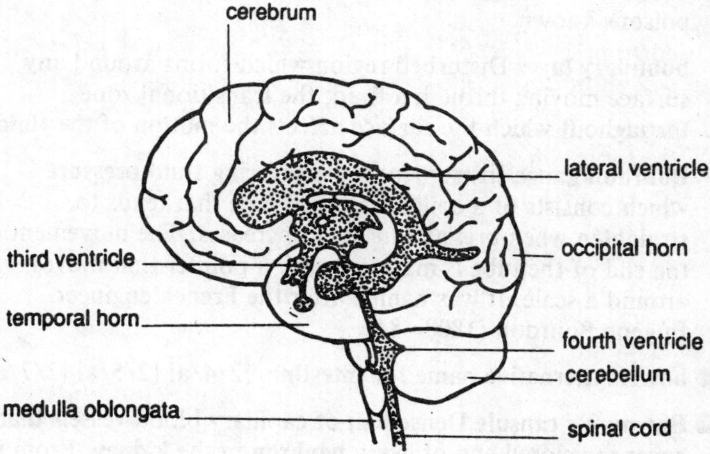

The human brain

■ **brain** Principal collection of **nerve cells** that form the anterior part of the central **nervous system**, consisting (in mammals) of a moist pinkish-grey mass protected by the bones of the cranium (skull). It receives, mostly via spinal nerves from the spinal cord but also via cranial nerves from organs in the head, sensory information through **afferent** sensory **neurones** carrying the impulses from sense organs, and it sends out instructions along **efferent** (motor) neurones to the effector organs (*e.g.* muscles). It is also the centre of intellect and memory so that behaviour can be based on past experience, and is responsible for the co-ordination of the whole body. *See also* **cerebellum**; **cerebrum**; **cortex**. [2/5/a]

brain stem Part of the brain at the end of the spinal cord, consisting of the midbrain, **medulla oblongata** and **pons**.

brainwave Pattern of electrical activity in the brain as revealed by an electroencephalograph. Alpha waves correspond to wakefulness with eyes closed, beta waves to wakefulness with eyes open and delta waves to deep sleep.

branched chain Side group(s) attached to the main chain in the molecule of an organic compound.

brass Alloy of **copper** and **zinc**.

breakdown diode Alternative name for **Zener diode**.

breastbone Alternative name for **sternum**.

■ **breathing** The physical actions of inhalation, gaseous exchange and exhalation. *See* **respiration**. [2/3/a]

breed Artificial subdivision of a species resulting from its domestication and selective breeding by human beings, to produce food animals (*e.g.* cattle, pigs, sheep), working animals (*e.g.* horses, dogs) or pets (*e.g.* cats).

■ **breeder reactor** **Nuclear reactor** that produces more fissile material (*e.g.* plutonium) than it consumes. [3/9/c]

Brewster's law The tangent of the angle of **polarization** of light by a substance is equal to the **refractive index** of the substance. It was named after the British physicist David Brewster (1781–1868).

brine Concentrated solution of common salt (**sodium chloride**).

British Thermal Unit (Btu) Amount of heat required to raise the temperature of 1 pound of water through 1 °F. 1 Btu = 1,055 joules.

Broca's area Speech centre of the **brain**. It was named after the French surgeon Paul Broca (1824–80).

bromide Binary compound containing **bromine**; a salt of hydrobromic acid.

bromide paper Photographic (light-sensitive) paper coated with an emulsion containing silver bromide, used for making black-and-white prints and enlargements.

■ **bromine** Br Dark red liquid non-metallic element in Group VIIA of the Periodic Table (the **halogens**), extracted from sea-water. It has a pungent smell. Its compounds are used in photography and as anti-knock additives to petrol. At. no. 35; r.a.m. 79.904. [4/8/a] [4/9/a]

bromine water Hypobromous acid (HBrO), made by dissolving **bromine** in water.

bromoform Alternative name for **tribromomethane**.

bronchiole Terminal air-conducting tube (1 mm in diameter) of mammalian **lungs**, arising from secondary subdivision of a **bronchus** and terminating in **alveoli**.

bronchitis Disorder caused by inflammation of the **bronchi** of the lungs resulting from infection or prolonged exposure to irritant chemicals such as tobacco smoke or pollutants. The airways to the lungs become narrowed and clogged with mucus, causing breathlessness, chest pain and a persistent cough.

bronchus One of the two air-carrying divisions of the **trachea** (windpipe) into the **lungs**. The bronchi become inflamed in the disorder **bronchitis**.

Brönsted-Lowry theory Concept of **acids** and **bases** in which an acid is defined as a substance with a tendency to lose a proton (H^+) and a base as a substance with a tendency to gain an electron. It was named after the British chemist Thomas Lowry (1874–1936) and the Danish chemist Johannes Brönsted (1879–1947), who proposed it independently. Alternative name: Lowry-Brönsted theory. *See also* **Lewis acid and base**.

bronze Alloy of **copper** and **tin**.

brown coal Alternative name for **lignite**.

Brownian movement Random motion of particles of a **solid** suspended in a **liquid** or **gas**, caused by collisions with molecules of the suspending medium. It was named after the British botanist Robert Brown (1773–1858). Alternative name: Brownian motion.

brown ring test Laboratory test for **nitrates** in solution. An acidic solution of iron (II) sulphate (ferrous sulphate) is added to the nitrate solution in a test-tube, and concentrated sulphuric acid carefully poured down the inside of the tube. A brown ring at the junction of the liquids indicates the presence of a nitrate.

Bryophyta Division of the plant kingdom that consists principally of the liverworts and mosses. In evolutionary terms, bryophytes are primitive plants which exhibit **alternation of generations**.

bryophyte Member of the plant division **Bryophyta**.

bubble chamber A container of pressurized liquid hydrogen at just above its normal boiling point, used to reveal the tracks of charged subatomic particles. As particles pass through the chamber, the pressure is suddenly lowered so that local boiling of the hydrogen produces a series of small bubbles along the particles' ionized tracks. *See also* **cloud chamber**.

bud *1*. In plants, an immature shoot that will bear leaves or flowers. *2*. In fungi and simple animals, a small outgrowth from a parent organism capable of detaching itself and living an independent existence.

■ **budding** *1*. Formation of **buds** from specialized cells on a plant shoot. *2*. **Grafting** of a bud onto a plant. *3*. Method of **asexual reproduction** employed by **yeasts** and simple animal organisms such as hydra; in plants it is termed gemmation. [2/7/a] [2/10/a]

buffer Solution that resists changes in **pH** on dilution or on the addition of **acid** or **alkali**.

bug Plant-sucking or blood-sucking insect of the order Hemiptera. The term is also used (incorrectly) for beetles, fleas and other insects.

bulb In botany, an underground perennating organ developed by some plants. It is usually formed from fleshy leaf bases, and has adventitious roots growing from a small piece of stem at its lower surface. Primarily a storage organ, it may also play a part in **vegetative propagation**.

bulk modulus Elastic modulus equal to the pressure on an object divided by its fractional decrease in volume. It is the ratio pV/v where p = intensity of stress, V = volume of the substance before pressure was applied, and v is the change in volume produced by the stress.

buna-S Alternative name for **styrene-butadiene rubber**.

Bunsen burner Gas burner that efficiently mixes air with the fuel gas, commonly used for heating in laboratories. It was named after the German chemist Robert Bunsen (1811–99).

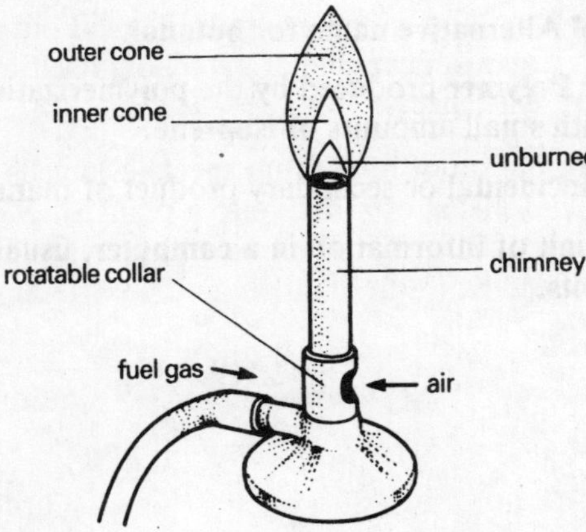

Bunsen burner

■ **buoyancy** Result of **upthrust** on an object floating or suspended in a fluid. *See* **Archimedes' principle**. [5/3/a]

burette Long vertical graduated glass tube with a tap, used for the addition of controlled and measurable volumes of liquids (*e.g.* in making **titrations** in **volumetric analysis**).

butadiene $CH_2=CHCH=CH_2$ **Hydrocarbon** gas with two double bonds, important in making **polymers** such as synthetic rubber.

butane C_4H_{10} Gaseous **alkane**, used as a portable supply of fuel.

butanedicarboxylic acid Alternative name for **adipic acid**.

butanedioic acid Alternative name for **succinic acid**.

butanol C_4H_9OH Colourless liquid **alcohol** that exists in four isomeric forms. Alternative name: butyl alcohol.

butyl alcohol Alternative name for **butanol**.

butyl rubber **Polymer** produced by the polymerization of isobutene with small amounts of **isoprene**.

by-product Incidental or secondary product of manufacture.

byte Single unit of information in a **computer**, usually a group of 8 **bits**.

C

cadmium Cd Silvery-white metallic element in Group IIB of the Periodic Table. It is used in **control rods** for nuclear reactors and as a corrosion-resistant electroplating on steel articles. Its compounds are used as yellow or red pigments. At. no. 48; r.a.m. 112.40.

caecum Pouch or pocket, such as one at the junction of the small and large intestines from which hangs the **appendix**.

caesium Cs Soft, reactive metallic element in Group IA of the Periodic Table (the **alkali metals**), a major fission product of uranium. It is used in photoelectric cells. The isotope caesium-137 is used in radiotherapy. At. no. 55; r.a.m. 132.905.

caesium clock **Atomic clock** used in the **SI unit** definition of the second.

■ **caffeine** White alkaloid with a bitter taste, obtained from coffee beans, tea leaves, kola nuts or by chemical synthesis. It is a diuretic and a stimulant to the central nervous system.

calamine Zinc ore whose main constituent is zinc oxide.

calcaneus Heel-bone; the major bone of the foot. Alternative name: calcaneum.

calciferol Fat-soluble **vitamin** formed in the skin by the action of sunlight, which controls levels of calcium in the blood. Alternative name: vitamin D.

■ **calcification** Depositing of calcium salts in body tissue, a normal process in the formation of **bones and teeth** but abnormal in the formation of **calculi** ('stones'). *See also* **cartilage**.

calcite Crystalline form of natural **calcium carbonate**. It is a major constituent of limestones and marbles, and as such is one of the most common minerals.

calcitonin Hormone secreted in vertebrates that controls the release of calcium from bone. In mammals it is secreted by the **thyroid gland**. Alternative name: thyrocalcitonin.

■ **calcium** Ca Silver-white metallic element in Group IIA of the Periodic Table (the **alkaline earths**). The fifth most abundant element on Earth, it occurs mainly in **calcium carbonate** minerals; it also occurs in bones and teeth. At. no. 20; r.a.m. 40.08. [4/9/a]

calcium carbide CaC_2 Compound of carbon and calcium made commercially by heating coal and lime to a high temperature. It reacts with water to produce **acetylene**, and was once used for this purpose in acetylene lamps. Alternative names: calcium acetylide, calcium ethynide, carbide.

■ **calcium carbonate** $CaCO_3$ White powder or colourless crystals, the main constituent of chalk, limestone and marble.

calcium chloride $CaCl_2$ White crystalline compound, which forms several **hydrates**, used to control dust, as a de-icing agent and as a refrigerant. The anhydrous salt is **deliquescent** and is employed as a **desiccant**.

calcium hydrogencarbonate $CaHCO_3$ White crystalline compound, stable only in solution and the cause of temporary **hardness of water**. Alternative name: calcium bicarbonate.

calcium hydroxide $Ca(OH)_2$ White crystalline powder which gives an alkaline aqueous solution known as limewater, used as a test for carbon dioxide (which turns it cloudy).

Alternative names: calcium hydrate, hydrated lime, caustic lime, slaked lime.

calcium oxide CaO White crystalline powder made commercially by roasting **limestone** (calcium carbonate). It is used to make **calcium hydroxide** (slaked lime), for treating acid soils, for making mortar and in smelting iron and other metals (to help to form **slag**). Alternative names: lime, quicklime.

calcium phosphate $Ca_3(PO_4)_2$ White crystalline solid which makes up the mineral component of bones and teeth, and occurs as the mineral **apatite**. It is produced commercially as bone ash and basic slag. Treated with sulphuric acid it forms the fertilizer known as superphosphate.

calcium silicate Ca_2SiO_4 White insoluble crystalline compound, present in various minerals and cements, and a component in the slag produced in a **blast furnace**.

calcium sulphate $CaSO_4$ White crystalline compound which occurs as the minerals anhydrite and (as the dihydrate) gypsum, used to make **plaster of Paris**. It is the cause of permanent **hardness of water**. It is used in making ceramics, paint, paper and sulphuric acid, and is the substance in blackboard 'chalk'.

calculus Abnormal hard accretion (a 'stone') of calcium salts and other compounds that may form in the kidneys and urinary tract, gall bladder, bile ducts or salivary glands.

calibration *1.* Measuring scale on a scientific instrument or apparatus. *2.* Method of putting a scale on a scientific instrument, usually by checking it against fixed quantities or standards.

californium Cf Radioactive element in Group IIIB of the

Periodic Table (one of the **actinides**), produced by alpha-particle bombardment of curium-242; it has several **isotopes**, with half-lives of up to 800 years. At. no. 98; r.a.m. 251 (most stable isotope).

calliper Measuring instrument that resembles a large pair of geometrical dividers; there are internal and external versions.

calomel Alternative name for **mercury (I) chloride**.

calomel half-cell Reference **electrode** of known potential consisting of mercury, mercury (I) chloride and potassium chloride solution. Alternative names: calomel electrode, calomel reference electrode.

Calor gas Portable fuel gas consisting mainly of **butane** and **propane**, stored under pressure in metal bottles.

Calorie (with a capital *C*) Amount of heat required to raise the temperature of 1kg of water by 1 °C at one atmosphere pressure; equal to 4.2 kilojoules. It is used as a unit of energy of food (when it is sometimes spelled with a small *c*). Alternative name: kilocalorie, large calorie. *See also* **calorie**.

calorie (with a small *c*) Amount of heat required to raise the temperature of 1g of water by 1 °C at one atmosphere pressure; equal to 4.184 joules. *See also* **Calorie**.

calorific value Quantity of heat liberated on the complete combustion of a unit weight or unit volume of fuel.

calorimeter Apparatus for measuring heat quantities generated in or evolved by materials in processes such as chemical reactions, changes of state and solvation.

calyx Outermost **whorl** of a flower composed of leaf-like **sepals**.

cambium Cellular tissue in which **secondary growth** occurs in the **stem** and **root** of a plant.

camphor Naturally-occurring organic compound with a penetrating aromatic odour. Alternative names: 2-camphanone, gum camphor.

■ **cancer** Disorder that results when body cells undergo unrestrained division (because of a breakdown in the normal control of cellular processes) to form a malignant **tumour**. Cancer may spread by **metastasis**, when tumour cells spread from their original position and invade other tissues. *See also* **carcinogen**.

candela Unit of luminous intensity.

candle power (cp) Luminous intensity expressed in candles.

cane-sugar Alternative name for **sucrose**.

canine tooth Pointed tooth in mammals (except most rodents and ungulates – *i.e.* herbivores), adapted for stabbing and tearing food. There are two canines in each jaw. Alternative name: eye-tooth.

Cannizzaro reaction In organic chemistry, the formation of an **alcohol** and an acid salt by the reaction between certain **aldehydes** and strong alkalis. It was named after the Italian chemist Stanislao Cannizzaro (1826–1910).

canonical form *See* **mesomerism**.

■ **capacitance** (*C*) Ratio of the charge on one of the conductors of a **capacitor** (there being an equal and an opposite charge on the other conductor) to the potential difference between them. The SI unit of capacitance is the farad. [5/9/b]

■ **capacitor** Device that can store charge and introduces **capacitance** into an electrical circuit. Alternative names: condenser, electrical condenser. [5/9/b]

capillarity In physics, a phenomenon – resulting from **surface tension** – that causes low-density liquids to flow along narrow (capillary) tubes or soak into porous materials (bécause of the cohesive nature of water molecules). Examples in biology include the movement of water between soil particles or up **xylem** vessels in a plant. Alternative name: capillary action.

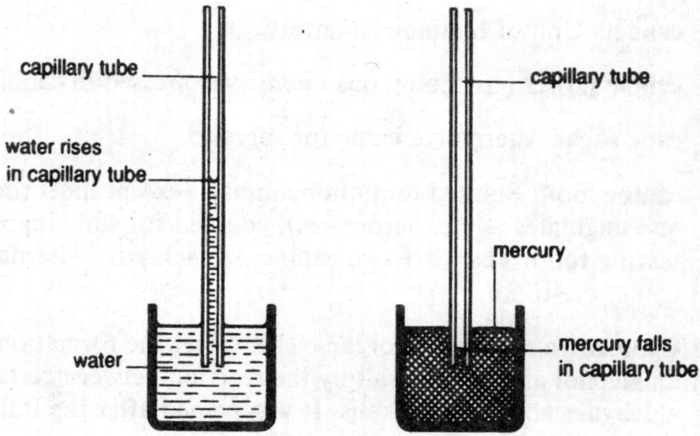

Capillarity causes liquids to rise or fall

■ **capillary** *1.* Any narrow tube, in which **capillarity** can occur. *2.* In biology, finest vessel of the **blood vascular system** in vertebrates. Large numbers of capillaries are present in **tissues**. Their walls are composed of a single layer of cells through which exchange of substances, *e.g.* **oxygen**, occurs between the tissues and **blood**. [2/5/a]

capillary action Alternative term for **capillarity**.

capillary tube Very fine tube up which water or other fluid rises by **capillarity**.

caprylic acid Alternative name for **octanoic acid**.

capsule *1*. In certain **bacteria**, a gelatinous extracellular envelope, which in some cases protects the cell. *2*. In mosses, structure in which spores are formed. *3*. In flowering plants, dry fruit that may liberate seeds in various ways (*e.g.* poppy). *4*. In anatomy, sheath of membrane that surrounds an organ or area of tissue; *e.g.* the synovial capsule that surrounds the moving parts of joints.

■ **carapace** Hard shield that covers the upper (dorsal) part of some crustaceans and chelonians (tortoises and turtles). In crustaceans, it is made of chitin and is part of the **exoskeleton**. In chelonians, it consists of bony plates which are fused together. [2/2/b] [2/3/b]

carbamide Alternative name for **urea**.

carbanion Transient negatively charged organic ion that has one more electron than the corresponding **free radical**.

carbene Organic radical that contains divalent carbon.

carbide *1*. Chemical compound consisting of carbon and another element; an **acetylide**. *2*. Alternative name for **calcium carbide**.

■ **carbohydrate** Compound of carbon, hydrogen and oxygen that contains a **saccharose** group or its first reaction product, and in which the ratio of hydrogen to oxygen is 2:1 (the same as in water). **Cellulose, starch** and all **sugars** are common carbohydrates. Digestible carbohydrates in the diet are a good source of energy.

carbolic acid Alternative name for the antiseptic **phenol**.

■ **carbon** C Non-metallic element in Group IVA of the Periodic Table which exists as several **allotropes** (including **diamond** and **graphite**). It occurs in all living things and its compounds are the basis of **organic chemistry**. It is the principal element in coal and petroleum. Its non-organic compounds include the oxides **carbon monoxide** (CO) and **carbon dioxide** (CO_2), **carbides** and **carbonates**. Carbon is used for making electrodes, brushes for electric motors, carbon fibres and in steel. Diamonds are used as gemstones and industrially as abrasives. Its isotope carbon-14 is the basis of **radiocarbon dating** (*see* **beta decay**). At. no. 6; r.a.m. 12.001. [4/8/d] [4/8/e]

carbonate Salt of **carbonic acid** (H_2CO_3), containing the ion CO_3^{2-} Carbonates commonly occur as minerals (*e.g.* **calcium carbonate**) and are readily decomposed by acids to produce carbon dioxide. *See also* **hydrogencarbonate**

carbonation Addition of **carbon dioxide** under pressure to a liquid. Carbonation is used in making fizzy drinks, and bottled and canned beer. Carbonated water is known as soda water.

carbon black Finely divided form of **carbon** obtained by the incomplete combustion or thermal decomposition of natural gas or petroleum oil.

■ **carbon cycle** Passage of **carbon** from the air (as **carbon dioxide**) to plants by photosynthesis (forming **sugars** and **starches**), then through the metabolism of animals, to decomposition products which ultimately return to the atmosphere in the form of carbon dioxide. [2/7/a] [2/8/d]

■ **carbon dating** *See* **radiocarbon dating**. [4/7/f]

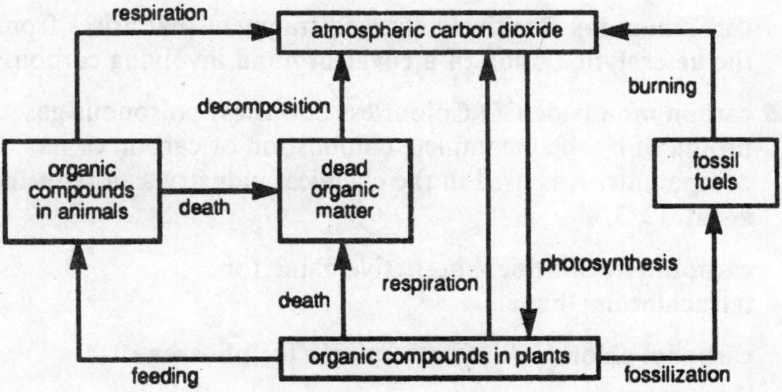

Carbon cycle

■ **carbon dioxide** CO_2 Colourless gas formed by the combustion of carbon and its organic compounds, by the action of acids on **carbonates**, and as a product of **fermentation** and **respiration**. It is a raw material of photosynthesis. It is used in fire extinguishers, in fizzy (carbonated) drinks, as a coolant in nuclear reactors and, as solid carbon dioxide (dry ice), as a refrigerant. The accumulation of carbon dioxide in the atmosphere creates the **greenhouse effect**. [2/8/d] [2/9/d]

carbon disulphide CS_2 Liquid chemical, used as a solvent for oils, fats and rubber and in paint-removers.

carbon fibre Synthetic carbon produced by charring any spun, felted or woven carbon-containing raw material at temperatures from 700 to 1,800 °C. It is used as reinforcement in plastic resins to make high-strength composites.

carbonic acid H_2CO_3 Acid formed by the combination of **carbon dioxide** and water. Its salts are **carbonates**.

carbonium ion Positively charged fragment that arises from the **heterolytic fission** of a **covalent bond** involving carbon.

■ **carbon monoxide** CO Colourless odourless poisonous gas produced by the incomplete combustion of carbon or its compounds. It is used in the chemical industry as a **reducing agent**. [2/3/c]

carbon tetrachloride Alternative name for **tetrachloromethane**.

carbonyl chloride Alternative name for **phosgene**.

carbonyl compound Chemical containing the radical =CO, formed when **carbon monoxide** combines with a metal (*e.g.* nickel carbonyl, $Ni(CO)_4$).

carbonyl group The group =CO, as in aldehydes, ketones and carboxyl compounds.

carborundum Form of the hard substance silicon carbide (SiC), used as an abrasive.

carboxyl group The organic group –COOH, characteristic of carboxylic acids. Alternative names: carboxy group, oxatyl group.

carboxylic acid Organic acid that contains the carboxyl group, –COOH; *e.g.* acetic (ethanoic) acid, CH_3COOH.

carburettor Part of an internal combustion engine in which air and liquid hydrocarbon fuel are mixed and vaporized.

carbylamine Alternative name for **isocyanide**.

■ **carcinogen** Substance capable of inducing a **cancer**. Most carcinogens are also **mutagens**, and this is thought to be their principal mode of action. [2/9/c]

carcinoma **Cancer** in epithelial tissue.

■ **cardiac** Relating to the **heart**. *See also* **cardiac muscle**. [2/4/a] [2/5/a] 2/7/a]

■ **cardiac muscle** Specialized **striated muscle**, found only in the heart, that continually contracts rhythmically and automatically without nervous stimulation (*i.e.* it is myogenic). [2/4/a] [2/5/a] 2/7/a]

cardiogram Alternative name for an **electrocardiogram**.

cardiovascular system **Heart** and the network of blood vessels that circulate **blood** around the body.

card punch Machine for punching coded sets of holes in punched cards, to be fed through a **card reader** for inputting **data** into a computer.

card reader Computer **input device** that reads **data** off punched cards.

carnallite Mineral consisting of a hydrated mixture of potassium and magnesium chlorides, mined as a source of potassium and magnesium chloride.

carnassial tooth Large **pre-molar** or **molar** tooth, found in **carnivores**, that is adapted for slicing food.

■ **Carnivora** Order of meat-eating mammals, characterized by the possession of **carnassial teeth** and large **canine teeth**, which includes badgers, bears, cats, dogs, weasels and wolves. [2/3/b] [2/4/b]

carnivore Meat-eating animal or a member of the **Carnivora**. *See also* **herbivore; omnivore.**

Carnot cycle Reversible cycle of four operations that occur in a perfect heat engine. These operations are isothermal expansion (*see* **isothermal process**), adiabatic expansion (*see* **adiabatic process**), isothermal compression and adiabatic compression. Carnot's principle states that the efficiency of a perfect heat engine depends on the temperature range in which it works, and not on the properties of the substances with which it works. It was named after the French physicist Nicolas Carnot (1796–1832).

Caro's acid An alternative name for peroxomonosulphuric (VI) acid, H_2SO_5 (persulphuric acid).

carotene Pigment that belongs to the **carotenoids**, a precursor of **retinol** (vitamin A).

carotenoid Member of a group of pigments found in *e.g.* carrots, which (like chlorophyll) can absorb light during **photosynthesis**. *See also* **carotene**.

carotid artery In vertebrates, either of two main **arteries** that carry blood from the heart to the head.

carotid body One of a pair of receptors located between the internal and external **carotid arteries**. It is sensitive to changes in the **carbon dioxide** and **oxygen** content of the blood, and sends impulses to the respiratory and blood-vascular centres in the brain to adjust these levels if necessary.

carpal Bone in the foot and wrist of **tetrapods**. There are 10–12 carpals in most animals; in human beings there are eight in each wrist.

■ **carpel** Female reproductive organ of a flowering plant, consisting of the **stigma, style** and **ovary**. Alternative name: gynaecium. [2/4/a] [2/7/a]

carpus The wrist, consisting of eight bones in human beings.

carrier *1.* Somebody who carries **pathogens** that cause disease, or carries genes that cause inherited disorders (*e.g.* haemophilia, colour blindness) and can pass them on to others, although not necessarily themselves having symptoms of the disorder. *2.* **Vector** (in medicine). *See also* **coenzyme; sex linkage.**

carrier wave In radio, the continuously transmitted radio wave whose amplitude or frequency is modulated by a signal representing the sound being broadcast. *See also* **amplitude modulation; frequency modulation.**

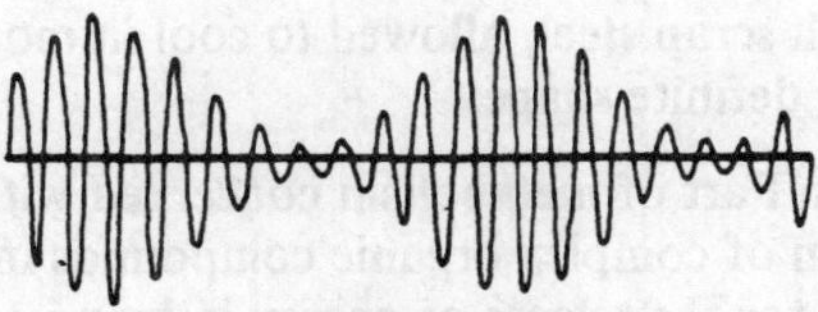

amplitude modulation

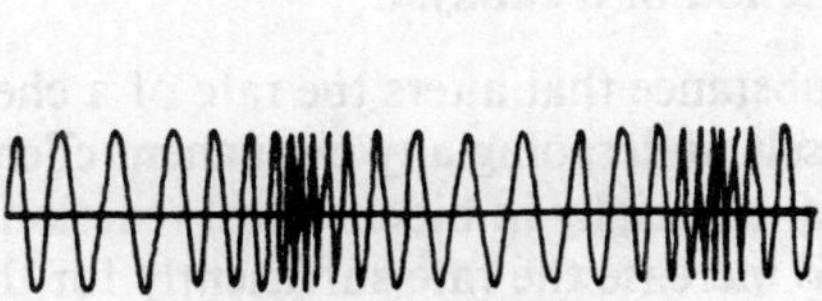

frequency modulation

A radio carrier wave is modulated

■ **cartilage** In animals that have a bony **skeleton**, pre-bone tissue occurring in young animals that becomes hard through **calcification**. It also occurs in some parts of the skeleton of adults, where it is structural (*e.g.* forming ear flaps and the larynx) or provides a cushioning effect during movement because it has a smoother surface and is softer than bone (*e.g.* intervertebral discs, cartilage at the ends of bones in joints). In some animals, *e.g.* sharks, rays and related fish (Chondrichthyes), the whole skeleton is composed of cartilage, even in the adult. [2/3/a] [2/5/a]

cartilaginous fish Fish that belongs to the class **Elasmobranchii** (Chondrichthyes); *e.g.* **sharks** and dogfish.

casein Protein that occurs in milk which serves to store **amino acids** as nutrients for the young of mammals. It is the principal constituent of cheese.

cast iron Hard brittle metal made from remelted **pig iron** mixed with scrap steel, allowed to cool in moulds so that it assumes a definite shape.

catabolism Part of **metabolism** concerned with the breakdown of complex organic compounds into simple molecules for the release of energy in living organisms. *See also* **anabolism**.

catalysis Action of a **catalyst**.

■ **catalyst** Substance that alters the rate of a chemical reaction without itself undergoing any permanent chemical change. Sometimes, especially in biochemical reactions, a catalyst is essential to increase the rate sufficiently for the reaction to be detectable. *See also* **enzyme**.

■ **catalytic converter** Device in the exhaust system of a vehicle with an internal combustion engine (petrol or diesel), designed to remove gases that contribute to atmospheric pollution. Using platinum or similar catalysts, it converts carbon monoxide to carbon dioxide, nitrogen oxides to nitrogen and unburned hydrocarbon fuel to water and carbon dioxide. [2/3/c]

catalytic cracking Process of breaking down long-chain **hydrocarbons** into more useful shorter-chain ones, employing a **catalyst**. It is used in an oil refinery to make fuels (*e.g.* petrol, kerosene) from **alkanes** of higher molecular weight.

catecholamine Member of a group of **amines** that include **neurotransmitters** (*e.g.* dopamine) and **hormones** (*e.g.* adrenalin).

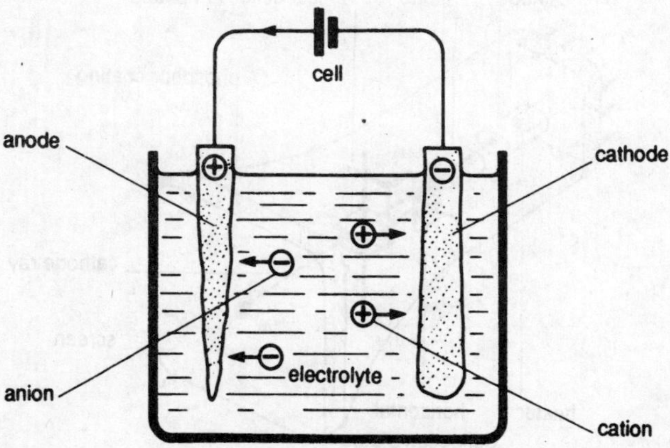

Cathode is the negative electrode

cathode Negatively-charged electrode in an **electrolytic cell** or **battery**.

cathode-ray oscilloscope (CRO) Apparatus based on a **cathode-ray tube** which provides a visible image of one or more rapidly varying electrical signals.

cathode rays Stream of **electrons** produced by the **cathode** (negative electrode) of an evacuated **discharge tube** (such as a **cathode-ray tube**).

cathode-ray tube Vacuum tube that allows the direct observation of the behaviour of cathode rays. It is used as the picture tube in television receivers and in cathode-ray oscilloscopes.

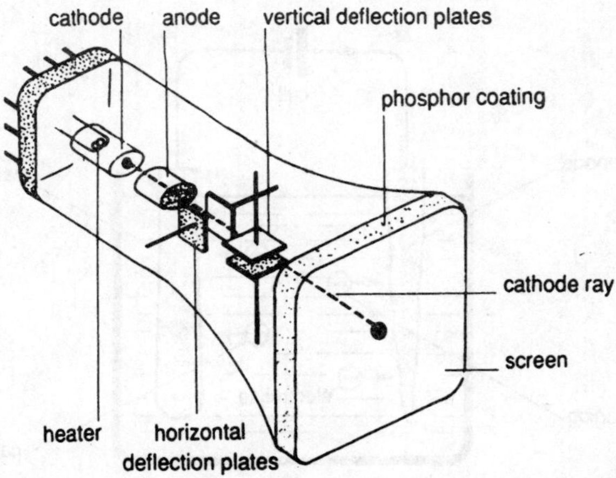

Cathode-ray tube

cation Positively-charged **ion**, which travels towards the **cathode** during **electrolysis**.

CAT scan Abbreviation of computerized axial tomography scan, a technique for producing X-ray pictures of cross-sectional 'slices' through the brain and other parts of the body.

cauda equina Sheaf of nerve roots that arise from the lower end of the spinal cord and serve the lower parts of the body.

caudal Relating to the tail.

caudal vertebra One of the set of **vertebrae** nearest to the base of the **spinal column**.

caustic Describing an **alkaline** substance that is corrosive towards organic matter.

Symbol on containers of caustic chemicals

caustic lime Alternative name for **calcium hydroxide**.

caustic potash Alternative name for **potassium hydroxide**.

caustic soda Alternative name for **sodium hydroxide**.

■ **cell** *1*. Fundamental unit of living organisms. It consists of a membrane-bound compartment, often microscopic in size, containing jelly-like protoplasm. In plants each cell is bounded by a wall composed of **cellulose**, with the cell membrane pressed closely to it because of the **turgor** of the cell. In animals each cell is bounded by a cell membrane only. In all cells the **nucleus** contains **nucleic acids** (*e.g.* DNA) essential for the synthesis of new **proteins**. Cells may be highly specialized for their particular function and grouped into **tissues**, *e.g.* muscle cells. *2*. Alternative name for an **electrolytic cell**.

cell body Part of a nerve cell that contains the cell **nucleus** and other cell components from which an axon extends.

cell differentiation Way in which previously undifferentiated **cells** change structurally and take on specialized roles during growth and development (*e.g.* becoming liver cells or bone cells).

■ **cell division** Splitting of cells in two (**mitosis**) by division of the **nucleus** and division of the **cytoplasm** after duplication of the cell contents. *See also* **meiosis**. [2/5/b]

cell membrane *See* **plasma membrane**.

cellulose Major **polysaccharide** (*i.e.* a carbohydrate) of plants found in cell walls and in some **algae** and **fungi**. It is composed of **glucose** units aligned in long parallel chains, and gives cell walls their strength and rigidity.

cellulose trinitrate Nitric acid **ester** of **cellulose**, used in **plastics**, lacquers and explosives. Alternative names: cellulose nitrate, nitrocellulose.

cell wall Wall that surrounds plant cells, consisting mainly of **cellulose**. A membrane encloses the cell's **protoplasm**, and the wall is secreted through the membrane. It defines how large the cell can get, gives the cell support and rigidity, and is important for regulating the water content and fabric of a plant's vascular system. *See also* **lignin**.

Celsius scale Temperature scale on which the **freezing point** of pure water is 0 °C and the **boiling point** is 100 °C. A degree Celsius is equal to a unit on the **kelvin** scale. To convert a Celsius temperature to kelvin, add 273.16 (and omit the degree sign). To convert a Celsius temperature to a Fahrenheit one, multiply by 9/5 and add 32. It is named after the Swedish astronomer Anders Celsius (1701–44).

cement Any bonding material. The cement used in building to make mortar and concrete consists of powdered roasted rock (*e.g.* Portland stone), which reforms a solid after absorbing water.

centigrade scale Former name for the **Celsius scale.**

central nervous system (CNS) Concentration of nervous tissue responsible for co-ordination of the body. In **vertebrates** it is highly developed to form the **brain** and **spinal cord**. The CNS processes information from the sense organs and effects a response, *e.g.* muscle movement.

central processor Heart of a **digital computer** which controls and co-ordinates all the other activities of the machine, and performs logical processes on **data** loaded into it according to **program** instructions it holds.

centre of curvature Geometric centre of a spherical mirror.

centre of mass Point at which the whole **mass** of an object may be considered to be concentrated. Alternative name: centre of gravity.

■ **centrifugal force** Apparent (but not real) outward force on any object moving in a circular path. *See also* **centripetal force**.

centrifuge Apparatus used for the separation of substances by **sedimentation** through rotation at high speeds, *e.g.* the separation of components of **cells**. The rate of sedimentation varies according to the size of the component.

centriole Cylindrical body present in most animal **cells**. During **mitosis** it forms the poles of the **spindle**.

■ **centripetal force** Force, directed towards the centre, that causes a body to move in a circular path. For an object of mass *m* moving with a speed *v* in a curve of radius of curvature *r*, it is equal to mv^2/r.

centromere Region on a **chromosome** that attaches to the **spindle** during **cell division**.

Cephalochordata Subphylum of the **Chordata** that consists of organisms with no vertebral column (*e.g.* amphioxus), although they do have a **notochord**.

cephalochordate Animal of the subphylum **Cephalochordata**.

cephalopod Member of the class **Cephalopoda**.

Cephalopoda Class of marine **molluscs** that includes cuttlefish, nautiloids, octopuses and squids (and extinct ammonites). Cephalopods have tentacles, highly developed eyes and nervous systems, and can move by forcing sea-water at high speed through their siphons.

cephalothorax Fused head and thorax (*e.g.* as in **arachnids**). Alternative name: prosoma.

■ **ceramics** Hard strong materials, produced by firing mixtures containing clay, which have important industrial applications, such as pottery and bricks. Glass and fused **silica** are sometimes also regarded as ceramics.

cerebellum Front part of hindbrain of **vertebrates**. It is important in balance and muscular coordination.

cerebral cortex Outer region of the **cerebral hemispheres** of the **brain** that contains densely packed **nerve cells** which are interconnected in a complex manner.

cerebral hemisphere In **vertebrates**, paired expansions of the anterior end of the forebrain. In mammals the forebrain is enormously enlarged by these expansions.

cerebrospinal fluid (CSF) Liquid that fills the cavities of the **brain**, which are continuous with each other and the central canal of the **spinal cord**. The fluid nourishes the tissues of the **central nervous system** (CNS) and removes its secretions.

cerebrum Part of forebrain which in higher **vertebrates** expands to form the **cerebral hemispheres**.

Cerenkov counter Method of measuring radioactivity by utilizing the effect of **Cerenkov radiation**.

Cerenkov radiation Light emitted when charged particles pass through a transparent medium at a velocity which is greater than that of light in the medium. It was named after the Soviet physicist Pavel Cerenkov (1904–).

cerium Ce Steel-grey metallic element in Group IIIB of the Periodic Table (one of the **lanthanides**). It is used in lighter flints, tracer bullets, catalytic converters for car exhausts and gas mantles. At. no. 58; r.a.m. 140.12.

cerumen Wax that forms in the **ear**.

cervical Relating to the neck or to the **cervix** of the womb.

cervical vertebra **Vertebra** in the neck region of the **spinal column**, concerned with movements of the head.

cervix Neck, usually referring to the neck of the uterus (womb) at the inner end of the vagina.

■ **CFC** Abbreviation of **chlorofluorocarbon**. [2/9/d]

c.g.s. unit Centimetre-gram-second unit, in general scientific use before the adoption of SI units.

Chaetognatha Phylum of invertebrate aquatic animals; the arrow-worms.

chain reaction Nuclear or chemical reaction in which the products ensure that the reaction continues (*e.g.* nuclear fission, combustion).

■ **chalk** Mineral form of **calcium carbonate**. Blackboard chalk is usually made from **calcium sulphate**.

chamber process Alternative name for **lead-chamber process**.

character In genetics, variation caused by a gene; an inherited trait. Alternative name: characteristic.

characteristic Typical feature (*e.g.* as in characteristics of living organisms). Alternative name: character.

charcoal Form of carbon made from incomplete burning of animal or vegetable matter.

■ **charge** *See* **electric charge**. [5/6/b] [5/9/b]

Charles's law At a given pressure, the volume of an ideal gas is directly proportional to its absolute temperature. It was named after the French physicist Jacques Charles (1746–1823).

chelate Chemical compound in which a central metal ion forms part of one or more organic rings of atoms. The formation of these compounds is useful in many contexts, *e.g.* medicine, in which chelating agents are administered to counteract poisoning by certain heavy metals. They can also act to buffer the concentration of metal ions (*e.g.* iron and calcium) in natural biological systems.

■ **chemical bond** Linkage between atoms or ions within a molecule. A chemical reaction and the input or output of energy are involved in the formation or destruction of a chemical bond. *See also* **covalent bond; ionic bond**. [4/8/d]

chemical combination Union of two or more chemical substances to form a different substance or substances.

chemical dating Determination of the age of an archaeological specimen by chemical analysis. *See also* **radiocarbon dating**.

chemical engineering Manufacture and operation of machines and plant necessary for industrial-scale chemical processing, often to produce other chemicals.

■ **chemical equation** Way of expressing a **chemical reaction** by placing the **formulas** of the **reactants** to the left and those of the **products** to the right, with an equality sign or directional arrows in between. The number of atoms of any particular element are the same on each side of the equation (when the reaction is in equilibrium). *E.g.* the equation for the reaction between two molecules of hydrogen and one molecule of oxygen to form two molecules of water is written as:

$$2H_2 + O_2 \rightarrow 2H_2O.$$ [4/9/b]

■ **chemical equilibrium** Balanced state of a **chemical reaction**, when the concentration of **reactants** and **products** remain constant. *See also* **equilibrium constant**. [4/7/c]

chemical potential Measure of the tendency of a **chemical reaction** to take place.

chemical reaction Process in which one or more substances react to form a different substance or substances.

chemical symbol Letter or pair of letters that stand for an **element** in chemical formulae and equations. *E.g.* the symbols of carbon, chlorine and gold are C, Cl and Au (from the Latin *aurum* = gold).

chemistry The study of elements and their compounds, particularly how they behave in **chemical reactions**.

chemoreceptor Cell that fires a nerve impulse in response to stimulation by a specific type of chemical substance; *e.g.* the taste buds on the tongue and olfactory bulbs in the nose contain chemoreceptors that provide the senses of taste and smell.

chemotaxis Response of organisms to chemical stimuli, *e.g.* the movement of **protozoa** towards nutrients.

chemotherapy Treatment of a disorder with drugs that are designed to destroy **pathogens** or cancerous tissue.

chemotropism Response of plants to chemical stimuli, *e.g.* the growth of the **pollen tube** towards the **ovary** in flowering plants.

chiasma *1.* Point along the **chromatid** of a homologous **chromosome** at which connections occur during crossing-over or exchange of genetic material in **meiosis**. *2.* Crossing-over point of the **optic nerves** in the brain.

Chile saltpetre Old name for impure **sodium nitrate**.

chimaera *1.* Genetic mosaic or organism composed of genetically different **tissues** arising *e.g.* from **mutation** or mixing of cell types of different organisms. It can be achieved by incorporating donor cells at an early stage of embryonic development of the recipient. Alternative name: chimera. *2.* Member of a group of deep-sea cartilaginous fish. Alternative name: ratfish.

chimera Alternative name for **chimaera** (sense *1.*).

chirality Property of a **molecule** that has a carbon atom attached to four different atoms or groups, and which can therefore exist as a pair of optically active **stereoisomers** (whose molecules are mirror images of each other). Commonly called 'handedness', chirality is significant in the biological activity of molecules, in some of which the right-handed version is active and the left-handed version is not, or vice versa (*e.g.* many **pheromones**).

chitin Polysaccharide (*i.e.* a **carbohydrate**) found in the **exoskeleton** of **arthropods**, giving it hard waxy properties.

chloral Alternative name for **trichloroethanal**.

chloral hydrate Alternative name for **trichloroethanediol**.

chlorate Salt of chloric acid ($HClO_3$).

chloride Binary compound containing **chlorine**; a salt of **hydrochloric acid** (HCl).

chlorination *1.* Reaction between **chlorine** and an organic compound to form the corresponding chlorinated compound. *E.g.* the chlorination of benzene produces chlorobenzene, C_6H_5Cl. *2.* Treatment of a substance with **chlorine**; *e.g.* to bleach or disinfect it.

■ **chlorine** Cl Gaseous non-metallic element in Group VIIA of the Periodic Table (the **halogens**), obtained by the electrolysis of sodium chloride (common salt). It is a green-yellow poisonous gas with an irritating smell, used as a disinfectant and bleach and to make chlorine-containing organic chemicals. At. no. 17; r.a.m. 35.453. [4/8/a] [4/9/a]

chlorine (I) oxide Alternative name for **dichlorine oxide**.

chlorite Salt of chlorous acid ($HClO_2$).

■ **chlorofluorocarbon** (CFC) **Fluorocarbon** that has chlorine atoms in place of some of the fluorine atoms. Chlorofluorocarbons and fluorocarbons have similar properties, and are used as aerosol propellants and refrigerants (although blamed for damage to the **ozone layer**. [2/9/d]

chloroform Alternative name for **trichloromethane**.

■ **chlorophyll** Green pigment found in photosynthetic cells (contained in chloroplasts) of green plants. It is the major light-absorbing pigment and the site of the first stage of photosynthesis. [2/3/d]

Chlorophyta Largest division of **algae**, consisting of the green algae (which possess chlorophyll). Unicellular or multicellular, freshwater and marine-living, they exhibit **sexual** and **asexual reproduction**.

■ **chloroplast** Organelle found in photosynthetic cells of plants, containing the green pigment **chlorophyll**. Subsequent production of **carbohydrates** from **photosynthesis** occurs in chloroplasts. [2/3/d]

chloroplatinic acid Solution of platinum chloride used in platinizing glass and ceramics.

chloroprene $CH_2 = CCLCH = CH_2$ Colourless liquid organic chemical used (through **polymerization**) to make artificial rubber. Alternative name: 2-chlorobuta-1,3-diene.

■ **cholesterol Sterol** found in animal tissues, a **lipid**-like substance that occurs in blood plasma, cell membranes, nerves and may form gallstones. High levels of cholesterol in the blood are connected to the onset of atherosclerosis, in which fatty materials are deposited in patches on artery walls and can restrict blood flow. Many **steroids** are derived from cholesterol.

choline $HOC_2H_4N(CH_3)_3OH$ Organic compound that is a constituent of the neurotransmitter **acetylcholine** and some **fats**. It is one of the B vitamins.

Chondrichthyes Alternative name for the fish class **Elasmobranchii.**

chondriosome Alternative name for **mitochondrion**.

■ **Chordata** Phylum of the animal kingdom that contains organisms with a **notochord, gill** slits and a hollow dorsal **nerve cord** at some stage of their development. Vertebrates are chordates, and thus the phylum includes mammals. [2/3/b] [2/4/b]

chordate Member of the phylum **Chordata**.

chorion *1*. Membrane that surrounds an implanted **blastocyst** and the embryo and foetus that develop from it in mammals, and the embryo in the eggs of reptiles and birds. *2*. Shell of an insect's egg.

chorionic gonadotrophin Hormone produced by the **placenta** during **pregnancy**. Its presence in the urine is the basis of many kinds of pregnancy testing.

chorionic villus sampling Testing during early pregnancy of small samples of tissue taken from the **chorion** for the presence of foetal abnormalities. *See also* **amniocentesis**.

choroid Layer of pigmented cells, rich in blood vessels, between the **retina** and **sclerotic** of the **eye**.

chromatic aberration *See* **aberration**.

■ **chromatid** One of two thread-like parts of a **chromosome**, visible during **prophase** of **meiosis** or **mitosis**, when the chromosome has duplicated. Chromatids are separated during **anaphase**. [2/8/b]

chromatin Basic **protein** that is associated with **eukaryotic chromosomes**, visible during certain stages of cell duplication.

■ **chromatography** Method of separating a mixture by carrying it in solution or in a gas stream through an absorbent. The separated substances may be extracted by **elution**. [4/5/b]

chromatophore *1*. **Chromoplast**. *2*. In some vertebrates, a cell with pigment granules, which on movement of the granules alters the animal's colour. *3*. Structure in **prokaryotes** in which the **photosynthetic** pigments are located.

chromite Mineral consisting of the oxides of **chromium** and **iron**. Alternative name: chrome iron ore.

■ **chromium** Cr Silver-grey metallic element in Group VIB of the Periodic Table (a **transition element**), obtained mainly from its ore chromite. It is electroplated onto other metals (particularly steel) to provide a corrosion-resistant decorative finish, and alloyed with nickel and iron to make stainless steels; its compounds are used in pigments and dyes. At. no. 24; r.a.m. 51.996.

chromophore Chemical grouping that causes compounds to have colour (*e.g.* the $-N=N-$ group in an **azo dye**).

chromoplast **Organelle** in a plant cell that contains pigments, *e.g.* chromoplasts in carrots contain **carotenoids**.

■ **chromosome** Structure within the **nucleus** of a cell that contains **protein** and the genetic **DNA**. Chromosomes occur in pairs in **diploid** cells (ordinary body cells); **haploid** cells (gametes or sex cells) have only one of each pair of chromosomes in their nuclei. The number of chromosomes varies in different **species**. During mitotic cell division, each chromosome doubles and the two duplicates separate into the two new daughter cells. [2/8/b]

chromosome map Diagram showing the positions of various **genes** on appropriate **chromosomes**.

chronic Describing a condition or disorder that is long-standing (and often difficult to treat). *See also* **acute**.

chrysalis Alternative name for **pupa**.

chyle Milky fluid resulting from the absorption of fats in the **lacteals** of the small intestine; it is removed by the **lymphatic system**.

chyme Partly digested food that passes from the stomach into the duodenum and small intestine.

cilia Plural of **cilium**.

ciliary body Ring of tissue that surrounds the lens of the eye. It generates the **aqueous humour** and contains ciliary muscles, which are used in **accommodation**.

ciliary muscle Muscle that controls the shape of the lens of the **eye** and thus achieves **accommodation**.

cilium (plural cilia) Small hair-like structure that moves rhythmically on the surface of a cell or the whole **epithelium**. Cilia usually cover the surface of a cell and cause movement in the fluid surrounding it. For example, ciliated cells line the respiratory tract, wafting mucus out of the bronchioles and bronchi. Cilia are also used for locomotion by some single-celled aquatic organisms. *See also* **flagellum.**

cinnamic acid $C_6H_5CH=CHCOOH$ White crystalline organic compound with a pleasant odour. Alternative name: 3-phenylpropenoic acid.

■ **circadian rhythm** Cyclical variation in the physiological, metabolic or behavioural aspects of an organism over a period of about 24 hours; *e.g.* sleep patterns. It may arise from inside an organism or be a response to a regular cycle of some external variation in the environment. Alternative term: diurnal rhythm. *See also* **biological clock.**

circuit breaker Safety device in an electric circuit that interrupts the current flow in the event of a fault. *See also* **fuse.**

■ **circulatory system** In animals, transport system that maintains a constant flow of **tissue fluid** in sealed vessels to all parts of body. *E.g.* in the **blood vascular system,** oxygen and food materials dissolved in blood diffuse into each cell; waste products, including carbon dioxide, diffuse out of the cells and into the blood. *See also* **lymphatic system.** [2/4/a] [2/5/a] [2/7/a]

cistron Functional unit of a **DNA** chain that controls **protein** manufacture.

■ **citric acid** $C_6H_8O_7$ Hydroxy-tri**carboxylic acid,** present in the juices of fruits. It is important in the energy-generating reactions in cells (Krebs cycle), and is much used as a flavouring.

citric acid cycle Alternative name for **Krebs cycle.**

Claisen condensation Chemical reaction in which two molecules combine to give a compound containing a **ketone** group and an **ester** group.

■ **class** In biological **classification**, one of the groups into which a phylum is divided, and which is itself divided into orders; *e.g.* Mammalia (mammals), Aves (birds). [2/3/b] [2/4/b]

classical physics Physics prior to the introduction of **quantum theory** and a knowledge of **relativity**.

■ **classification** In biology, the placing of living organisms into a series of groups according to similarities in structure, physiology, genotype, and other characteristics. The smallest group is the **species**. Similar species are placed in a genus, similar genera are grouped into families, families into orders, orders into classes, classes into phyla (or divisions in plants), and phyla into **kingdoms**. There may also be intermediate groups such as suborders, subclasses and subphyla. Modern classification is usually intended to reflect degrees of evolutionary relationship, although not all experts agree on single classification schemes for animals or for plants. *See also* **binomial nomenclature**. [2/3/b] [2/4/b]

clathrate Chemical structure in which one atom or molecule is 'encaged' by a structure of other molecules, and not held by **chemical bonds**. *See also* **chelate**.

■ **clavicle** In vertebrates, the anterior bone of the ventral side of the shoulder girdle. Alternative name: collarbone. [2/3/a]

cleavage *1.* In biology, a series of **mitotic** divisions of a fertilized **ovum**. *2.* In biochemistry, the splitting of chemical bonds (*e.g.* protease enzymes cleave peptide bonds from proteins to release amino acid residues).

climax community Complex but stable plant community that can perpetually regenerate itself under prevailing environmental conditions. It is the final stage of plant **succession**. Alternative name: climax vegetation.

■ **clinical thermometer** Type of (mercury) **thermometer** used for taking body temperature. [2/7/a]

clinostat Horizontal or vertical turntable used to rotate plants so as to counteract the effect of a stimulus (*e.g.* gravity) that normally acts in only one direction.

clitoris Mass of erectile tissue in female mammals, the equivalent of the **penis** in males, situated in front of the opening of the **vagina**.

cloaca *1*. In most non-mammalian vertebrates, the single posterior opening to the body. Into it open the anus, reproductive ducts and urinary ducts. *2*. Terminal part of the intestine in some invertebrates, *e.g.* sea cucumbers.

■ **clone** One of many descendants produced by **vegetative propagation** from one original plant seedling, or **asexual reproduction** or **parthenogenesis** from a single animal. Members of a clone have identical genetic constitution. [2/10/a]

clotting factor Protein structure, *e.g.* thrombin and fibrinogen, that induces blood **coagulation** when a blood vessel is broken.

■ **cloud chamber** Apparatus for observing the tracks of subatomic particles. It consists of a chamber containing a liquid (such as ethanol) and its saturated vapour. Sudden expansion of the chamber (by retracting one wall) lowers the temperature, so that droplets of liquid condense from the now supersaturated vapour onto ions along the tracks of the particles. *See also* **bubble chamber**.

Cnidaria Phylum of invertebrate animals, which includes corals, hydra, jellyfish and sea-anemones. They were formerly classified as **Coelenterata**.

cnidarian Animal that is a member of the phylum **Cnidaria**.

CNS Abbreviation of **central nervous system**.

■ **coagulation** *1*. Process by which bleeding is arrested. Thrombin is produced in the absence of antithrombin from the combination of prothrombin and calcium ions. This interacts with soluble fibrinogen to precipitate it as the insoluble blood protein fibrin, which forms a mesh of fine threads over the wound. Blood cells become trapped in the mesh and form a clot. *2*. Irreversible setting of protoplasm on exposure to heat or poison. *3*. Precipitation of colloids, *e.g.* proteins, from solutions. [2/5/a]

■ **coal** Black mineral consisting mainly of **carbon**, used as a fuel and source of organic chemicals, and for making **coal gas**, **coal-tar**, **coke** and (by **hydrogenation**) oil. It is the remains of plants from the Carboniferous and Permian periods that have been subjected to high pressures underground. *See also* **peat**. [2/4/c]

coal gas Fuel gas made by the **destructive distillation** of coal in closed iron retorts, which yields **coke** and **coal-tar** as by-products. The gas consists mainly of hydrogen and methane, with some carbon monoxide.

coal-tar Thick black oily liquid, consisting mainly of **aromatic compounds**, obtained as a by-product of **coal gas** manufacture. *See also* **tar**.

■ **cobalt** Co Silver-white magnetic metallic element in Group VIII of the Periodic Table, used in alloys to make cutting tools and magnets. The radioactive isotope Co-60 is used in **radiotherapy**. At. no. 27; r.a.m. 58.9332. [2/1/c]

COBOL Acronym of Common Business Oriented Language, a computer programming language designed for commercial use.

■ **coccus** Spherical-shaped **bacterium**. Cocci may join together to form clumps (staphylococci) or chains (streptococci). [2/5/a]

■ **coccyx** Bony structure in primates and amphibians, formed by fusion of tail vertebrae; *e.g.* in human beings it consists of three to five vestigial vertebrae at the base of the spine. [2/5/a]

■ **cochlea** Spirally coiled part of the inner **ear** in mammals. It translates sound-induced vibrations into nerve impulses that travel along the auditory nerve to the brain, where they are interpreted as sounds. [2/4/a] [2/5/a] [2/7/a]

Cockcroft-Walton generator High-voltage **direct current** generator used for accelerating nuclear particles to high speeds. It was named after the British scientist John Cockcroft (1897–1967) and the Irish physicist Ernest Walton (1903–).

codeine Pain-killing drug (a narcotic analgesic), the methyl derivative of **morphine**.

co-dominant Describing a gene condition in which neither **allele** is **dominant** or **recessive** (*e.g.* A, B and AB blood groups).

coefficient Number or parameter that measures some specified property of a given substance; *e.g.* coefficient of friction, coefficient of viscosity.

■ **Coelenterata** Major group of invertebrate animals, formerly given the status of a phylum but now divided into Cnidaria and Ctenophora. Coelenterates have a body made up of two layers of cells, **ectoderm** and **endoderm**, which are separated by a middle layer of jelly, the **mesoglea**. They are symmetrical, and have a single body cavity with one opening at the mouth. The nervous system is a diffuse network, and they have neither excretory nor blood systems. Common coelenterates include jellyfish, corals and hydra. [2/4/b]

coelenterate Member of the **Coelenterata**.

coelom *1*. Body cavity of **triploblastic** animals (*e.g.* vertebrates), formed from the **mesoderm**. It contains coelomic fluid, in which colourless **phagocytic** corpuscles as well as the gut are suspended. Coelomic fluid keeps the body moist for respiration and the corpuscles keep the space free from **bacteria**. *2*. Body cavity of an insect.

coenocyte Multinucleate mass of **protoplasm** formed by the division of the **nucleus** from a cell that has only a single nucleus, *e.g.* as in **fungi** and **algae**.

coenzyme Organic compound essential to catalytic activities of **enzymes** without being utilized in the reaction. Coenzymes usually act as carriers of intermediate products, *e.g.* ATP.

coherent units System of units in which the desired units are obtained by multiplying or dividing base units, with no numerical constant involved. *See also* **SI units**.

■ **cohesion** Attraction between similar particles (atoms or molecules of the same substance); *e.g.* between water molecules to create **surface tension**. *See also* **adhesion**. [5/2/a]

coke Brittle grey-black solid containing about 85 per cent carbon, made by roasting coal in a limited supply of air. It is used as a smokeless fuel and source of carbon (*e.g.* in smelting metal ores).

cold-blooded Alternative name for **poikilothermic**.

Coleoptera Beetles, a major order of **insects**. *See* **beetle**.

collagen Fibrous protein connective tissue that binds together bones, ligaments, cartilage, muscles and skin.

collarbone Alternative name for the **clavicle**. [2/3/a] [2/4/a]

■ **colloid** Form of matter that consists of extremely small particles, about 10^{-4} to 10^{-6} mm across, so small that they remain suspended and dispersed in a medium such as air or water. Common colloids include aerosols (*e.g.* fog, mist), emulsions (*e.g.* milk) and gels (*e.g.* gelatin, rubber). Mucus may also be colloidal, and clay consists of mainly colloidal-sized soil particles. A non-colloidal substance is termed a crystalloid. [4/9/a,b]

■ **colon** *1*. Large intestine of vertebrates, in which the main function is the absorption of water from **faeces**. *2*. In insects, the wide posterior part of the hind gut. [2/4/a] [2/5/a] [2/7/a]

colonization Process by which species begin to live in new **habitats**, usually through natural means or sometimes through introduction by human activity.

colony *1*. Bacterial growth on a solid medium that forms a visible mass. *2*. Growth of a group of individual plants of one species that invade new ground. *3*. Collection of individuals that live together with some degree of interdependence (*e.g.* social insects such as some ants, bees and wasps).

colorimeter Instrument for measuring the colour intensity of a medium such as a coloured solution (which can be related to concentration and therefore provide a method of quantitative analysis). The technique is termed colorimetry.

colostrum Yellowish milky fluid secreted from the mammalian breast immediately before and after childbirth. It contains more **antibodies** and **leucocytes** (and less fat and carbohydrates) than true milk, which follows within a few days.

■ **colour** Visual sensation or perception that results from the adsorption of light energy of a particular **wavelength** by the cones of the retina of the eye. There are two or more types of cone, each of which are sensitive to different wavelengths of light. The brain combines nerve impulses from these cones to produce the perception of colour. The colour of an object thus depends on the wavelength of light it reflects (other wavelengths being absorbed) or transmits. [5/9/a] [5/9/c]

■ **colour blindness** Inability to distinguish between certain colours. It is a **congenital** abnormality that affects 6% of human males and 1% of females. The most common defect is red-green colour blindness, which results in observation of both colours in grey, blue or yellow, depending on the amount of yellow and blue present in the light. [2/9/c]

commensalism Close relationship between two plant or animal organisms or communities from which one usually benefits more than the other (which is nevertheless unharmed by the association); *e.g.* a sea anemone living on the shell of a hermit crab is carried around and benefits from bits of food that float away from the crab while it is feeding. *See also* **parasitism**; **symbiosis**.

communications satellite Man-made object that orbits the Earth, used for relaying radio, television and telephone signals. Many such satellites are in **geostationary orbits**.

community In biology, collection of interacting but different species that occupy the same **habitat**.

■ **commutator** Device for altering or reversing the direction of an electric current, *e.g.* on a d.c. electric motor. [5/7/c]

compact Describing a condition that results from physical pressure (*e.g.* a compact soil holds less water and air than a non-compact one).

compass Apparatus for finding direction (parallel to the Earth's surface), usually by alignment with the Earth's magnetic field. In its simplest form, it consists of a magnetized needle pivoted horizontally at its centre. Such magnetic compasses are affected by nearby magnetic materials and electrical equipment, and have to be corrected for the continuous slow movement of the position of the north magnetic pole. A gyrocompass overcomes these problems by using a **gyroscope** with its axis permanently aligned with north.

■ **competition** In biology, the struggle within a **community** between organisms of the same or different species for survival. *See also* **natural selection**. [2/4/d] [2/6/a]

compiler Computer **program** that converts a source language into **machine code** (readable by the computer).

complex compound Alternative name for a **coordination compound**.

complex ion **Cation** bonded by means of a **coordinate bond**.

compost Gardening term for a collection of leaves, weeds and other organic debris that is left to decay and then used as fertilizer.

■ **compound** Substance that consists of two or more **elements** chemically united in definite proportions by weight. *E.g.* sodium chloride (common salt), NaCl, is a compound of the alkali metal sodium and the halogen gas chlorine. Alternative name: chemical compound. [4/6/a]

compound eye One of the paired eyes of most adult

arthropods (*e.g.* insects), consisting of many units (called ommatidia) each with its own lens. A fly, for example, can detect movement very easily as an image moves from one ommatidium to the next.

compound pendulum Vertical bar or rod pivoted at the top and with a weight near the bottom. If its centre of mass is a distance h from the pivot and the **radius of gyration** is k, the period of swing t is given by $t = 2\pi[(h^2 + k^2)/hg]^{1/2}$, where g is the acceleration of free fall.

Compton effect Reduction in the energy of a photon as a result of its interaction with a free electron. Some of the photon's energy is transferred to the electron. It was named after the American physicist Arthur Compton (1892–1962).

computer Electronic device that can accept **data**, apply a series of logical operations to it (obeying a **program**), and supply the results of these operations. *See* **analog computer**; **digital computer**.

concave Describing a surface that curves inwards (as opposed to one that is convex).

■ **concentration** Strength of a mixture or solution. Concentrations can be expressed in very many ways; *e.g.* parts per million (for traces of a substance), percentage (*i.e.* parts per hundred by weight or volume), gm or kg per litre of solvent or per litre of solution, moles per litre of solution (**molarity**), moles per kg of solvent (**molality**), or in terms of normality (*see* **normal**). *See also* **solubility**.

conception Fertilization of a human egg, usually taken to mean also **implantation** in the womb of the resultant **zygote** (which goes on to develop into an embryo and foetus).

condensation *1.* Change of a gas or vapour into a liquid or solid by cooling. *2.* Chemical reaction in which two or more small molecules combine to form a larger molecule, often with the elimination of a simple substance such as water. It is used to make **polymers**.

condensation pump Alternative name for **diffusion pump**.

■ **condensation reaction** Chemical reaction in which two or more small molecules combine to form a larger one, often with the elimination of a simpler substance, usually water; *e.g.* acetaldehyde (ethanal), CH_3CHO, condenses with hydroxylamine, H_2NOH, to form an oxime, $CH_3CH{=}NOH$. [4/8/b]

■ **condenser** *1.* In electricity, an alternative name for a **capacitor**. *2.* In optics, a lens or mirror that concentrates a light source.

■ **conditioned reflex** Animal's response to a neutral stimulus that learning has associated with a particular effect; *e.g.* a laboratory rat may learn (be conditioned) to press a lever when hungry because it associates this action with receiving food. [2/2/c]

conductance Ability to convey energy as heat or electricity. Electrical conductance, measured in siemens, is the reciprocal of **resistance**. Alternative name: conductivity.

■ **conduction band** Energy range in a **semiconductor** within which **electrons** can be made to flow by an applied **electric field**.

conduction, electrical Passage of electricity by materials.

conduction, thermal Transmission of heat by materials.

■ **conductor** *1.* Material that allows heat to flow through it by **conduction**. *2.* Material that allows electricity to flow through it; a conductor has a low **resistance**. [4/6/a]

cone *1.* Light-sensitive nerve cell present in the **retina** of the eye of most vertebrates; it can detect colour. *See also* **rod**. *2.* Reproductive structure of **gymnosperms** (*e.g.* pines and other conifers).

configuration *1.* In chemistry, the arrangement in space of the atoms in a molecule. *2.* In atomic physics, the arrangement of electrons about the nucleus of an atom. Alternative name: electron configuration.

conformation Particular shape of a molecule that arises through the normal rotation of its atoms or groups about single bonds.

congenital Dating from birth or before birth. Congenital conditions may be caused by enviromental factors or be inherited.

conifer Plant that is a member of the order **Coniferales**.

■ **Coniferales** Large order of cone-bearing, usually evergreen shrubs or trees; conifers. They include pines and firs. Larch is the only European conifer that is deciduous. [2/2/b] [2/2/c]

conjugated Describing an organic compound that has the single and double triple bonds; *e.g.* buta-1,3-diene, $H_2C=CH\text{-}CH=CH_2$.

conjugation In biology, form of reproduction that involves the permanent or temporary union of two isogametes, *e.g.* in certain green **algae**. In **protozoa**, two individuals partly fuse, exchanging nuclear materials. When separated, each cell divides further to give new individuals or, more usually, uninucleate **spores** (which eventually develop into full adults).

connective tissue Strong **tissue** that binds **organs** and tissues together. It consists of a **glycoprotein** matrix containing

collagen in which cells, fibres and vessels are embedded. The most widespread connective tissue is **areolar tissue**.

■ **conservation of energy** Law that states that in all processes occurring in an isolated system the energy of the system remains constant. *See also* **thermodynamics, laws of**. [3/5/c] [3/10/b]

■ **conservation of mass** Principle which states that the products of a purely chemical reaction have the same total mass as the reactants.

■ **conservation of momentum** Law which states that the total momentum of two colliding objects before impact is equal to their total momentum after impact. [5/10/a]

constant Quantity that remains the same in all circumstances. *E.g.* the acceleration of free fall is a physical constant; in the expression $2y = 5x^2$, the numbers 2 and 5 are constants (and x and y are variables).

constant composition, law of In any given chemical compound, the same **elements** are always combined in the same proportions by mass.

■ **contact process** Industrial process for the manufacture of **sulphuric acid**, involving the catalytic **oxidation** of **sulphur dioxide** to **sulphur trioxide**. [4/4/b]

continuous stationery Long length of fan-folded paper with sprocket holes down each side for transporting it through a computer printer. The paper may be perforated to facilitate tearing off the sprocket holes and for tearing the paper into single sheets after printing.

■ **continuous wave** **Electromagnetic wave** of constant **amplitude**. [5/9/a]

contraception Avoidance of **conception**, by means of a method or device that is designed to prevent male sperm from reaching the female egg (ovum) during or soon after intercourse, or by means of **hormones** (the contraceptive pill) that suppress **ovulation** by interfering with the woman's **menstrual cycle**.

contractile vacuole Membranous sac within a single-celled organism (*e.g.* amoebae and other protozoans) which fills with water and suddenly contracts, expelling its contents from the cell. It carries out **osmoregulation** and excretion.

control Part of a scientific experiment in which the experimental conditions are checked against standard (*i.e.* non-experimental) ones.

■ **control rod** Length of **neutron**-absorbing material, *e.g.* boron, cadmium, that is a good absorber of **thermal neutrons**, used to control the rate of fission **(chain reaction)** in a **nuclear reactor** by being moved in or out of the core. [4/8/d]

■ **convection** Transport of **heat** by the movement of the heated substance (usually a fluid such as air or water). [3/6/c]

convergent evolution Tendency of species that live in a single uniform environment to develop similar characteristics.

converging lens **Lens** capable of bringing light to a focus; a convex lens.

coordinate bond Type of **covalent bond** that is formed by the donation of a **lone pair of electrons** from one atom to another. Alternative name: dative bond.

coordination compound Chemical compound that has **coordinate bonds**; *e.g.* potassium ferricyanide, $K_3Fe(CN)_6$. Alternative names: complex, complex compound.

coordination number Number of nearest neighbours of an atom or an ion in a chemical compound.

Copepoda Subclass of **Crustacea,** many of which are minute parasitic marine animals. They also include the non-parasitic, freshwater *Daphnia*, commonly known as the water flea.

copepod Crustacean that is a member of the subclass **Copepoda**.

copolymer Polymer built up from two or more different kinds of **monomers,** *e.g.* the hard plastic ABS (acrilonitrile-butadiene-styrene).

■ **copper** Cu Reddish metallic element in Group IB of the Periodic Table (a **transition element)** which occurs as the free metal (native) and in various ores, chief of which is chalcopyrite. The metal is a good conductor of electricity and is used for making wire, pipes and coins. Its chief alloys are **brass** and **bronze**. Its compounds are used as pesticides and pigments. Copper is an important **trace element** in many plants and animals. At. no. 29; r.a.m. 63.546.

copper (I) Alternative name for cuprous in copper compounds.

copper (II) Alternative name for cupric in copper compounds.

copper (II) carbonate $CuCO_3$ Green crystalline compound which occurs (as the **basic salt)** in the minerals azurite and malachite. It is also a component of verdigris, which forms on copper and its alloys exposed to the atmosphere.

copper (II) chloride $CuCl_2$ Brown covalently bonded compound, which forms a green crystalline dihydrate. It is used in fireworks to give a green flame and to remove sulphur in the refining of **petroleum**. Alternative name: cupric chloride.

copper (I) oxide Cu_2O Insoluble red powder, used in rectifiers and as a **pigment**. Alternative names: copper oxide, cuprite, cuprous oxide, red copper oxide.

copper (II) oxide CuO Insoluble black solid, used as a **pigment**. Alternative names: copper oxide, cupric oxide.

copper (II) sulphate $CuSO_4$ White **hygroscopic** compound, which forms a blue crystalline pentahydrate, used as a wood preservative, fungicide, dyestuff and in **electroplating**. Alternative names: blue vitriol, cupric sulphate.

coral Substance containing calcium carbonate that is secreted by various marine organisms (*e.g.* Anthozoa) for support and habitation.

cordite Smokeless propellant explosive prepared from nitrocellulose (**cellulose trinitrate**) and nitroglycerin (**glyceryl trinitrate**).

core Element in a computer **memory** consisting of a piece of magnetic material that can retain a permanent positive or negative electric charge until a current passes through it (when it changes polarity).

■ **cork** Layer of dead cells on the outside of plant stems and roots. It is impermeable to air and water, and so protects the living cells inside against water loss and physical injury. In the cork oak this layer is exceptionally thick and can be stripped off, to be used for making bottle stoppers ('corks') or heat-resistant insulating material. Alternative name: phellem. *See also* **lenticel**. [2/4/a]

cork cambium *See* **phellogen**.

■ **corm** Underground organ developed by some plants. It is formed from **stem** tissue, and has adventitious roots growing from its lower surface. Primarily a storage organ, it may also play a part in **vegetative propagation** by producing buds that develop into new plants. [2/4/a] [2/5/a] [2/7/a]

■ **cornea** Transparent connective tissue at the front surface of the **eye** of vertebrates, overlying the **iris**. Together with the **lens**, it focuses incoming light onto the **retina**. [2/4/a] [2/5/a] [2/7/a]

cornification Alternative name for **keratinization**.

■ **corolla** Inner part of a flower, consisting of petals. It is often coloured. [2/4/a] [2/5/a] [2/7/a]

coronary vessels Arteries and **veins** that carry the blood supplying the heart muscle in vertebrates.

corpus callosum Thick bundle of nerve fibres in the middle of the **brain** that connects the two cerebral hemispheres.

■ **corpus luteum** Yellow body formed in the **ovary** of female mammals which produces the hormone **progesterone**. It develops from the **Graafian follicle** after **ovulation**. If **fertilization** does not occur, the corpus luteum degenerates. [2/5/a]

■ **corrosion** Gradual chemical breakdown, often **oxidation** of metals, by air, water or chemicals. [4/7/b]

corrosive sublimate Alternative name for **mercury (II) chloride**.

cortex Outer layer of a structure, *e.g.* the rind of plant parenchyma tissue surrounding the vascular cylinder in the stem and roots of plants, or the outer layer of cells of the **adrenal gland** or **brain** of mammals. *See also* **medulla**.

corticosteroid Hormone secreted by the **adrenal cortex**, which controls sodium and water **metabolism** as well as **glycogen** formation.

corticotrophin Alternative name for **adrenocorticotrophic hormone** (ACTH).

■ **cortisone Hormone** isolated from the **adrenal cortex**, used in the treatment of severe hay fever, rheumatoid arthritis and other inflammatory conditions. [2/7/a]

corundum Hard mineral consisting of aluminium oxide (Al_2O_3), whose coloured crystalline forms are ruby and sapphire. It is used as an abrasive (*e.g.* as in emery).

■ **cosmic radiation** Radiation of extremely short wavelength (about 10^{-15} m) from outer space. *See also* **cosmic rays**. [4/7/f]

■ **cosmic rays** Radiation consisting of high-energy particles from outer space. Primary rays consist of **protons** and nuclei as well as other subatomic particles. Collision with nitrogen and oxygen atoms in the atmosphere generates secondary cosmic rays, consisting of elementary particles and **gamma-rays**. [4/7/f]

costal Concerning the ribs.

cotyledon *1*. Leaf that forms part of the embryo of seeds, usually lacking chlorophyll (until exposed to light after germination). Monocotyledons have one and dicotyledons have two cotyledons in each seed. In certain plants, *e.g.* peas and beans, these are food-storage organs. Cotyledons of many plants appear above ground, develop chlorophyll and synthesize food material by **photosynthesis**. *2*. One of the leaves of the embryo in flowering plants. *3*. Leaf developed by a young fern plant.

■ **coulomb** (C) SI unit of electric charge, defined as the quantity of electricity transported by a current of 1 ampere in 1 second. It was named after the French physicist Charles Coulomb (1736–1806). [5/9/b]

counter *1*. In physics, electronic apparatus for detecting and counting particles, usually by making them generate pulses of electric current; the actual counting circuit is a **scaler**. *2*. In computing, any device that accumulates totals (*e.g.* of repeated program loops or cards passing through a punched **card reader**).

■ **couple** In physics, a pair of equal and parallel forces acting in opposite directions upon an object. This produces a turning effect (**torque**) equal to one of the forces times the distance between them. [5/7/b]

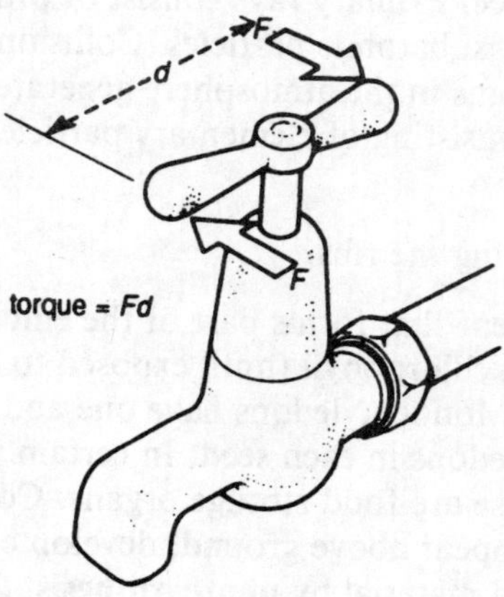

A simple tap uses a couple

coupling In physics, connection between two oscillating systems.

■ **covalent bond** Chemical bond that results from the sharing of a pair of **electrons** between two atoms. *See also* **coordinate bond; ionic bond**. [4/8/d] [4/8/e]

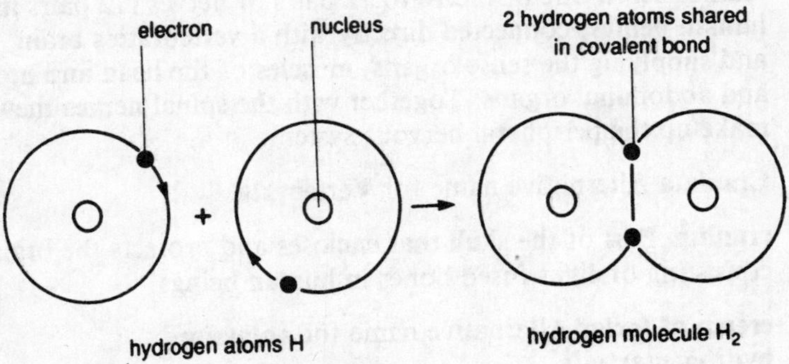

Formation of a covalent bond

covalent compound Chemical compound in which the atoms are covalently bonded.

covalent radius Effective radius of an atom involved in a **covalent bond**.

Cowper's gland One of a pair of small glands located below the **prostate gland** and connected to the **urethra**; it produces fluid for **semen**. It was named after the British anatomist William Cowper (1666–1709).

Coxsackie virus Member of a group of **viruses** that cause inflammatory diseases in human beings. It was named after the city in New York state where it was first found.

■ **cracking** Industrial process, usually employing heat or a **catalyst**, in which heavy **hydrocarbons** are broken down into lower boiling point fractions, *e.g.* in making petrol from crude oil. It is used in **reforming**. [4/8/b] [4/10/d]

cranial Relating to the **cranium** and **brain**.

cranial nerve One of the 10 to 12 pairs of nerves (12 pairs in human beings) connected directly with a vertebrate's brain and supplying the sense organs, muscles of the head and neck and abdominal organs. Together with the spinal nerves they make up the peripheral nervous system.

Craniata Alternative name for **Vertebrata**.

cranium Part of the **skull** that encloses and protects the brain, consisting of eight fused bones in human beings.

cream of tartar Alternative name for **potassium hydrogentartrate**.

creatine $NH_2C(NH)N(CH_3)CH_2COOH$ White crystalline amino acid present in muscle, where it plays an important role in muscle contraction. It is broken down to **creatinine**.

creatinine $C_4H_7N_3O$ Heterocyclic crystalline solid formed by the breakdown of **creatine** and excreted in urine.

creosote Colourless oily fluid distilled from wood tar; used as a disinfectant. *See also* **creosote oil**.

creosote oil Dark brown distillation product of **coal-tar**; used as a wood preservative. Alternative names: coal-tar creosote; creosote.

cresol Alternative name for **methylphenol**.

critical angle In optics, smallest angle of incidence at which **total internal reflection** occurs.

critical mass Minimum amount of fissile material required to maintain a nuclear **chain reaction** (*see* **nuclear fission**).

critical pressure Pressure necessary to condense a substance at its **critical temperature**.

critical state Conditions of temperature and pressure at which the liquid and gas phases become one phase – *i.e.* they have the same density.

critical temperature *1.* Temperature above which a gas cannot be liquefied (no matter how high the pressure).
2. Temperature at which a magnetic material loses its magnetism. Alternative name: Curie point.

critical volume Volume of one unit of mass of a substance at its **critical pressure** and **critical temperature**.

CRO Abbreviation of **cathode-ray oscilloscope**.

Cro-Magnon man Earliest recognized form of modern man. Cro-Magnons are known to have lived in Europe 30–40,000 years ago, but may have appeared in Africa or Asia at an earlier date.

■ **crossing over** Exchange of material between homologous **chromosomes** during **meiosis** (cell division). It is the mechanism that alters the pattern of genes in the chromosomes of offspring, giving the genetic variation associated with **sexual reproduction**. [2/8/b]

■ **Crustacea** Large class of **arthropods** which includes barnacles, crabs, lobsters, prawns and woodlice. Characterized by having two pairs of **antennae**, most crustaceans are aquatic. [2/3/b] [2/4/b]

crustacean Member of the class **Crustacea**.

cryogenics Branch of physics that involves low temperatures and their effects.

cryolite Na_3AlF_6 Pale grey mineral found mainly in Greenland, used as flux in the electrolytic extraction of **aluminium**. Alternative name: sodium aluminium fluoride.

crystal Solid that has a definite regular shape. Crystals of the same substance have the same specific shapes, reflecting the way its component atoms or ions are regularly arranged in a lattice. *See also* **liquid crystal**.

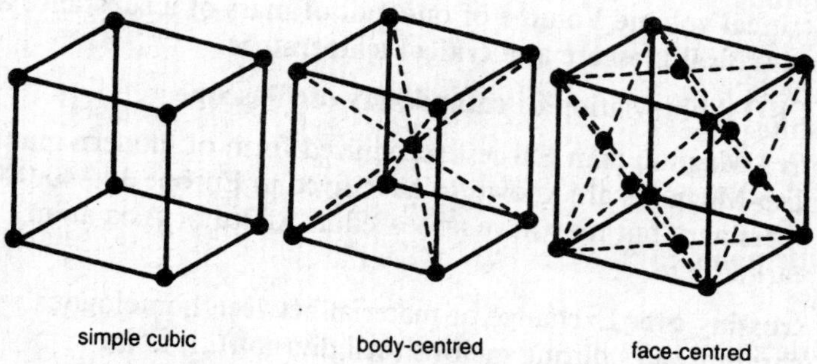

Types of cubic crystal structures

crystal lattice Regular arrangement of atoms or ions in a **crystal**.

crystalline Having the form of crystals, as opposed to being **amorphous**.

crystallization Process in which crystals form from a molten mass or solution.

crystallography Study of crystals.

crystalloid Substance that is not a **colloid** and can therefore pass through a semipermeable membrane.

CS gas $C_6H_4ClCH=C(CN)_2$ White irritant organic compound used as a tear gas for riot control. Alternative name: (2-chlorobenzylidene)-malononitrile.

cupric Bivalent copper. Alternative names: copper (II), copper (2+).

cuprite Mineral form of **copper (I) oxide**.

cuprous Monovalent copper. Alternative names: copper (I), copper (1+).

curie (Ci) Measure of radioactivity. 1 Ci = 3.700×10^{10} disintegrations per second. It was named after the Polish-born French scientist Marie Curie (1867–1934). It has been replaced in **SI units** by the **becquerel**.

Curie point Temperature above which a magnetic material loses its **magnetism**. Alternative name: critical temperature.

Curie's law In a **paramagnetic** substance, the **magnetic susceptibility** is inversely proportional to the **absolute temperature**.

curium Cm Radioactive element in Group IIIA of the Periodic Table (one of the **actinides**), made by the alpha-particle bombardment of plutonium-239. It has several **isotopes**, with half-lives up to 1.7 × 107 years. At. no. 96; r.a.m. (most stable isotope) 247.

■ **current, electric** Flow of electrons along a conductor, measured in **amperes**. [5/5/b]

cusp In anatomy, pointed part of a tooth.

cuticle *1.* In plants, deposit of waxy waterproof material that forms a continuous layer over aerial parts, broken only by **stomata**. It provides protection against water loss and injury. *2.* In animals, protective layer of hard material that covers many arthropods (*e.g.* crustaceans, insects), where it may form the **exoskeleton**. [2/4/a] [2/5/a]

cutin Mixture resulting from oxidation and condensation of fatty acids which is deposited on or in the outer layer of cell walls (cuticle) of plants.

cyanamide NH=C=NH. Colourless crystalline compound, used as an industrial source of **ammonia**.

cyanide Compound containing the cyanide ion (CN^-); a salt of hydrocyanic acid (HCN). All cyanides are highly poisonous.

cyanide, organic *See* **nitrile**.

Cyanobacteria Alternative name for the **Cyanophyta**.

cyanoferrate *See* **ferricyanide**; **ferrocyanide**.

cyanogen NCCN Colourless highly poisonous gas made by the action of acids on cyanides.

Cyanophyta Division of single-celled, photosynthetic, **prokaryotic** organisms. Cyanophytes sometimes join together in colonies, and are most commonly found in water. Alternative names: blue-green algae; Cyanobacteria; Cyanophyceae.

cyanophyte Member of the **Cyanophyta**.

cybernetics Science that employs control systems resembling those of living things for mechanisms and electronic systems, (*e.g.* in building industrial robots).

cycle One of a repeating series of similar changes; *e.g.* in a wave motion or vibration. One cycle is equal to the period of the motion; the number of cycles per unit time is its frequency. A frequency of 1 cycle per second = 1 **hertz**.

cyclic Describing the molecule of any (usually organic) compound whose atoms form a ring. **Benzene** and the **cycloalkanes** are simple cyclic compounds. *See also* **acyclic**; **alicyclic**.

cycloalkane Saturated **cyclic** hydrocarbon whose molecule has a ring of carbon atoms, general formula C_nH_{2n}; *e.g.* cyclohexane, C_6H_{12}.

cyclonite $(CH_2NNO_2)_3$ Powerful explosive derived from **hexamine**. Alternative names: hexogen, RDX.

cyclopentadiene C_5H_6 Colourless liquid cyclic hydrocarbon, with a five-membered ring containing two carbon-carbon double bonds. It readily forms the cyclopentadienyl ion $C_5H_5^-$, present in compounds such as **ferrocene**.

cyclotron Machine for accelerating atomic particles to high speeds. Particles follow a spiral path in a magnetic field between two D-shaped electrodes.

cystine $(SCH_2CH(NH_2)COOH)_2$ Dimeric form of the **amino acid** cysteine, found in keratin.

cytochrome Respiratory pigment found in organisms that use **aerobic respiration**.

cytokinin Plant **hormone** that stimulates the division of plant cells.

cytology Study of **cells**.

cytoplasm **Protoplasm** of a cell other than that of the **nucleus**.

cytosine Colourless crystalline compound, derived from **pyrimidine**. It is a major constituent of **DNA** and **RNA**.

cytotoxic Describing a drug that destroys or prevents the replication of cells, used in **chemotherapy** to treat **cancer**.

D

dalton Alternative name for **atomic mass unit**.

Dalton's atomic theory Theory that states that matter consists of tiny particles (atoms), and that all the atoms of a particular **element** are exactly alike, but different from the atoms of other elements in behaviour and mass. The theory also states that chemical action takes place as a result of attraction between atoms, but it fails to account satisfactorily for the volume relationships that exist between combining gases. It was proposed by the British scientist John Dalton (1766–1844).

Dalton's law of partial pressures In a mixture of gases, the pressure exerted by one of the component gases is the same as if it alone occupied the total volume.

Daniell cell Electrolytic cell that consists of a zinc **half-cell** and a copper half-cell, usually arranged as a zinc **cathode** and a copper **anode** dipping into an **electrolyte** of dilute sulphuric acid.

daraf Unit of the elastance of an electrical component; it is the reciprocal of **capacitance** (the word is **farad** backwards).

■ **Darwinism** Theory of **evolution** which states that living organisms arise in their different forms by gradual change over many generations, and that this process is governed by **natural selection**. It was proposed by the British naturalist Charles Darwin (1809–82). [2/9/c]

data Collection of information, often referring to results of a statistical study or to information supplied to, processed by or provided by a computer (excluding the **program**).

data bank Large file of computer **data**, usually accessible (to many users) by direct access.

data base *1.* Organized collection of **data** that is held on a computer, where it is regularly updated and can easily be accessed. *2.* **Applications program** that controls and makes use of a data base.

data transmission Transfer of **data** between outstations and a central **computer** or between different computer systems.

dative bond Alternative name for a **coordinate bond**.

daughter cell One of the two cells produced when a plant or animal cell divides by **mitosis**.

daughter element One of the elements produced when an atom divides by **nuclear fission**.

daughter nucleus New atomic nucleus produced when the nucleus of a radio-isotope decays.

Davy lamp Alternative name for **safety lamp**.

d.c. Abbreviation of **direct current**.

■ **DDT** Abbreviation of dichlorophenyltrichloroethane. This compound was once much used as an insecticide, but is now banned in most countries because of its high toxicity and the fact that it can pass through food chains without degrading. [2/7/d] [2/9/d] [2/10/e]

deaminase **Enzyme** that catalyses the removal of an **amino group** ($-NH_2$) from an organic molecule.

deamination Enzymatic removal of an **amino group** ($-NH_2$) from a compound. The process is important in the breakdown of **amino acids** in the liver and kidney. Ammonia formed by deamination is converted to **urea** and excreted.

de Broglie wavelength Wavelength of a wave that is associated with the motion of a particle. For a particle of mass *m* moving with speed *v*, the wavelength of the wave is given by $\lambda = h/mv$, where h is **Planck's constant**. It is named after the French physicist Louis de Broglie (1892–1987).

debug To remove a bug from a computer system.

Debye-Hückel theory Theory that explains variations from the ideal behaviour of **electrolytes** in terms of inter-ionic attraction, and assumes that electrolytes in solution are completely dissociated into charged ions. It was named after the Dutch physicists Peter Debye (1884–1966) and Erich Hückel (1896–).

deca- Prefix denoting 10 times in the metric system.

decane $C_{10}H_{22}$ Colourless liquid hydrocarbon; an **alkane** and tenth member of the methane series.

decant Carefully pour away the liquid above a precipitate, once it has settled.

decapod Member of the crustacean order **Decapoda**.

Decapoda *1*. Large order of **crustaceans** that includes crabs, lobsters, crayfish and shrimps. There are more than 8,000 species of decapods, most of which are marine. *2*. Sub-order of **cephalopods** that includes squids and cuttlefish.

■ **decay** *1*. Natural breakdown of an organic substance; **decomposition**. *2*. Breakdown, through **radioactivity**, of a radioactive substance. Such decay is typically exponential. *See also* **half-life**. [2/2/d]

■ **decibel** Unit used for comparing power levels (on a logarithmic scale); one-tenth of a bel. It is commonly used in comparisons of sound intensity. *See also* **bel**.

■ **deciduous** Shedding leaves at certain seasons of the year. The process occurs most commonly in autumn for plants in temperate regions. *See also* **evergreen**. [2/1/b] [2/2/b] [2/2/c]

decimal system Number system that uses the base 10; *i.e.* it uses the digits 1 to 9 and 0.

declination Alternative term for **magnetic declination**.

decomposer Saprophyte organism that breaks down organic materials into simple molecules, *e.g.* **fungi**. [2/2/d]

■ **decomposition** *1.* Breaking down of a chemical compound into its component parts, often brought about by heating. *2.* Rotting of a dead organism, often brought about by bacteria or fungi. [4/5/c]

decrepitation *1.* Crackling sound that is often produced when crystals are heated. It is caused by internal stresses and fracturing. *2.* Fragmentation of powder particles that results from heating.

defect Irregularity in the ordered arrangement of atoms, ions or electrons in a **crystal**.

■ **deficiency disease** Disorder brought about by the lack of a certain food substance in the diet, *e.g.* a vitamin, mineral or amino acid. [2/9/a]

■ **deforestation** Loss of forest trees by felling or by the action of erosion or **acid rain**. Because it results in there being fewer trees to convert atmospheric carbon dioxide into oxygen (by **photosynthesis**), deforestation on a large scale contributes to the **greenhouse effect**. *See also* **afforestation**. [2/6/d]

degaussing Process of demagnetizing a metal object, *e.g.* a ship, by encircling it with an **electric field**.

deglutition Another term for swallowing.

degradation *1.* In chemistry, conversion of a molecule into simpler components. *2.* In physics, the irreversible loss of energy available to do work.

degree *1.* Unit of difference used in temperature scales. *2.* Unit derived by dividing a circle into 360 segments, used to measure angles and describe direction; it is subdivided into minutes and seconds (of arc). Both types of degrees have the symbol °.

degree of freedom One of several variable factors, *e.g.* temperature, pressure and concentration, that must be made constant for the condition of a system at equilibrium to be defined.

dehiscene Opening to liberate seeds; in particular, the fruits of a plant, *e.g.* poppy.

dehiscent Term that describes seeds and berries that burst open when ripe or mature.

dehydrating agent Alternative name for a **desiccant**.

■ **dehydration** Loss or removal of water which, in a living organism, results in a reduction of tissue fluid, possibly to a harmful level. [2/7/a]

dehydrogenase **Enzyme** that catalyses the removal of hydrogen from a compound, and thus causes the compound's **oxidation**.

deionization Method of purifying or otherwise altering the composition of a solution using **ion exchange**.

deliquescence Gradual change undergone by certain substances that absorb water from the atmosphere to become first damp and then aqueous solutions.

delocalization Phenomenon that occurs in certain molecules, *e.g.* benzene. Some of the electrons involved in chemical bonding the atoms are not restricted to one particular bond, but are free to move between two or more bonds. The electrical conductivity of metals is due to the presence of delocalized electrons.

delta metal Tough alloy of copper and zinc that contains a small amount of iron.

delta ray Electron that is ejected from an atom when it is struck by a high-energy particle.

demagnetization Removal of the **magnetism** from a ferromagnetic material (*e.g.* by using a diminishing alternating current field or merely by striking it).

deme Any distinct population of interbreeding organisms that share a set of characteristics. A deme is a sub-division of a species that is much more localized but less differentiated than a **subspecies**. *See also* **race**; **variety**.

demodulation Process by which information is extracted and sorted from the **carrier wave** of a radio broadcast. *See* **amplitude modulation**; **frequency modulation**.

denature To unfold the structure of the **polypeptide** chain of a **protein** by exposing it to a higher temperature or extremes of **pH**. This results in the loss of biological activity and a decrease in solubility.

■ **dendrite** *1.* In biology, an elongated process from a **nerve cell** that links it with another nerve cell. Alternative name: dendron. *2.* In crystallography, a branching **crystal**, such as occurs in some rocks and minerals.

dendrochronology Science of dating based on annual growth rings in trees. Variations in climate produce variations in ring width, and researchers have been able to establish patterns of these variations in certain regions that go back as much as 3,000 years.

dendron Thread-like outgrowth of a **nerve cell** (neurone) that carries impulses away from the cell body. Alternative name for **dendrite**. *See also* **axon**.

■ **denitrification** Process that occurs in organisms, *e.g.* bacteria in soil, which breaks down **nitrates** and **nitrites**, with the liberation of **nitrogen**. [2/2/d]

■ **density** *1*. Mass of a unit volume of a substance. For an object of mass m and volume V, the density d is m/V. It is commonly expressed in units gm cm^{-3} (although the SI unit is kg m^{-3}). *2*. Number of items in a defined surface area or volume (*e.g.* charge density, population density).

dental caries Decay of teeth caused by bacteria that live on food debris in plaque. The bacteria produce acid (as a waste product) that dissolves calcium salts in the tooth **enamel**.

■ **dental formula** Notation that shows the number of each kind of tooth possessed by a **mammal**. The number of teeth in one side of the jaw is given, with the number in the upper jaw before that in the lower jaw, and in the order incisors, canines, premolars, molars; *e.g.* the human formula is 2/2 1/1 2/2 3/3.

dentine Layer that occurs under the **enamel** of teeth. It is similar to bone, but harder, and is perforated by thin extensions from tooth-forming cells.

■ **deoxyribonucleic acid** (DNA) Nucleic acid that is usually referred to by its abbreviation. *See* **DNA**. [2/10/b]

deoxyribonucleotides Compounds that are the fundamental units of **DNA** (deoxyribonucleic acid). Each **nucleotide** contains a nitrogenous **base**, a pentose **sugar** and **phosphoric acid**. The four bases characteristic of deoxyribonucleotides are **adenine, guanine, cytosine** and **thymine**.

depolarization Removal or prevention of electrical polarity or **polarization**.

depression of freezing point Reduction of the **freezing point** of a liquid when a solid is dissolved in it. At constant pressure and for dilute solutions of a non-volatile solvent, the depression of the freezing point is directly proportional to the concentration of the solutes.

derivative In chemistry, a compound (usually organic) obtained from another compound.

Dermaptera Order of winged insects that consists of the earwigs. There are about 1,800 species world-wide.

■ **dermis** Layer of skin that is the innermost of the two main layers. It is composed of **connective tissue** and contains blood, lymph vessels, sensory nerves, hair follicles, sweat glands and some muscle cells. *See also* **epidermis**. [2/4/a] [2/5/a]

Dermoptera Small sub-order of the **insectivores** that includes the 'flying lemur', the most highly developed of all gliding mammals.

DERV Acronym of Diesel Engine Road Vehicle; another name for **diesel fuel**.

desalination Removal of salts (mainly sodium chloride) from seawater to produce water pure enough for drinking, irrigation and use in steam turbines and water-cooled machinery. Methods employed include **distillation, ion exchange, electrodialysis** and **molecular sieves.**

desiccant Substance that absorbs water and can therefore be used as a drying agent or to prevent **deliquescence** (*e.g.* anhydrous calcium chloride, silica gel), often used in a **desiccator**. Alternative name: dehydrating agent.

desiccator Apparatus used for drying substances or preventing **deliquescence**, *e.g.* a closed glass vessel containing a **desiccant**.

desk top publishing (DTP) Technique that uses a **microcomputer** linked to a **word processor** (with access to various type fonts and justification programs) and a laser **printer** to produce multiple copies of a document that rivals conventional printing in quality. The addition of a scanner allows the introduction of simple graphics (illustrations).

desmid Member of the Desmidaceae family of unicellular green **algae**, usually found in unpolluted freshwater habitats and **plankton**.

desorption *1.* Reverse process to **adsorption**. *2.* Removal of an adsorbate from an **adsorbent** in **chromatography**.

desquamation Shedding of the skin, as when a snake or lizard sloughs its skin.

destructive distillation Heating of a solid (such as coal, wood) in a closed vessel to a high temperature until it decomposes and the products of decomposition (*e.g.* gas, tar) can be carried off as vapour and possibly condensed.

■ **detergent** Substance that is used as a cleaning agent. Detergents are particularly useful for cleaning because they lower **surface tension** and emulsify fats and oils, allowing them to go into solution with water without forming a scum with any of the substances that cause **hardness of water**. Soaps act in a similar way, but form an insoluble scum in hard water. Detergents may, however, be a source of **pollution** in rivers and lakes.

determinant *1.* In medicine and biology, region or regions of an **antigen** molecule required for its 'recognition' (binding) by a particle or **antibody**. The selective nature of this molecular interaction confers specificity on the immune reaction of the antibody-producer. *2.* Also in biology, a factor that transmits inherited characteristics, *e.g.* a **gene**.

deuterated compound Substance in which ordinary hydrogen has been replaced by **deuterium**.

■ **deuterium** D_2 One of the three **isotopes** of **hydrogen**, having one neutron and one proton in its atomic nucleus. Its oxide, deuterium oxide, is also known as heavy water. It has a relative atomic mass of 2.0141. Alternative name: heavy hydrogen. *See also* **tritium**.

deuterium oxide D_2O Chemical name of **heavy water**.

deuteron Positively charged particle that is composed of one **neutron** and one **proton**; it is the nucleus of a **deuterium** atom. Deuterons are often used to bombard other particles inside a **cyclotron**.

deviation Error in a compass reading caused by nearby magnetic disturbances.

Dewar flask Double-walled container that has a vacuum between the two walls in order to reduce heat transmission by **conduction** or **convection**. The walls are usually silvered to minimize transmission by radiation. It was named after the British scientist James Dewar (1842–1923). Alternative name: vacuum flask.

dew point Temperature to which air must be cooled in order for it to become saturated with water vapour, which then condenses as mist or dew. Alternative name: dew temperature.

dextrin Polysaccharide of intermediate chain length produced from the action of **amylases on starch**.

dextronic acid Alternative name for **gluconic acid**.

dextrorotatory Describing an **optically active** compound that causes the plane of polarized light to rotate in a clockwise direction. It is indicated by the prefix (+)- or *d*-.

dextrose Alternative name for **glucose**.

■ **diabetes** Disorder caused by a lack of a hormone (**insulin**), which is normally secreted by the islets of Langerhans in the pancreas to control the levels of sugar (glucose) in the blood. Full name: diabetes mellitus. [2/9/c]

diakinesis Phase of **cell division** that occurs at the final stage of **prophase** of the first division in **meiosis**. During this phase the **chromosomes** become short and thick, forming more **chiasmata**, the **nucleoli** and nuclear membrane disappear, and the **spindle** appears for the process of division.

dialysed iron Colloidal solution of iron (III) hydroxide ($Fe(OH)_3$). It is a red liquid, used in medicine.

dialysis Separation of **colloids** from **crystalloids** using selective diffusion through a semipermeable membrane. It is the process by which globular proteins can be separated from low-molecular-weight solutes, as in filtering ('purifying') blood in an artificial kidney machine: the membrane retains protein molecules and allows small solute molecules and water to pass through.

diamagnetism Phenomenon in which magnetic susceptibility is negative, *i.e.* repelled by a magnet.

2,8-diaminoacridine methochloride Alternative name for **acriflavine**.

■ **diamond** Colourless crystalline natural **allotrope** of **carbon**, the hardest mineral. Gem-quality diamonds are used in jewellery, and industrial ones for cutting tools and precision instruments such as watches. In practice, most diamonds are not pure, and so have some degree of colour. *See also* **allotropy**. [4/8/d]

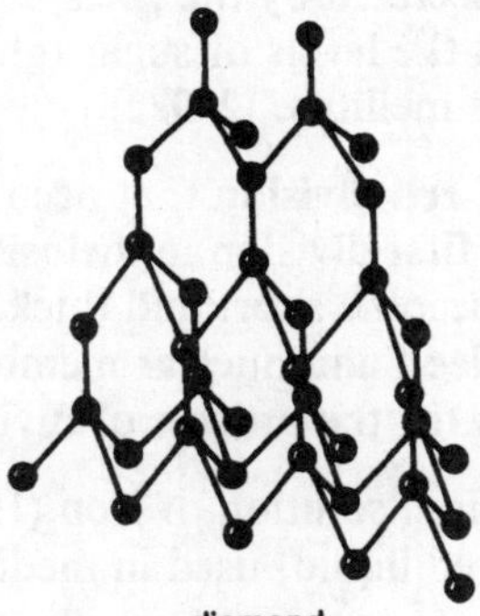

Atomic structure of a diamond

diapause Pause in the development of an individual insect that may occur at any stage of growth – egg, larva or pupa. Diapause is usually induced by an adverse seasonal change, and the organism postpones development until conditions become more favourable.

■ **diaphragm** *1.* In anatomy, a sheet of **muscle** present in mammals, located below the **lungs**. It is attached to the body wall at the sides, and separates the **thorax** from the **abdomen**. During **respiration** the muscle contracts and relaxes, so forming an important part of the mechanism for filling and emptying the lungs. *2.* In human biology, a birth-control device fitted over the entrance to the uterus to prevent the entry of sperm (alternative name: Dutch cap). [2/4/a] [2/5/a] *3.* In physics, a thin membrane that vibrates in response to, or to produce, sound waves, *e.g.* the cone in a loudspeaker.

diaspore Structure that functions as a means of dispersal for plant and fungus species, *e.g.* a **seed** or a **spore**.

diastase Alternative name for **amylase**.

diastema Gap in the jaw of mammals (usually **herbivores**) where there are no teeth. It permits manipulation of leafy food by the animal's tongue.

diastole Phase of the **heartbeat** in which the heart undergoes relaxation and refills with blood from the veins. The term also applies to a contractile vacuole in a cell when it refills with fluid. *See also* **systole**.

diatom Alternative name for a member of the **Bacillariophyta**.

diatomic Describing a molecule that is composed of two identical atoms, *e.g.* O_2, H_2 and Cl_2. *See also* **atomicity**.

diatomite Alternative name for **kieselguhr**.

diazo compound Organic compound that contains two adjacent nitrogen atoms, but only one attached to a carbon atom. Formed by **diazotization**, diazo compounds are very important in synthesis, being the starting point of various dyes and drugs.

diazonium compound Organic compound of the type $RN_2^+X^-$, where R is an **aryl group**. The compounds are colourless solids, extremely soluble in water, used for making

azo dyes. Many of them (particularly the nitrates) are explosive in the solid state.

diazotization Formation of a **diazo compound** by the interaction of sodium nitrite, an inorganic acid, and a primary aromatic amine at low temperatures.

■ **dibasic** Describing an **acid** that contains two replaceable hydrogen atoms in its molecules; *e.g.* carbonic acid, H_2CO_3, sulphuric acid, H_2SO_4. A dibasic acid can form two types of **salts**: a normal salt, in which both hydrogen atoms are replaced by a metal or its equivalent (*e.g.* sodium carbonate, Na_2CO_3), and an **acid salt**, in which only one hydrogen atom is replaced (*e.g.* sodium hydrogensulphate (bisulphate), $NaHSO_4$).

dibromoethane $C_2H_4Br_2$ Volatile liquid that exists in two isomeric forms. It is added to petrol to remove lead. Alternative name: ethylene dibromide.

dicarboxylic acid Organic acid that contains two **carboxyl groups**.

dichlorine oxide Cl_2O Yellowish-red gas which dissolves in water to produce **hypochlorous acid** (HClO) and explodes to give chlorine and oxygen when heated. Alternative name: chlorine (I) oxide.

dichotomous Of plants, divided into two equal branches.

dichroism Property of a few substances that makes them transmit some colours and reflect others, or which display certain colours when viewed from one angle and different colours when viewed from another.

dichromate Salt containing the dichromate (VI) ion ($Cr_2O_7^{2-}$), an **oxidizing agent**; *e.g.* potassium dichromate, $K_2Cr_2O_7$. Alternative name: bichromate.

■ **dicotyledon** Flowering plant that has two seed leaves, broad-veined leaves and stems with **vascular bundles**. *See also* **monocotyledon**. [2/4/d]

dielectric Nonconductor of electricity in which an electric field persists in the presence of an inducing field. A dielectric is the insulating material in a **capacitor**.

dielectric constant Alternative name for relative **permittivity**.

dielectric strength Property of an **insulator** that enables it to withstand electric stress without breaking down.

dielectrophoresis Movement of electrically polarized particles in a variable **electric field**.

Diels-Alder reaction Addition reaction in organic chemistry in which a 6-membered ring system is formed without elimination of any compounds. It was named after the German chemists Otto Diels (1876–1954) and Kurt Alder (1902–58).

diesel fuel Type of liquid fuel used in a diesel engine, consisting of **alkanes** (boiling range 200–350 °C) obtained from **petroleum**. Alternative name: **DERV**.

diet Food that is eaten according to certain criteria. *See also* **balanced diet**.

1,1-diethoxyethane Alternative name for **acetal**.

differentiator Analog computer device whose (variable) output is proportional to the time differential of the (variable) input.

■ **diffraction** Bending of the path of a beam (*e.g.* of light or electrons) at the edge of an object. [5/10/c]

■ **diffraction grating** Optical device that is used for producing **spectra**. It consists of a sheet of glass or plastic marked with closely spaced parallel lines (as many as 10,000 per centimetre). The spectra are produced by a combination of **diffraction** and **interference**. [5/10/c]

diffusion Process by which gases or liquids mix together – *e.g.* in gas exchange between plant leaves and air. *See also* **active transport**; **dialysis**; **osmosis**.

■ **diffusion of gases** Phenomenon by which gases mix together, reducing any concentration gradient to zero; *e.g.* in gas exchange between plant leaves and air. [4/6/e]

■ **diffusion of light** Spreading or scattering of light. [5/10/c]

diffusion of solutions Free movement of **molecules** or **ions** of a dissolved substance through a solvent, resulting in complete mixing. *See also* **osmosis**.

diffusion pump Apparatus used to produce a high vacuum. The pump employs mercury or oil at low vapour-pressure which carries along in its flow molecules of a gas from a low pressure established by a backing pump. Alternative names: condensation pump, vacuum pump.

■ **digestion** Breakdown of complex substances in food by **enzymes** in the **alimentary canal** to produce simpler soluble compounds, which pass into the body by **absorption** and **assimilation**. Ultimately **carbohydrates** (*e.g.* starch, sugar) are broken down to glucose, **proteins** to amino acids, and **fats** to fatty acids and glycerol. [2/4/a] [2/5/a] [2/7/a]

digit In biology, a finger or toe, or other analogous structure.

digital/analog converter Device that converts digital signals into continuously variable electrical signals for use by an **analog computer**.

digital computer Computer that operates on **data** supplied and stored in digital or number form.

digital display Display that shows readings of a measuring machine, clock, etc. by displaying numerals.

digitalis Potent **alkaloid** that is extracted from plants of the genus *Digitalis* (foxgloves). It is used in medicine as a heart stimulant.

dihydrate Chemical (a **hydrate**) whose molecules have two associated molecules of **water of crystallization**; *e.g.* sodium dichromate, $Na_2Cr_2O_7.2H_2O$.

2,3-dihydroxybutanedioic acid Alternative name for **tartaric acid.**

dihydroxypurine Alternative name for **xanthine.**

dihydroxysuccinic acid Alternative name for **tartaric acid.**

dilate To widen; to produce **dilation.**

■ **dilation** In biology and medicine, the widening or expansion of an organ, opening, passage or vessel (*e.g.* of the **cervix** during birth). Alternative term: dilatation. [2/7/a]

diluent Solvent used to reduce the strength of a **solution.**

dilute *1.* To reduce the strength of a **solution** by adding water or other solvent. *2.* Describing a solution in which the amount of **solute** is small compared to that of the **solvent.**

dilution Process that involves the lowering of concentration.

dimension Power to which a fundamental unit is raised in a derived unit; *e.g.* acceleration has the dimensions of $[LT^{-2}]$, *i.e.* +1 for length and −2 for time, equivalent to length divided by the square of time.

dimensional analysis Prediction of the relationship of quantities. If an equation is correct the **dimensions** of the quantities on each side must be identical. It is an important way of checking the validity of an equation.

dimer Chemical formed from two similar **monomer** molecules.

dimethylbenzene Alternative name for **xylene.**

dimorphism Existence of two distinct forms of an organism; *e.g.* sexual dimorphism in some animals, aerial and submerged leaves of some aquatic plants.

dinitrogen oxide N_2O Colourless gas made by heating ammonium nitrate and used as an anaesthetic. Alternative names: nitrogen oxide, nitrous oxide, dental gas, laughing gas.

dinitrogen tetroxide N_2O_4 Colourless solid which melts at 9 °C to give a pale yellow liquid. It is used as an oxidant, *e.g.* in rocket fuel.

dinosaur Extinct reptile which existed during the **Mesozoic** era. Dinosaurs were a successful and diverse group that dominated the terrestrial environment of Earth for 140 million years. Climatic changes induced by continental shifts are thought by many to have caused their extinction about 65 million years ago at the end of the **Cretaceous** period, although other theories give different reasons.

■ **diode** *1.* Electron tube (valve) containing two electrodes, an anode and a cathode. *2.* **Rectifier** made up of a **semiconductor** crystal with two terminals.

dioecious Describing plants in which the male and female reproductive organs are borne on different individuals or parts.

diol Organic compound containing two **hydroxyl groups** and having the general formula $C_nH_{2n}(OH)_2$. Diols are thick liquids or crystalline solids, and some have a sweet taste. Ethane-1,2-diol (ethylene glycol) is the simplest diol, widely used as a solvent and as an antifreeze agent. Alternative names: dihydric alcohol, glycol.

■ **dioptre** Unit that is used to express the power of a lens. It is the reciprocal of the **focal length** of the lens in metres. The power of a convergent lens with a focal length of one metre is said to be +1 dioptre. The power of a divergent lens is given a negative value.

dioxan $(CH_2)_2O_2$ Colourless liquid cyclic **ether**. It is inert to many reagents and frequently used in mixtures with water to increase the solubility of organic compounds such as alkyl halides. Alternative name: 1,4-dioxan.

dioxin $C_{12}H_4Cl_4O_2$ Highly toxic by-product of organic synthesis, used as a defoliant and herbicide, which in even small doses can cause allergic skin reactions.

dip Angle measured in a vertical plane between the direction of the Earth's **magnetic field** and the horizontal. Alternative name: magnetic dip.

diploblastic Describing an animal in which the body wall consists of two cellular layers: the outer **ectoderm** and the inner **endoderm**, sometimes separated by a middle **mesoglea** layer. No organs develop from the latter, but only from the ectoderm and endoderm, as in *e.g.* coelenterates.

■ **diploid** In a cell or organism, describing the existence of **chromosomes** in homologous pairs, *i.e.* twice the **haploid** number ($2n$). It is characteristic of all animal cells except **gametes**. In lower plants exhibiting **alternation of generations**, the **sporophyte** is diploid and the **gametophyte** is haploid. [2/8/b]

diplotene In **meiosis**, the stage in late prophase when the pairs of **chromatids** begin to separate from the **tetrad** formed by the association of homologous **chromosomes. Chiasmata** can be seen at this stage.

dipole *1*. Pair of equal and opposite electric charges at a (short) distance from each other. *2*. Simple type of radio aerial (antenna) consisting of a pair of horizontal or vertical metal rods in line, the signal being picked up at their adjacent ends.

dipole moment Product of one charge of a **dipole** and the distance between the charges.

■ **direct current** (d.c.) **Electric current** that flows always in the same direction (as opposed to **alternating current**). [5/5/b]

direct dye Dye that does not require a **mordant**.

directive body Alternative name for **polar body**.

disaccharide Sugar with molecules that consist of two **monosaccharide** units linked by **glycoside** bonds; *e.g.* **sucrose, maltose, lactose**.

discharge *1*. High-voltage 'spark' (current flow) between points of large potential difference (*e.g.* lightning). *2*. Removal of the charge between the plates of a **capacitor** by allowing current to flow out of it. *3*. Removal of energy from an **electrolytic cell** (battery or accumulator) by allowing current to flow out of it. *4*. Process by which **ions** are converted to neutral atoms at an electrode during **electrolysis** (by gain or loss of electrons).

discharge tube Evacuated or gas-filled tube with sealed-in **electrodes** between which a high-voltage electric **discharge** takes place. *See also* **Geissler tube**.

disintegration Break-up of an atomic nucleus either through bombardment by subatomic particles or through radioactive **decay**.

disintegration constant Probability of a radioactive **decay** of an atomic nucleus per unit time. Alternative names: decay constant, transformation constant.

disk Magnetic disc used to record data in computers. *See also* **floppy disk; hard disk.**

diskette Alternative name for a **floppy disk**.

dislocation Imperfection in a **crystal** lattice.

disodium oxide Alternative name for **sodium monoxide**.

disodium tetraborate Alternative name for **borax**.

dispersal Means by which a plant's seeds are scattered; *e.g.* by the wind, stuck to the fur of animals, eaten (with fruit) by birds and even by flowing water.

■ **dispersion** Splitting of an electromagnetic radiation (*e.g.* visible light) into its component **wavelengths** when it passes through a medium (because different wavelengths undergo different degrees of **diffraction** or **refraction**). [5/10/c]

displacement reaction Alternative name for **substitution reaction**.

display Short name for a **liquid-crystal display** (LCD) or a **visual display unit** (VDU).

dissociation Temporary reversible chemical decomposition of a substance into its component atoms or molecules, which often take the form of **ions**. *E.g.* it occurs when most ionic compounds dissolve in water.

dissociation constant Equilibrium constant of a **dissociation** reaction, and therefore a measure of the affinity of atoms or molecules in a compound.

dissolving Making a **solution**, usually by adding a solid **solute** to a liquid **solvent**.

distance ratio Alternative name for **velocity ratio**.

distillate Condensed liquid obtained by **distillation**.

■ **distillation** Method for purification or separation of liquids by heating to the **boiling point**, condensing the vapour, and collecting the distillate. Formerly, the method was used to produce distilled water for chemical experiments and processes that required water to be much purer than in the mains water supply. In this application distillation has been largely superseded by ion exchange.

distilled water *See* **distillation**.

distortion Change from the ideal shape of an object or image, or in the form of a wave pattern (*e.g.* an electrical signal).

disulphuric acid Alternative name for **oleum**.

dithionate Salt derived from **dithionic acid** ($H_2S_2O_6$). Alternative name: hyposulphate.

dithionic acid $H_2S_2O_6$ Strong acid that decomposes slowly in concentrated solutions and when heated. Alternative name: hyposulphuric acid.

dithionite Name that is given to any of the **salts** of **dithionous acid**, all of which are strong reducing agents.

dithionous acid $H_2S_2O_4$ Strong but unstable acid that is found only in solution.

diurnal During the day, daily (*e.g.* a diurnal animal is active by day).

divalent Capable of combining with two atoms of hydrogen or their equivalent. Alternative name: bivalent.

divergent evolution Alternative name for **adaptive radiation**.

diverging lens Lens that spreads out a beam of light passing through it, often a **concave** lens.

division *1.* In biological **classification,** one of the major groups into which the plant kingdom is divided. The members of the group, although often quite different in form and structure, share certain common features, *e.g.* bryophytes include the mosses and liverworts. Divisions are divided into classes, often with an intermediate subdivision. The equivalent of a division in the animal kingdom is a **phylum**. *2.* In biology, the formation of a pair of **daughter cells** from a parent cell (*see* **cell division**).

dl-form Term indicating that a mixture contains the **dextrorotatory** and the **laevorotatory** forms of an optically active compound in equal molecular proportions.

D-lines Pair of characteristic lines in the yellow region of the spectrum of sodium, used as standards in spectroscopy.

■ **DNA** Abbreviation of deoxyribonucleic acid, the long thread-like molecule that consists of a double helix of polynucleotides (combinations of a sugar, organic bases and phosphate) held together by **hydrogen bonds**. DNA is found chiefly in **chromosomes** and is the material that carries the hereditary information of all living organisms (although most, but not all, viruses have only RNA, ribonucleic acid). [2/10/b]

DNA hybridization Technique in which **DNA** from one species is induced to undergo base pairing with DNA or **RNA** from another species to produce a hybrid DNA (a process known as annealing).

dodecanoic acid Alternative name for **lauric acid**.

dodecylbenzene $C_6H_5(CH_2)_{11}CH_3$ Hydrocarbon of the **benzene** family, important in the manufacture of **detergents**.

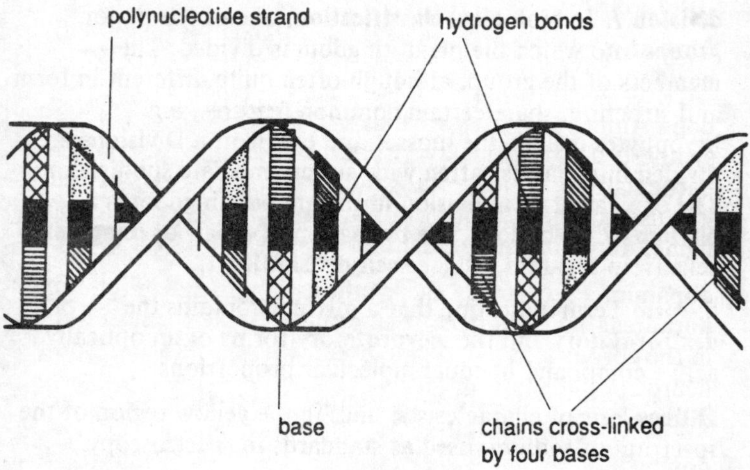

Structure of DNA

dolomite $CaCO_3.MgCO_3$ Naturally occurring calcium-magnesium carbonate, named after a mountain range in which it is found. Alternative name: pearl spar.

domain In a **ferromagnetic** substance, a microscopic region in which atomic **magnetic moments** can be aligned to give it permanent **magnetism**.

■ **dominant** In a **heterozygous** organism, describing the **gene** that prevents the expression of a **recessive** allele in a pair of homologous **chromosomes**. Thus the **phenotype** of an organism with a combination of dominant and recessive genes is similar to that with two dominant alleles. [2/7/b] [2/9/c]

■ **donor** *1*. In chemistry, an atom that donates both electrons to form a **coordinate bond**. *2*. In medicine, person or animal that donates blood, tissue or organs for use by another person or animal. *3*. In physics, an element that donates electrons to form an *n*-type **semiconductor**; *e.g.* antimony or arsenic may be donor elements for germanium or silicon. *See also* **doping**.

dopa Amino acid derivative that is a precursor in the synthesis of **dopamine** and is laevorotatory (L-dopa). Found particularly in the adrenal gland and in some types of beans, it is used in the treatment of Parkinson's disease. Alternative name: dihydroxyphenylalanine.

dopamine Precursor in the synthesis of **adrenalin** and **noradrenalin** in animals. It is found in highest concentration in the corpus striatum of the brain, where it functions as a neurotransmitter. Low levels are associated with Parkinson's disease in human beings.

■ **doping** Addition of impurity atoms to *e.g.* germanium or silicon to make them into **semiconductors**. Doping with an element of valency 5 (*e.g.* antimony, arsenic, phosphorus) donates electrons to form an *n*-type semiconductor; doping with an element of valency 3 (*e.g.* aluminium, boron, gallium) donates 'holes' to form a *p*-type semiconductor.

■ **Doppler effect** Phenomenon in which the wavelength of a wave (electromagnetic or sound) changes because its source is moving relative to the observer. If the source is approaching the observer, the wavelength decreases and the frequency increases (*e.g.* sound rises in pitch). If the source is receding, the wavelength increases and the frequency decreases (*e.g.* sound falls in pitch, or light shifts towards the red end of the spectrum). The effect is used in astronomy (called red shift) and in Doppler **radars**, which distinguish between moving objects and stationary ones. It was named after the Austrian physicist Christian Doppler (1803–53).

dormancy Period of minimal metabolic activity of an organism or reproductive body. It is a means of surviving a period of adverse environmental conditions, *e.g.* cold or drought. Examples of some dormant structures are spores, cysts and perennating organs of plants. Environmental factors such as day length and temperature control both the onset and ending of dormancy. Dormancy may also be prompted and terminated by the action of hormones, *e.g.* abscinic acid and **gibberellins**, respectively. *See also* **aestivation**; **hibernation**.

dorsal Describing the upper surface of an organism. In vertebrates this is the surface nearest to the backbone. In plants the dorsal surface of a leaf is the upper surface, usually with the thicker **cuticle**.

dosimeter Instrument used for measuring the dose of **radiation** received by an individual or an area.

■ **double bond Covalent bond** that is formed by sharing two pairs of electrons between two atoms. [4/8/d]

double decomposition Reaction between two dissolved ionic substances (usually salts) in which the reactants 'change partners' to form a new soluble salt and an insoluble one, which is precipitated. *E.g.* solutions of sodium chloride (NaCl) and silver nitrate ($AgNO_3$) react to form a solution of sodium nitrate ($NaNO_3$) and a precipitate of silver chloride (AgCl).

■ **double recessive** Homozygous condition in which two **recessive alleles** of a particular **gene** are at the same locus on a pair of **homologous chromosomes**, so that the recessive form of the gene is expressed in the **phenotype**. [2/7/b]

double refraction Phenomenon shown by certain crystals (*e.g.* calcite) that splits an incident ray of light into two refracted rays (termed ordinary and extraordinary rays) polarized at right-angles to each other. Alternative name: birefringence.

■ **Down's syndrome** Abnormal chromosomal condition caused by the presence of an extra **autosomal** chromosome 21. It is characterized by abnormal physical development and mental retardation. Alternative names: mongolism, trisomy 21. [2/9/c]

drug Chemical (often a biochemical) that has a stimulating, narcotic or healing effect on the body. *See also* **addiction**; **analgesic**; **antibiotic**.

drupe Fleshy fruit covered by **epicarp** and containing one or more seeds surrounded by a hard stony wall, the **endocarp**. Drupes with one seed include plums and cherries; many-seeded drupes include holly and elder fruits. Blackberries and raspberries are collections of small drupes or drupelets. Alternative name: pyrenocarp. *See also* **berry**.

■ **dry cell** Electrolytic cell containing no free liquid **electrolyte**. A moist paste of ammonium chloride (NH_4Cl) often acts as electrolyte. Dry cells are used in batteries for torches, portable radios, etc.

dry ice Solid **carbon dioxide**.

DTP Abbreviation for **desk top publishing**.

ductile Describing a substance that exhibits **ductility**.

ductility Property of a metal that allows it to be drawn out into a wire; *e.g.* copper and silver are very ductile metals.

■ **ductless gland** Alternative name for **endocrine gland**. [2/7/a]

■ **duodenum** First section of the **small intestine** which is mainly secretory in function, producing digestive **enzymes**. It also

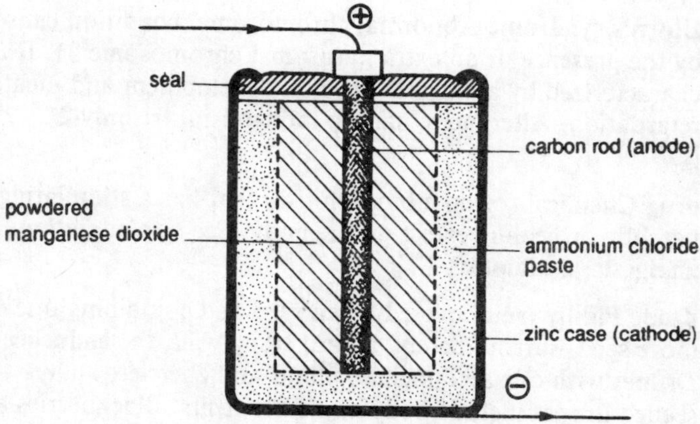

Construction of a dry cell battery

receives pancreatic juice from the **pancreas** and bile from the gall bladder. [2/4/a] [2/5/a] [2/7/a]

Duralumin Strong **alloy** of aluminium containing 4%copper and traces of magnesium, manganese and silicon, much used in the aerospace industry.

dura mater In vertebrates, the connective tissue containing blood vessels that surrounds the brain and spinal cord.

dynamic isomerism Alternative name for **tautomerism**.

dynamics Branch of **mechanics** that deals with the actions of forces on objects in motion.

dynamite High explosive consisting of **nitroglycerine** (and sometimes other explosives) absorbed into the earthy mineral kieselguhr. It was invented by the Swedish chemist Alfred Nobel (1833–96).

dynamo Device for converting mechanical energy into electrical energy in the form of direct current (d.c.). It consists of conducting coils that rotate between the poles of a powerful magnet.

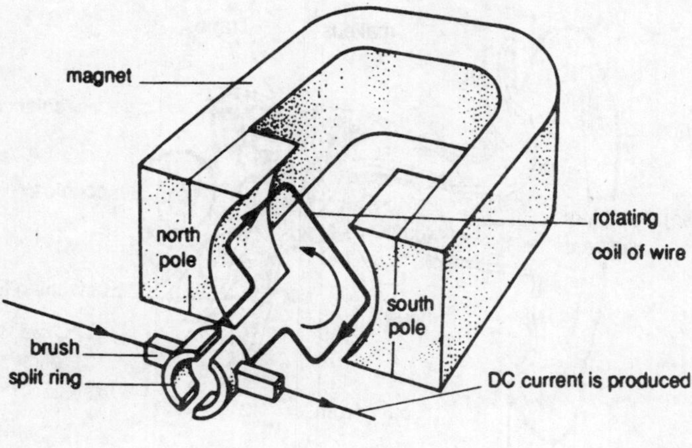

Principle of the dynamo

dyne Force that gives an object of mass 1 gram an acceleration of 1 cm^{-2}. The SI unit of force is the **newton,** equal to 10^5 dynes.

dysprosium Dy Silvery metallic element in Group IIIB of the Periodic Table (one of the **lanthanides**), used to make magnets and nuclear reactor **control rods**. At. no. 66; r.a.m. 162.50.

E

■ **ear** One of a pair of hearing and balance sensory organs situated on each side of the head of vertebrates. In mammals it consists of three parts: the **outer**, **middle** and **inner ear**. [2/4/a] [2/5/a] [2/7/a]

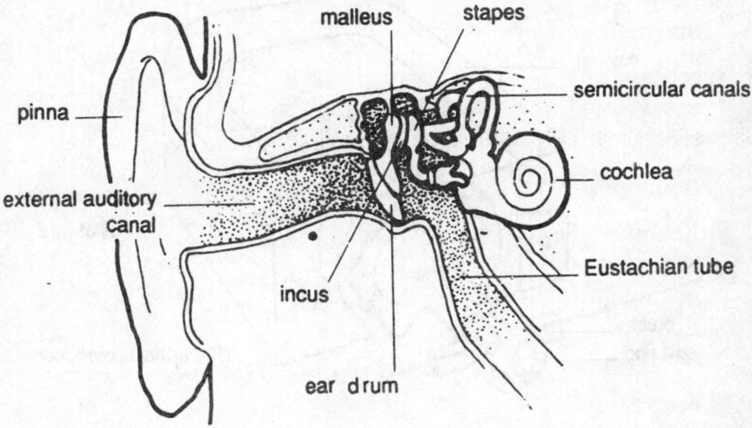

The human ear

ear drum Membrane at the inner end of the auditory canal of the **outer ear** which transmits sound vibrations to the **ear ossicles** of the **middle ear**. Alternative name: tympanum.

ear ossicle Small bone found in the **middle ear** of vertebrates. In mammals there are three in each ear, which transmit sound waves from the **ear drum** to the oval window (fenestra ovalis), which vibrates fluid in the **inner ear**, causing an impulse to travel via the **auditory nerve** to the brain. The three mammalian ossicles are the **malleus, incus** and **stapes**. Amphibians, reptiles and birds have only one ear ossicle, the columella auris.

■ **earth** In electric circuits, a connection to a piece of metal that is in turn linked to the Earth. It has the effect of preventing any earthed apparatus from retaining an **electric charge**.

Earth's magnetism *See* **dip**; **magnetic north**; **magnetic pole**; **magnetic storm**.

ecdysis Alternative name for **moulting** of an animal's **exoskeleton** to facilitate growth (*e.g.* as do many insects and crustaceans). *See also* **desquamation**.

echinoderm Member of the phylum **Echinodermata**.

■ **Echinodermata** Phylum of marine invertebrates that are characterized by chalky plates embedded in the skin and the possession of **tube feet**, powered by a water vascular system. Echinoderms include starfish, brittle stars, sea-cucumbers, sea-lilies and sea-urchins. [2/4/b]

■ **echo** Sound or **electromagnetic radiation** that is reflected or refracted, so that it is delayed and received as a signal distinct from that directly transmitted and it apparently comes from a different direction. [5/4/c] [5/5/a]

echolocation Method of estimating the location of something by transmitting high-frequency sounds and detecting their echoes. It is the basic principle underlying the operation of **sonar**.

echo sounder Device for estimating the depth of sea beneath a vessel by measuring the time taken for an ultrasound impulse to reach the sea bed and for its **echo** to return to a receiver. *See also* **sonar**.

eclipse Interception of the light of one heavenly body by another. In a solar eclipse, the light of the Sun is blocked out by the intervention of the Moon. In a lunar eclipse, the Earth moves between the Moon and the Sun so that the Moon reflects no sunlight and does not shine. When the light-emitting body is not totally obscured, it is termed a partial eclipse (as opposed to a total eclipse).

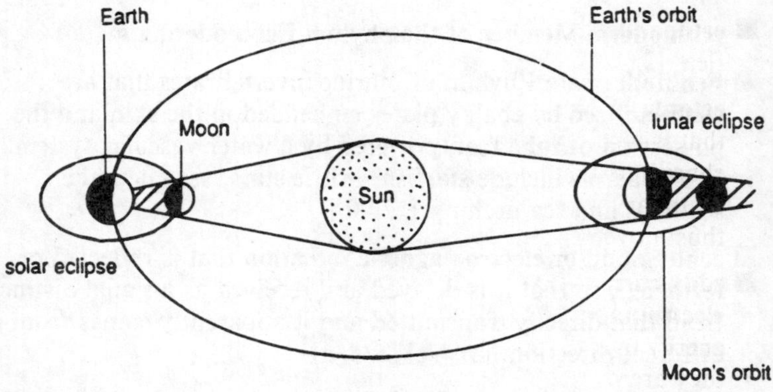

Lunar and solar eclipses

■ **ecological niche** Position that a particular species occupies within an **ecosystem**. The term both describes the function of a species in terms of interactions with other species, *e.g.* feeding behaviour, and defines the physical boundaries of the environment occupied by the species. Bats, for instance, are said to occupy an airborne niche. [2/4/d] [2/7/d]

ecology Study of interaction of organisms between themselves and with their physical environment.

ecosphere Planet looked at from an ecological point of view. Earth is the only known ecosphere.

ecosystem Natural unit that contains living and non-living components (*e.g.* a **community** and its **environment**) interacting and exchanging materials, and generally balanced as a stable system; *e.g.* grassland or rainforest.

ectoderm Outermost **germ layer** of the embryo of a **metazoan** which develops into tissues of the **epidermis**, *e.g.* skin, hair, sense organs, enamel and in lower animals the **nephridia**. Alternative name: epiblast.

■ **ectoparasite Parasite** that lives on the outside of its **host**; *e.g.* flea. [2/4/d] [2/6/a]

ectoplasm Non-granulated jelly-like outer layer of **cytoplasm** that is located below the **plasma membrane**. It is characteristic of most amoeboid animal cells, in which at its boundary with **plasmasol** it aids cytoplasmic streaming, and thus movement; *e.g.* in amoeboid protozoa and leucocytes.

■ **eddy current** Electric current within a conductor caused by **electromagnetic induction**. Such currents result in losses of energy in electrical machines (*e.g.* a transformer, in which they are overcome by laminating the core), but are utilized in induction heating and some braking systems. [5/7/c]

Edison accumulator Alternative name for **nickel-iron accumulator**.

EDTA Abbreviation of ethylenediaminetetracetic acid. It is a white crystalline organic compound, used generally as its sodium salt as an analytical reagent and antidote for heavy-metal poisoning, when it forms **chelates.**

effective resistance Total alternating current resistance of a conductor of electricity.

effector Tissue or organ that responds to a nervous stimulus (*e.g.* **endocrine gland**, **muscle**).

efferent Leading away from, as applied to vessels, fibres and ducts leading from **organs.** *See also* **afferent.**

effervescence Evolution of bubbles of a gas from a liquid.

efficiency Ratio of usable energy output to energy input of a machine, generally expressed as a percentage. For a simple machine it is the ratio of the **mechanical advantage** (force ratio) to the **velocity ratio** (distance ratio).

efflorescence Property of certain crystalline salts that lose **water of crystallization** on exposure to air and become powdery; *e.g.* sodium carbonate decahydrate (washing soda), $Na_2CO_3.10H_2O$. It is the opposite of **deliquescence.**

effort In a simple machine (*e.g.* lever, pulley) the force that is applied to move a load. The ratio of the load to the effort is the **mechanical advantage** (force ratio).

effusion Passage of gases under pressure through small holes.

egg *See* **ovum.**

egg membrane Thin protective membrane that surrounds the fertilized **ovum** of animals. It is secreted by the **oöcyte** and the **follicle** cells. *See also* **chorion.**

ego Psychological term for the aspect of personality concerned with rationality and common sense. *See also* **id.**

einsteinium Es Radioactive metallic element in Group IIIB of the Periodic Table (one of the **lanthanides**). It has several **isotopes**, with half-lives of up to 2 years. At. no. 99; r.a.m. (most stable isotope) 254.

Elasmobranchii Class of animals that includes cartilaginous fish (sharks, skates and rays). Alternative name: **Chondrichthyes**.

■ **elasticity** Property of a material that makes it resume its original shape after a deforming force acting upon it is removed. But beyond the **elastic limit**, permanent deformation occurs. [4/4/a]

elastic limit A stress applied to an elastic material produces a proportional strain (*see* **Hooke's law**) up to a point, after which the slightest increase in stress produces a large strain (the material becomes plastic) and permanent deformation. That point is the elastic limit.

elastic modulus Ratio of stress to strain for an elastic material below its **elastic limit**. *See also* **bulk modulus; Hooke's law; rigidity modulus; Young's modulus**.

elastomer Substance that returns to its original shape when stress is removed. *See* **elasticity**.

E-layer Alternative name for the **Heaviside layer**.

electrical condenser Alternative name for **capacitor**.

electrical line of force Line radiating from an **electric field**.

electric arc Luminous discharge produced by the passage of high-voltage electricity between two **electrodes**.

■ **electric charge** Excess or deficiency of electrons in an object, giving rise to an overall positive or negative electric charge respectively. [5/9/a]

■ **electric current** Drift of electrons through a conductor in the same direction, usually because there is a **potential difference** across it. [5/8/b]

electric displacement Electric charge per unit area, in coulombs per square metre (C m^{-2}). Alternative name: electric flux density.

■ **electric field** Region surrounding an **electric charge** in which a charged particle is subjected to a force. [5/7/d]

■ **electric flux** Lines of force that make up an **electric field**. [5/7/d]

■ **electricity** *1*. Branch of science that is concerned with all phenomena caused by static or dynamic electric charges. *2*. Supply of electric current. [5/7/c]

electric meter Instrument for measuring the consumption of mains electricity (in terms of watt-hours, or units).

■ **electric motor** Device that converts electrical energy into mechanical energy. A simple electric motor consists of a current-carrying coil that rotates in the magnetic field between the poles of a permanent magnet. *See also* **induction motor; linear motor**. [5/7/c]

electric polarization Difference between the displacement of charge and the electric field strength in a **dielectric**.

electrocardiogram (ECG) Record of the electrical activity of the heart produced by an electrocardiograph machine. Alternative name: cardiogram.

electrocardiograph Machine that uses electrodes taped to the body to produce **electrocardiograms**.

electrochemical series List of metals arranged in order of their **electrode potentials**. A metal will displace from their

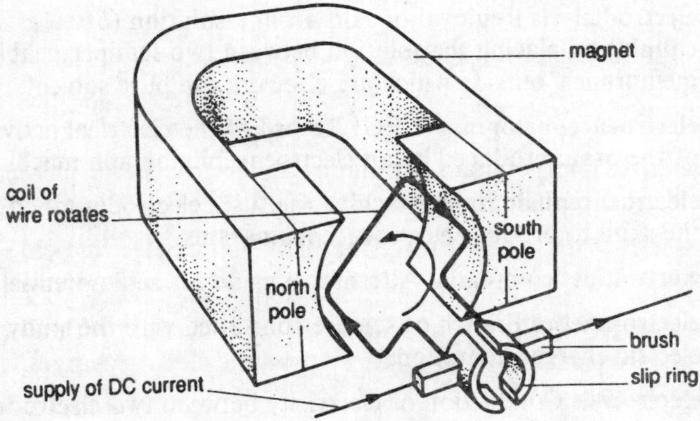

Principle of the d.c. electric motor

salts metals lower down in the series. Alternative name: electromotive series.

■ **electrochemistry** Branch of science that is concerned with the study of electrical chemical energy, such as the effects of electric current on chemicals (particularly **electrolytes**) and the generation of electricity by chemical action (as in an **electrolytic cell**). [4/6/d]

electrode *1.* Conducting plate (anode or cathode) that collects or emits electrons from an **electrolyte** during **electrolysis**. *2.* Conducting plate in an **electrolytic cell** (battery), discharge tube or vacuum tube.

electrodeposition Deposition of a substance from an **electrolyte** on to an electrode, as in **electroplating**.

electrode potential Electric potential developed by a substance in equilibrium with a solution of its ions.

electrodialysis Removal of salts from a solution (often a **colloid**) by placing the solution between two semipermeable membranes, outside which are electrodes in pure solvent.

electroencephalogram (EEG) Record of the electrical activity of the brain produced by an electroencephalograph machine.

electroencephalograph Machine that uses electrodes taped to the skull to produce **electroencephalograms**.

electrokinetic potential Alternative name for **zeta potential**.

electrokinetics Branch of science concerned with the study of electric charges in motion.

■ **electrolysis** Conduction of electricity between two **electrodes**, through a solution of a substance (or a substance in its molten state) containing **ions** and accompanied by chemical changes at the electrodes. *See also* **electroplating**. [4/10/c]

electrolysis, Faraday's law of Mass of a given element liberated during **electrolysis** is directly proportional to the amount of electricity consumed. When the same quantity of electricity is passed through different electrolytes, the masses of different substances liberated are directly proportional to the relative atomic weights of the substances divided by their ionic charges.

■ **electrolyte** Substance that in its molten state or in solution can conduct an electric current. [4/6/d]

electrolytic capacitor Electrolytic cell in which a thin film of non-conducting substance has been deposited on one of the electrodes by an electric current.

electrolytic cell Apparatus that consists of **electrodes** immersed in an **electrolyte**.

electrolytic dissociation Partial or complete reversible decomposition of a substance in solution or the molten state into electrically charged **ions**.

electrolytic rectifier Rectifier that consists of two **electrodes** and an **electrolyte**, in which the current flows in one direction only.

■ **electrolytic refining** Method of obtaining pure metals by making the impure metal the **anode** in an **electrolytic cell** and depositing the pure metal on the **cathode**. [4/5/c]

electrolytic separation Method of separating metals from a solution by varying the applied potential according to the **electrode potentials** of the metals.

electromagnet Temporary magnet consisting of a current-carrying coil of wire wound on a ferromagnetic core. It is the basis of many items of electrical equipment, *e.g.* electric bells and lifting magnets.

■ **electromagnetic induction Electromotive force** (e.m.f.) produced in a conductor when it is moved in a **magnetic field**. It is the working principle of an electrical generator (*e.g.* **dynamo**). It can give rise to a **back e.m.f.** and **eddy currents**. [5/7/c]

■ **electromagnetic interaction** Interaction between electrically charged **elementary particles**. [4/8/d]

electromagnetic radiation Energy that results from moving electric charges and travels in association with electric and magnetic fields, *e.g.* radio waves, heat rays, light and X-rays, which form part of the **electromagnetic spectrum**.

■ **electromagnetic spectrum** Range of frequencies over which **electromagnetic radiations** are propagated. In order of increasing frequency (decreasing wavelength) it consists of radio waves, microwaves, infra-red radiation, visible light, ultraviolet radiation, X-rays and gamma-rays. [5/9/a]

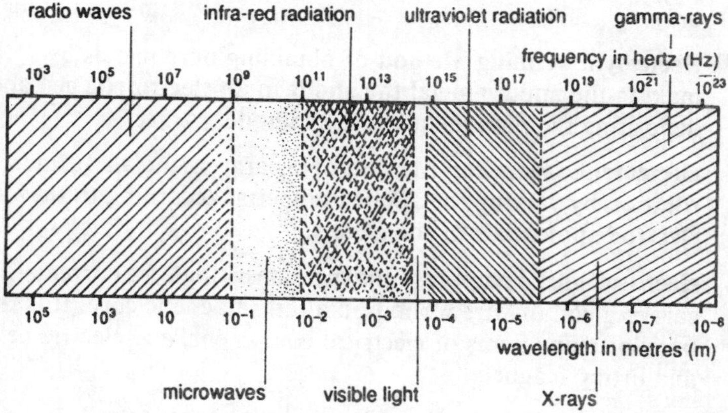

Electromagnetic spectrum

electromagnetic units (EMU) System of electrical units based on a unit magnetic pole.

electromagnetic wave Wave formed by electric and magnetic fields, *i.e.* of **electromagnetic radiation.** Such waves do not require a medium in which to propagate, and will travel in a vacuum.

electromagnetism Combination of an **electric field** and a magnetic field, their interaction with stationary or moving **electric charges,** and their study and application. It therefore applies to light and other forms of **electromagnetic radiation**, as well as to devices such as electromagnets, electric motors and generators.

electrometallurgy Electrical methods of processing metals. Such methods are used in industry, *e.g.* for **electrodeposition**, **electrolytic refining** and the separation of metals from their ores.

electromotive force (e.m.f.) Potential difference of a source of electric current, such as an **electrolytic cell** (battery) or generator. Often it can be measured only at equilibrium (when there is no current flow). Alternative name: voltage.

electromotive series Alternative name for **electrochemical series**.

■ **electron** Fundamental negatively-charged subatomic particle (radius 2.81777×10^{-15} m; rest mass 9.109558×10^{-31}; charge 1.602192×10^{-19}). Every neutral atom has as many orbiting electrons as there are **protons** in its **nucleus**. A flow of electrons constitutes an electric current. [4/8/d]

electron affinity Energy liberated when an **electron** is acquired by a neutral atom.

electron capture *1*. Formation of a negative ion through the capture of an **electron** by a substance. *2*. Transformation of a **proton** into a **neutron** in the nucleus of an atom (accompanied by the emission of X-rays) through the capture of an orbital electron, so converting the element into another with an atomic number one less.

electron charge (e) Charge on an electron, a fundamental physical constant equal to 1.602102×10^{-19} coulombs.

■ **electron configuration** *See* **configuration**. [4/8/d]

electron-deficient compound Compound in which there are insufficient electrons to form two-electron **covalent bonds** between all the adjacent atoms *e.g.* boranes.

electron density Density of **electric charge**.

electron diffraction Method of determining the arrangement of the atoms in a solid, and hence its crystal structure, by the **diffraction** of a beam of **electrons**.

electron donor Alternative name for **reducing agent**.

electronegativity Power of an atom in a molecule to attract electrons. For elements arranged in the Periodic Table, it increases up a group and across a period.

electron gun Electrode assembly for producing a narrow beam of electrons, as used, *e.g.*, in **cathode-ray tubes**.

electronics Branch of science concerned with the study of electricity in a vacuum, in gases and in **semiconductors**.

electron lens Arrangement of **electrodes** or of permanent or **electromagnets** used to focus or divert beams of electrons in the same way as an optical lens modifies a beam of light, as in, *e.g.*, an **electron microscope**.

■ **electron microscope** Instrument that uses a beam of electrons from an **electron gun** to produce magnified images of extremely small objects, beyond the range of an optical microscope.

electron multiplier Alternative name for **photomultiplier**.

electron octet *See* **octet**.

electron optics Study of the control of free electrons by curved electric and magnet fields, particularly the use of such fields to focus and deflect beams of electrons.

electron probe microanalysis (EPM) Quantitative analysis of small amounts of substances by focusing a beam of electrons on to a point on the surface of the sample so that characteristic X-ray intensities are produced.

electron radius (r_e) Radius of an electron, a fundamental physical constant equal to 2.81777×10^{-15} m.

electron rest mass (m_e) Mass of an electron, a fundamental physical constant equal to 9.10908×10^{-31} kg.

electron-spin resonance (ESR) Branch of microwave spectroscopy in which radiation of measurable frequency and wavelength is used to supply energy to protons.

electron transport Process found mainly in **aerobic respiration** and **photosynthesis** that provides a source of energy in the form of **ATP**. Hydrogen atoms are used in this system and taken up by a hydrogen carrier, *e.g.* **FAD**; the **electrons** of the hydrogen pass along a chain of carriers which are in turn reduced and oxidized. This is coupled to the formation of ATP. The hydrogen atoms together with oxygen eventually form water.

electron volt (eV) General unit of energy equal to work done on an electron when it passes through potential gradient of 1 volt.

electrophile Electron-deficient ion or molecule that attacks molecules of high electron density. *See also* **nucleophile**.

electrophilic addition Chemical reaction that involves the addition of a molecule to an **unsaturated** organic compound across a double or triple bond.

electrophilic reagent Reagent that attacks molecules of high electron density.

electrophilic substitution Reaction that involves the substitution of an atom or group of atoms in an organic compound. An **electrophile** is the attacking substituent.

electrophoresis Movement of charged **colloid** particles in a solution placed in an **electric field**.

■ **electrophorus** Device for producing charges by electrostatic induction. [5/9/b]

■ **electroplating** Deposition of a thin coating of a metal using **electrolysis**. The object to be plated is the **cathode**, and the plating metal is the **anode**. Metal ions are stripped from the anode, pass through the electrolyte, and are deposited on the cathode.

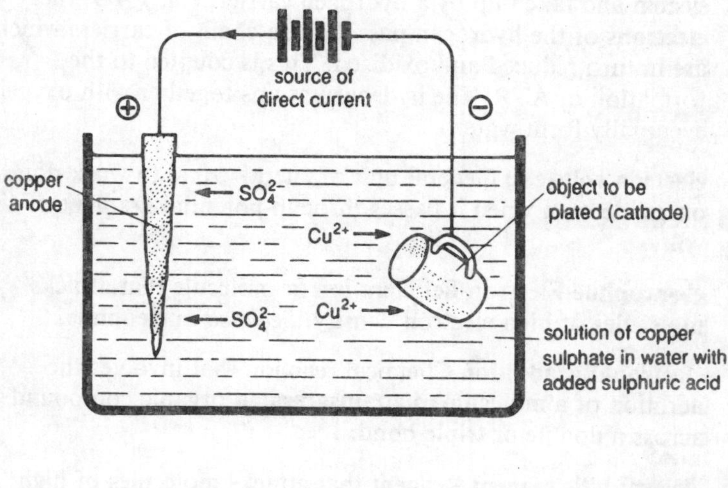

Simple electroplating bath

electropositive Tending to form positive ions; having a deficiency of electrons.

■ **electroscope** Instrument for detecting electric charges or gaseous ions. [5/9/b]

■ **electrostatic field** Electric field associated with stationary electric charges. [5/9/b]

electrostatic printer Machine that prints by heat-fusing finely powdered carbon to paper on which characters have been 'imprinted' as patterns of electrostatic charge. *See also* **xerography**.

electrostatics Branch of electricity concerned with the study of electrical charges at rest.

electrostatic units (ESU) System of electrical units based on the force exerted between two electric charges.

electrostriction Change in the dimensions of a **dielectric** that is caused by the reorientation of molecules when an **electric field** is applied.

electrotype Printing plate made by the electrodeposition of copper in a mould (often made from paper or plastic that has been pressed onto metal type).

■ **electrovalent bond** Alternative name for **ionic bond**. [4/8/d] [4/8/e]

electrovalent crystal Crystal in which the **ions** are linked by a bond resulting from electrostatic attraction. Alternative name: ionic crystal.

electrum Naturally occurring **alloy** of gold and up to 40 per cent silver that looks superficially like pure gold.

■ **element** *1.* Substance consisting of similar **atoms** of the same atomic number. It cannot be decomposed by chemical action to a simpler substance. Alternative name: chemical element. *See also* **isotope**; **Periodic Table**. *2*. In physics, one of several lenses in a compound lens, or one of several components in an electrical circuit. *3*. General term for the high-resistance coil in an electric fire or heater. [4/7/e]

■ **elementary particle** Subatomic particle not known to be made up of simpler particles. [4/7/f]

elevation of boiling point Rise in the boiling point of a liquid caused by dissolving a substance in the liquid.

Elinvar Alloy of chromium, iron and nickel, used for making watch hairsprings.

eluate Solution obtained from **elution**.

eluent Solvent used for **elution**, the mobile phase.

elution Removal of an adsorbed substance by washing the adsorbent with a solvent (eluent). The technique is used in some forms of **chromatography**.

■ **embryo** *1*. In animals, organism formed after cleavage of the **zygote** before hatching or birth. A maturing embryo is often termed a **foetus**. *2*. In lower plants, structure that develops from the zygote of **bryophytes** and **pteridophytes**, and in higher plants is the **seed** before germination.

embryology Study of **embryos**, their formation and development.

embryo sac *1*. In lower plants, female **gametophyte** containing the **eggs**, synergids and polar and antipodal **nuclei**. *2*. In flowering plants, large oval cell in which egg fertilization and embryonic development occur.

emery Naturally occurring form of **aluminium oxide** (corundum) containing oxides of iron and silica as impurities. It is used as an abrasive.

e.m.f Abbreviation for **electromotive force.**

emission spectrum Spectrum obtained when the light from a luminous source undergoes dispersion and is observed directly.

empirical formula Chemical formula that shows the simplest ratio between atoms of a molecule. *E.g.* glucose, molecular formula $C_6H_{12}O_6$, and acetic (ethanoic) acid, $C_2H_4O_2$, both have the same empirical formula, CH_2O. *See also* **molecular formula; structural formula**.

■ **emulsion Colloidal** suspension of one liquid dispersed in another *(e.g.* milk).

■ **enamel** White protective calcified outer coating of the crown of the **tooth** of a vertebrate. It is produced by epidermal cells and consists almost entirely of calcium salts bound together by **keratin** fibres. It is extremely hard and takes much longer to decay in an old skull than the bone. The calcium salts are readily attacked by the acid produced by bacteria in **plaque**, causing **dental caries**. [2/1/c]

encephalin One of two peptides which are natural **analgesics**, produced in the brain and released after injury. The encephalins have properties similar to **morphine**. Alternative names: endorphin, enkephalin.

■ **endemic** Describing a disease that continually occurs among people or animals in a particular region. *See also* **epidemic**; **pandemic**. [2/7/c]

■ **endocarp** Inner of the (usually) three layers of a fruit, which may be a hard stone, as in a **drupe**. *See also* **epicarp**; **pericarp**. [2/3/a]

■ **endocrine gland** Ductless organ or discrete group of cells that synthesize **hormones** and secrete them directly into the bloodstream. Such glands include the **pituitary**, **pineal**, **thyroid**, **parathyroid** and **adrenal glands**, the **gonads** and **placenta** (in mammals), **islets of Langerhans** (in the pancreas) and parts of the **alimentary canal**. Their function is parallel to the **nervous system**, that of regulation of responses in animals, but it is much slower than a nervous response. Alternative name: ductless gland. [2/7/a]

endocrinology Study of structure and function of **endocrine glands** and the roles of their **hormones** as the chemical messengers of the body. [2/7/a]

endoderm Innermost **germ layer** of the **zygote** of a **metazoan**, which eventually develops into the **gastrula** wall as well as the lining of its archenteron canal and its derivatives, *e.g.* **liver** and **pancreas**. In birds and mammals it also forms the **yolk** sac and **allantois**.

endodermis Innermost layer of the **cortex** of plant tissue that surrounds the **vascular tissue**. It consists of a single layer of cells which controls the movement of water and minerals between the cortex and the **stele**.

endogenous Produced or originating within an organism.

endolymph In vertebrates, fluid that fills the cavity of the **middle**, **inner** and the **semicircular canals** of the **ear**.

■ **endoparasite** **Parasite** that lives inside the body of its **host**, *e.g* fluke, malaria parasite, tapeworm. [2/4/d]

endoplasm Central portion of **cytoplasm**, surrounded by the **ectoplasm** and containing **organelles**. Alternative name: plasmasol.

endoplasmic reticulum (ER) Structure that occurs in cells in the form of a flattened membrane-bound sac of cell **organelles**, continuous with the outer **nuclear membrane**.

When covered with **ribosomes** it is termed rough ER; in their absence smooth ER. Its main function is the synthesis of **proteins** and their transport within or to the outside of the cell. In liver cells ER is involved in detoxification processes and in **lipid** and **cholesterol** metabolism. In association with the **Golgi apparatus**, ER is involved in **lysosome** production.

endorphin One of a group of **peptides** that are produced by the **pituitary gland** and which act as painkillers in the body.

endoscope Tubular optical device, perhaps using **fibre optics**, that is inserted into a natural orifice or a surgical incision to study organs and tissues inside the body.

■ **endoskeleton Skeleton** that lies inside the body of an organism, *e.g.* the bony skeleton of **vertebrates**. *See also* **exoskeleton**. [2/3/a]

endosperm Food-storage **tissue** that surrounds the developing **embryo** in monocotyledonous seed plants, providing nourishment.

endospore Tough asexual **spore** that is formed by some bacteria and some fungi to resist adverse conditions.

endothelium Tissue formed from a single layer of cells found lining spaces and tubes within the body, *e.g.* lining the **heart** in vertebrates. *See also* **epithelium**.

endothermic Describing a process in which heat is taken in; *e.g.* in many chemical reactions. *See also* **exothermic**.

end point Point at which a chemical reaction is complete, such as the end of a **titration**. *See also* **volumetric analysis**.

■ **energy** *1.* Capacity for doing work, measured in joules. Energy takes various forms: *e.g.* **kinetic energy**, **potential energy**, electrical energy, chemical energy, **heat**, **light** and

sound. All forms of energy can be regarded as being aspects of kinetic or potential energy; *e.g.* heat energy in a substance is the kinetic energy of that substance's molecules. *2.* Fuel or power source (*e.g.* in such expressions as alternative energy, nuclear energy). [5/9/a]

■ **energy level** The energy of **electrons** in an atom is not continuously variable, but has a discrete set of values, *i.e.* energy levels. At any instant the energy of a given electron can correspond to only one of these levels. *See* **Bohr theory**. [4/8/d]

■ **energy transfer** Process by which chemical energy in the form of food is converted to heat energy and other forms in living organisms. *See* **biomass**; **food chain**; **pyramid of numbers**; **trophic level**. [2/3/d]

engine Machine for converting one form of energy into another form, or for producing a **mechanical advantage**. In many engines, combustion of a fuel converts its chemical energy into usable mechanical energy. Alternative name: motor.

enol Organic compound that contains the group C=CH(OH); the alcoholic form of a **ketone**.

■ **enrichment** *1.* In microbiology, isolation of a particular type of organism by enhancing its growth over other organisms in a mixed population. *2.* In the nuclear industry, processing an ore or fuel to increase the proportion of a required fissionable **isotope**; *e.g.* increasing the amount of uranium-235 in uranium fuel rods for nuclear reactors.

enterokinase **Enzyme** in blood that helps to bring about clotting.

enthalpy (*H*) Amount of heat energy a substance possesses, measurable in terms of the heat change (ΔH) that accompanies a chemical reaction carried out at constant pressure. In any system, $H = U + pV$, where U is the internal energy, p the pressure and V the volume.

entropy (S) In **thermodynamics**, quantity that is a measure of a system's disorder, or the unavailability of its energy to do work. In a reversible process the change in entropy is equal to the amount of energy adsorbed divided by the absolute temperature at which it is taken up.

■ **environment** All the conditions in which an organism lives, including the amount of light, acidity (pH), temperature, water supply and presence of other (competing) organisms. [2/5/c]

■ **enzyme Protein** that acts as a **catalyst** for the chemical reactions that occur in living systems. Without such a catalyst most of the reactions of metabolism would not occur under the conditions that prevail. Most enzymes are specific to a particular **substrate** (and therefore a particular reaction) and act by activating the substrate and binding to it.

■ **ephemeral** *1.* In botany, plant with a short life-cycle; germination to seed-production may occur several times in one year. *See also* **annual**; **biennial**; **perennial**. *2.* In zoology, animal with a short life-cycle; a member of the insect order Ephemeroptera (*e.g.* mayflies). [2/3/d]

epiblast Alternative name for **ectoderm**.

epicarp The outer of the (usually) three layers of a fruit, which may be a skin, rind or hard shell. *See also* **endocarp**; **pericarp**.

■ **epidemic** Describing a disease that, for a limited time, affects many people or animals in a particular region. *See also* **endemic**; **pandemic**. [2/7/c]

■ **epidemiology** Study of diseases as they affect the population, including their incidence and prevention. [2/7/c]

■ **epidermis** Layer of cells at the surface of a plant or animal. In plants and some invertebrates, it forms a single protective layer, often overlaid by a **cuticle** which is impermeable to water. In vertebrates, it forms the skin and is composed of several layers of cells, the outermost becoming keratinized (*see* **keratinization**). [2/4/a] [2/5/a] [2/7/a]

epididymis Long coiled tube in the **testes** of some vertebrates through which **sperm** from the **seminiferous tubules** pass, before going into the **vas deferens** and to the exterior.

epigamic Describing animal characteristics that are attractive to the opposite sex during courtship; *e.g.* in birds, the colour of feathers and types of songs.

epigeal *1.* Describing a type of seed germination in which the **cotyledons** appear above the ground. *See also* **hypogeal**. *2.* Describing animals that live above ground, as opposed to underground.

■ **epiglottis** Valve-like flap of **cartilage** in mammals that closes the opening into the **larynx**, the glottis, during swallowing. [2/4/a] [2/5/a] [2/7/a]

epinephrine Alternative name for **adrenalin**.

epiphysis *1.* Growing end of a bone, at which cartilage is converted to solid bone. *2.* Alternative name for **pineal gland**.

■ **epiphyte** Plant that grows on another plant but does not feed on it (*i.e.* it is not a parasite); *e.g.* various lichens and mosses. Epiphytes use other plants for support, and absorb water from the air. *See also* **saprophyte**. [2/6/a]

epithelium Animal lining **tissue** of varying complexity, whose main function is protective. It may be specialized for a particular organ; *e.g.* squamous epithelium lines capillaries and is permeable to molecules in solution, glandular epithelium contains cells that are secretory.

■ **epithermal neutron Neutron** that has energy of between 10^{-2} and 10^2 electron volts (eV); a neutron having energy greater than that associated with thermal agitation.

epoxide Organic compound whose molecules include a three-membered oxygen ring (a cyclic ether).

epoxy resin Synthetic polymeric **resin** with **epoxide** groups. Such resins are used in surface coatings and as adhesives. Alternative name: epoxide resin.

Epsom salt Alternative name for **magnesium sulphate**.

equation *See* **chemical equation** and the following article.

equation of motion Five parameters can describe the motion of an object moving in a straight line: initial velocity (v_1), final velocity (v_2), acceleration (a), distance travelled (s) and time taken (t). These give rise to five equations, each containing only four of the parameters:

$$s = \tfrac{1}{2}t(v_1 + v_2)$$
$$s = v_1 t + \tfrac{1}{2}at^2$$
$$s = v_2 t - \tfrac{1}{2}at^2$$
$$v_2 = v_1 + at$$
$$v_2^2 = v_1^2 + 2as$$

equation of state Any formula that connects the volume, pressure and temperature of a given system, *e.g.* **van der Waals' equation**.

equation of time Difference between apparent time (time given by a clock) and mean solar time (sundial time).

equilibrium *1.* An object is in **equilibrium** when the forces acting on it are such that there is no tendency for the object to move. *2.* State in which no change occurs in a system if no change occurs in the surrounding environment (*e.g.* chemical equilibrium).

equilibrium constant (K_c) Concentration of the products of a chemical reaction divided by the concentration of the reactants, in accordance with the chemical equation, at a given temperature.

equimolecular mixture Mixture of substances in equal molecular proportions.

equivalence point Theoretical **end point** of a **titration**. *See also* **volumetric analysis**.

equivalent proportions, law of When two **elements** both form chemical compounds with a third element, a compound of the first two contains them in the relative proportions they have in compounds with the third one. *E.g.* Carbon combines with hydrogen to form methane, CH_4, in which the ratio of carbon to hydrogen is 12:4; oxygen also combines with hydrogen to form water, H_2O, in which the ratio of oxygen to hydrogen is 16:2. Carbon and oxygen form the compound carbon monoxide, CO, in which the ratio of carbon to oxygen is 12:16. Alternative name: law of reciprocal proportions.

equivalent weight Number of parts by mass of an element that can combine with or displace one part by mass of hydrogen.

erase head Part of a tape recorder, video recorder or computer input/output device that erases recorded signals (data), on tape or disk, before the **write head** records new ones.

erbium Er Metallic element in Group IIIB of the Periodic Table (one of the **lanthanides**), used in making lasers for medical applications. Its pink oxide is used as a pigment in ceramics. At. no. 68; r.a.m. 167.26.

erg Energy transferred when a force of 1 **dyne** moves through 1 cm, equivalent to 10^{-7} joules.

ergosterol White crystalline **sterol**. It occurs in animal **fat** and in some micro-organisms. In animals it is converted to vitamin D_2 by ultraviolet radiation.

erosion Gradual removal of something by natural means; *e.g.* of soil by the action of wind and rain, or of tissue from the cervix of the womb during childbirth.

erythrocyte Red **blood cell**. It contains **haemoglobin** and carries **oxygen** around the body. In mammals, erythrocytes have no **nuclei**. *See also* **leucocyte**.

escape velocity Velocity that an object at a given point requires to escape from a particular **gravitational field**; *e.g.* a rocket leaving Earth's gravity needs an escape velocity of 11,200 m s^{-1}. The velocity must be such that the **kinetic energy** of the object is greater than its **potential energy** resulting from the gravitational field. Alternative name: escape speed.

essential amino acid Any **amino acid** that cannot be manufactured in some vertebrates, including human beings. These acids must therefore be obtained from the diet. They are as follows: arginine, histidine, isoleucine, leucine, lysine, methionine, phenylalanine, threonine, tryptophan and valine.

essential fatty acid Any **fatty acid** that is required in the diet of mammals because it cannot be synthesized. They include linoleic acid and γ-linolenic acid, obtained from plant sources.

essential oil Volatile oil with a pleasant odour, obtained from various plants. Such oils are widely used in perfumery.

ester Compound formed when the hydrogen atom of the **hydroxyl group** in an oxygen-containing acid is replaced by an **alkyl group**, as when a **carboxylic acid** reacts with an **alcohol**; *e.g.* acetic (ethanoic) acid, CH_3COOH, and ethanol (ethyl alcohol), C_2H_5OH, react to form the ester ethyl acetate (ethanoate), $CH_3COOC_2H_5$.

esterification Formation of an **ester**, generally by reaction between an **acid** and an **alcohol**.

ethanal Alternative name for **acetaldehyde**.

ethanal trimer Alternative name for **paraldehyde**.

ethanamide Alternative name for **acetamide**.

ethane C_2H_6 Gaseous **alkane** which occurs with **methane** in natural gas.

ethanedioic acid Alternative name for **oxalic acid**.

ethanediol $HOCH_2CH_2OH$ Syrupy liquid **glycol** (dihydric alcohol), used for making polymers and as an antifreeze. Alternative names: ethylene glycol, glycol.

ethane-propane rubber (EPR) Synthetic rubber prepared by **polymerization** of **ethane** and **propane**.

ethanoate Alternative name for **acetate**.

ethanoic acid Alternative name for **acetic acid**.

ethanol C_2H_5OH Colourless liquid **alcohol**. It is the active constituent of alcoholic drinks; it is also used as a fuel and in the preparation of esters, ethers and other organic compounds. Alternative names: ethyl alcohol, alcohol.

ethanoyl Alternative name for **acetyl group**.

ethanoyl chloride Alternative name for **acetyl chloride**.

ethene $CH_2=CH_2$ Colourless gas with a sweetish smell, important in chemical synthesis. Alternative name: ethylene.

ether Organic compound that has the general formula ROR′, where R and R′ are **alkyl groups**. The commonest, diethyl oxide (diethyl ether, or simply ether), $(C_2H_5)O_2$, is a useful volatile solvent formerly used as an anaesthetic.

ethology Scientific study of animal behaviour in the wild.

ethyl acetate Alternative name for **ethyl ethanoate**.

ethyl alcohol Alternative name for **ethanol**.

ethylbenzene Alternative name for **styrene**.

ethyl carbamate Alternative name **urethane**.

ethylene Alternative name for **ethene**.

ethylene dibromide Alternative name for **dibromoethane**.

ethylene glycol Alternative name for **ethanediol**.

ethylene tetrachloride Alternative name for **tetrachloroethene**.

ethyl ethanoate Colourless liquid **ester** with a fruity smell, produced by the reaction between ethanol (ethyl alcohol) and ethanoic (acetic) acid, used as a solvent and in medicine. Alternative name: ethyl acetate.

ethyne Alternative name for **acetylene**.

ethynide Alternative name for **acetylide**.

■ **etiolation** Phenomenon that occurs in green plants grown without light, which appear yellow due to lack of formation of **chlorophyll** and are abnormally long-stemmed; the leaves also become reduced. [2/5/a] [2/3/d]

etiology Alternative spelling of **aetiology**.

eubacteria Largest group of **bacteria**, containing the most commonly encountered forms that inhabit soil and water. The group contains Gram-positive bacteria and green photosynthetic bacteria.

eucaryote Alternative name for **eukaryote**.

eugenics Theory and practice of improving the human race through genetic principles. This can range from the generally discredited idea of selective breeding programmes to counselling of parents who may be carriers of harmful genes.

eukaryote Cell with a certain level of complexity. Eukaryotes have a **nucleus** separated from the **cytoplasm** by a nuclear membrane. Genetic material is carried on **chromosomes** consisting of **DNA** associated with **protein**. The cell contains membrane-bounded organelles, *e.g.* **mitochrondria** and **chloroplasts**. All organisms are eukaryotic except for **bacteria** and **cyanophytes**, which are **prokaryotes**. Alternative name: eucaryote.

eukaryotic Describing or relating to a **eukaryote**.

europium Eu Silvery-white metallic element in Group IIIB of the Periodic Table (one of the **lanthanides**), used in nuclear reactor **control rods**. At. no. 63; r.a.m. 151.96.

Eustachian tube Channel that connects the **middle ear** with the **pharynx** at the back of the throat in mammals and some other vertebrates. It ensures that the air pressure on each side of the ear drum is equal. It was named after the Italian anatomist Bartolomeo Eustachio (1520–74).

■ **eutectic mixture** Mixture of substances in such proportions that no other mixture of the same substances has a lower freezing point.

Eutheria Placental mammals, a subclass that includes all mammals except the **monotremes** and **marsupials**.

eutrophic Describing a lake or other body of water that is well supplied with nutrients. *See also* **oligotrophic**.

eutrophication Large increase in nutrients in lakes and other fresh water, leading to overgrowth of **algae** and other plants, with consequent decrease of oxygen and depletion of fish stocks. It may be caused by run-off of agricultural fertilizers or by pollution.

evaluation Review of *e.g.* an experiment to try to improve on the method used so that it becomes more accurate or efficient.

evaporation Process by which a liquid changes to its vapour. It can occur (slowly) at a temperature below the boling point, but is faster if the liquid is heated and fastest when the liquid is boiling. [4/6/e]

■ **evergreen** Plant that possesses leaves throughout all the seasons, *e.g.* pines and firs. The leaves are shed, but only after several years and then not all at once. *See also* **deciduous**. [2/1/b] [2/2/b]

■ **evolution** Successive altering of **species** through time. Evolutionary theory states that the origin of all species is through evolution, and thus they are related by descent. *See also* **Darwinism**; **natural selection**. [2/9/c]

excess electron Electron that is added to a **semiconductor** from a **donor** impurity. *See also* **doping**.

excitation Addition of energy to a system, such as an atom or nucleus, causing it to transfer from its **ground state** to one of higher energy.

excitation energy Energy required for **excitation.**

excited state Energy state of an atom or molecule that is higher than the **ground state,** resulting from **excitation.** *See also* **energy level.**

exclusion principle Alternative name for the **Pauli exclusion principle.**

■ **excretion** Removal of waste products of **metabolism,** carried out by elimination from the body or storage in insoluble form. Products of protein metabolism are the main substances liberated (in the form of **urea** or **uric acid**). The chief organs of excretion in vertebrates are the **kidneys.** [2/7/a]

exhalation The action of breathing out.

■ **exocrine gland** Gland that discharges secretions into ducts, *e.g.* salivary glands. *See also* **endocrine gland.** [2/7/a]

exogamy Outbreeding. *See also* **inbreeding.**

exogenous Originating outside an organism, organ or cell. The term may refer to such things as substances (*e.g.* nutrients) or stimuli (*e.g.* light).

■ **exoskeleton Skeleton** located on the external part of the body; *e.g.* in arthropods the exoskeleton is impregnated with **chitin,** which gives the animal protection. *See also* **endoskeleton.** [2/3/a]

■ **exothermic** Describing a process in which heat is evolved. [4/6/c]

expansion of gas Increase in volume of an ideal gas is at the rate of $^1/_{273}$ of its volume at 0 °C for each degree rise in temperature. *See* **Charles' law.**

expansivity Increase in size of a substance per unit temperature rise. Linear expansivity relates to increase in length of a solid, superficial expansivity to increase in area of a solid, and volume expansivity to increase in volume of a solid, liquid or gas (*see* **expansion of gas**). Alternative name: thermal expansion.

explosion Rapid production of heat and pressure from a chemical or nuclear reaction, accompanied by the evolution of large volumes of gas and a destructive shock wave.

exponential growth Growth that occurs, *e.g.* in cultures of micro-organisms, in which a population of cells increases in numbers logarithmically.

■ **extracellular** External to a cell; in a multicellular organism, extracellular tissue may still be within the organism. *See also* **intracellular**. [2/5/a] [2/7/a]

■ **extrinsic semiconductor Semiconductor** that has its conductivity increased by the introduction of tiny, but controlled, amounts of certain impurities. *See* **doping**.

■ **eye** Sense organ for detecting light. Its structure varies among organisms; *e.g.* the ocellus found in some invertebrates is simple. The vertebrate eye is complex in comparison, as are the **compound eyes** of adult insects. In the human eye, light passes through the transparent cornea and lens, which together focus it on the light-sensitive retina at the back of the eye. The amount of light entering is controlled by the pupil, whose size can be changed by the iris. Ciliary muscles stretch the lens to change its shape for focusing on near or far objects. A liquid (aqueous humour) between the cornea and the lens, and a jelly-like fluid (vitreous humour) between the lens and the retina keep the eyeball in shape. [2/4/a] [2/5/a] [2/7/a]

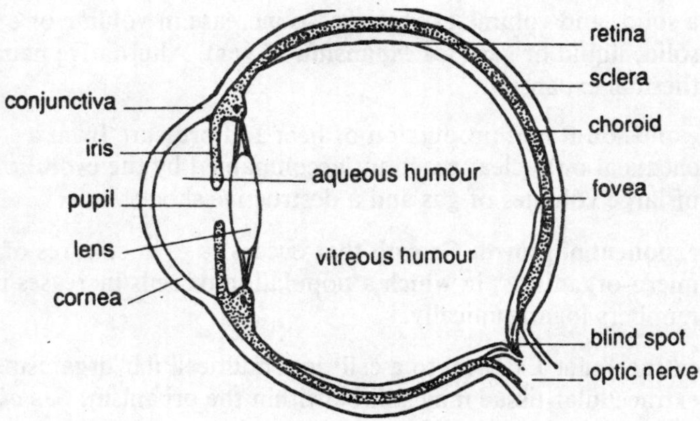

The human eye

eyespot Primitive light-sensitive structure found in many unicellular organisms; it contains **carotenoid** pigment.

■ **eye tooth** Alternative name for **canine tooth**. [2/5/a]

F

face-centred cube Crystal structure that is cubic with an **atom** or **ion** at the centre of each of the six faces of the cube in addition to the eight at its corners.

facilitated diffusion Mode of transport through a **membrane** that involves carrier molecules in the membrane, which eases the transport of a specific substance but does not involve the use of energy; *e.g.* the uptake of glucose by erythrocytes (red blood cells). The transport system can become saturated with the transported substance, in contrast to simple diffusion.

F-actin *See* **actin**.

FAD Abbreviation of **flavin adenine dinucleotide**.

faeces Undigested food that is eliminated from the **alimentary canal** via the **anus** after water and useful salts have been absorbed by the **colon**.

Fahrenheit scale Temperature scale on which the freezing point of water is 32 °F and the boiling point 212 °F. It was named after the German physicist Gabriel Fahrenheit (1686–1736). *See also* **Celsius scale**.

Fajans' rules Set of rules that state when **ionic bonds** are likely in a chemical compound (as opposed to covalent bonds). They were named after the Polish chemist Kasimir Fajans (1887–1975).

Fallopian tube One of a pair of tubes that in female mammals conducts **ova (eggs)** from an **ovary** to the **uterus** (womb) by ciliary action. **Fertilization** can occur when **sperm** meet eggs in the tube. It was named after the Italian anatomist Gabriel Fallopius (1523–62). Alternative name: oviduct.

fall-out Radioactive substances that fall to Earth from the atmosphere after a nuclear explosion.

false fruit Fruit in which the structure is formed from an enlarged receptacle; *e.g.* strawberry. Alternative name: pseudocarp.

false pregnancy Alternative name for **pseudopregnancy**.

family *1.* In biological **classification**, one of the groups into which an order is divided, and which is itself divided into genera; *e.g.* Canidae (dogs). *2.* A group consisting typically of parents and their children or offspring; *e.g.* a herd of elephants or pride of lions often consists of one family group. [2/3/b] [2/4/b]

farad (F) Unit of electrical **capacitance**, defined as the capacitance that, when charged by a potential difference of 1 volt, carries a charge of 1 coulomb.

faraday (F) Unit of electric charge equal to the quantity of charge that during **electrolysis** liberates one gram equivalent of an element. It has the value 9.65×10^4 coulombs per gram-equivalent. It was named after the British scientist Michael Faraday (1791–1867). *See also* **farad**.

Faraday cage Shield, commonly made of metal wire, used to protect apparatus or equipment from external electric fields.

Faraday constant (*F*) Fundamental physical constant, the electric charge carried by one mole of singly charged ions or electrons, equal to 9.6487×10^4 coulombs per mole. It is the product of the **Avogadro constant** and the **electron charge**.

Faraday effect Rotation of the plane of vibration of a beam of polarized light passing through a substance such as glass, in the direction of an applied magnetic field.

Faraday's laws of electrolysis *1.* The amount of chemical decomposition that takes place during **electrolysis** is proportional to the electric current passed. *2.* The amounts of substances liberated during electrolysis are proportional to their chemical equivalent weights.

■ **Faraday's laws of electromagnetic induction** *1.* An induced **electromotive force** is established in an electric circuit whenever the magnetic field linking that circuit changes. *2.* The magnitude of the induced electromotive force in any circuit is proportional to the rate of change of the **magnetic flux** linking the circuit. [5/7/c]

farina *1.* Starch or flour. *2.* Alternative name for **pollen**.

fascicle Alternative name for **vascular bundle**.

fast neutron Neutron produced by **nuclear fission** that has lost little of its energy and travels too fast to produce further fission and sustain a **chain reaction** (unlike a slow, or thermal, neutron).

fat *See* **fats and oils**.

fatigue *1.* In biology, inability of an organ or organism to function to its full capacity that results from overactivity. *2.* In physics, permanent weakness in a substance that results from stresses placed upon it.

fats and oils Naturally occurring **esters** (of **glycerol** and **fatty acids**) that are used as energy-storage compounds by animals and plants. They are hydrocarbons and members of a larger class of naturally occurring compounds called **lipids**.

fatty acid Monobasic **carboxylic acid**, an essential constituent of **fats and oils**. The simplest fatty acids are formic (methanoic) acid, HCOOH, and acetic (ethanoic) acid, CH_3COOH.

fatty degeneration Disease of tissue caused by poisoning or lack of oxygen, in which droplets of fat form within cells.

feather Outgrowth of a bird's skin made of **keratin**. There are various types, including contour feathers, flight feathers and down, which traps air and acts as heat insulation. [2/3/b]

feature Noticeable external aspect of an organism's appearance.

feedback Process in which a system or device is controlled or modified as a result of its activity. In positive feedback, the activity is increased; in negative feedback, it is decreased.

Fehling's solution Test reagent consisting of two parts: a solution of copper (II) sulphate and a solution of potassium sodium tartrate and sodium hydroxide. When the two solutions are mixed, an alkaline solution of a soluble copper (II) complex is formed. In the presence of an **aldehyde** or **reducing sugar**, a pink-red precipitate of copper (I) oxide forms. It was named after the German chemist Hermann Fehling (1812–85). *See also* **Fehling's test**.

Fehling's test Test for an **aldehyde** group or **reducing sugar**, indicated by the formation of copper (I) oxide as a pink-red precipitate with **Fehling's solution**. *See also* **Benedict's test**.

feldspar Any of a large group of igneous crystalline rocks, chiefly silicates of aluminium with potassium, sodium and calcium. Feldspar is used in making porcelain, tiles and glazes. Alternative name: felspar.

felspar Alternative name for **feldspar**.

■ **femur** *1.* In four-legged (tetrapod) vertebrates, the thigh bone. *2.* In insects, the segment of the leg nearest the body. [2/3/a]

fermentation Energy-producing breakdown of organic compounds by micro-organisms (in the absence of oxygen); *e.g.* the breakdown of **sugar** by yeasts into ethanol, carbon dioxide and organic acids. Fermentation is a type of **anaerobic respiration**.

fermium Fm Radioactive element in Group IIIB of the Periodic Table (one of the **actinides**). It has several **isotopes**, with half-lives of up to 95 days. At. no. 100; r.a.m. (most stable isotope) 257.

ferric Trivalent iron. Alternative names: iron (III), iron (3+).

ferricyanide $[Fe(CN)_6]^{3-}$ Very stable complex ion of iron (III). A solution of the potassium salt gives a deep blue precipitate (Prussian blue) in the presence of iron (II) (ferrous) ions. Alternative name: hexacyanoferrate (III).

ferrimagnetism Property of certain compounds in which the **magnetic moments** of neighbouring ions align in anti-parallel fashion (*e.g.* **ferrites**).

ferrite Non-conducting ceramic material that exhibits ferrimagnetism; general formula MFe_2O_4, where M is a divalent metal of the **transition elements**. Ferrites are used to make powerful magnets in radars and other high-frequency electronic apparatus, such as computer memories.

ferrocene $C_{10}H_{10}Fe$ Orange organometallic compound, whose molecules consist of an iron atom 'sandwiched' between two molecules of **cyclopentadiene**. Alternative name: dicyclopentadienyliron.

ferrocyanide $[Fe(CN)_6]^{4-}$ Very stable complex ion of iron (II). Alternative name: hexacyanoferrate (II).

ferromagnetism Property of certain substances that in a magnetizing field have induced magnetism (because of aligned **magnetic moments**), which persists when the field is removed and they become permanent magnets. The magnetized regions are called **magnetic domains**. Examples of ferromagnetic materials include iron, cobalt and their alloys.

ferrous Bivalent iron. Alternative names: iron (II), iron (2+).

fertilization Fusion of specialized sex cells or **gametes** which are **haploid** to form a single cell, a **diploid zygote**. It occurs in **sexual reproduction**; *e.g.* in vertebrates the **ovum** (female gamete) is fertilized by the **sperm** (male gamete). Fertilization may be internal or external. See also **pollination**.

■ **fertilizer** Substance used to increase the fertility of soil. Natural, or organic, fertilizers consist of animal or plant residues and are usually called manures. Artificial, or inorganic, fertilizers supply nitrogen (in the form of compounds such as ammonium nitrate, ammonium sulphate and ammonium phosphate), and sometimes also phosphorus and potassium (potash). They are specified in terms of their NPK (nitrogen, phosphorus, potassium) content. [2/5/c]

Feulgen's test Test for the presence of **DNA** (deoxyribonucleic acid); *e.g.* in tissue sections by staining, which gives a purple colour. It was named after the German chemist R. Feulgen (1884–1955).

fibre Thread-like structure. Natural fibres include wool and other **protein**-containing animal products, and plant fibres (*e.g.* cotton) consisting mainly of **cellulose**. The cellulose fibres in food, often referred to as roughage, are an important part of a **balanced diet**.

fibre optics Branch of optics that uses bundles of pure glass fibres within straight or curved 'pipes', along which light travels as it is internally reflected. A modulated light signal can carry much more data (*e.g.* computer data, telephone signals, television channels) than a wire of similar dimensions.

fibrin Meshwork of fibres formed from **fibrinogen**, which creates a **blood** clot in a wound where blood is exposed to air.

fibrinogen Soluble **plasma protein** found in blood which, after triggering of chemical factors when platelets in blood are exposed to air by wounding, is converted to **fibrin** as part of the blood-clotting mechanism.

fibroblast Long flat cell found in **connective tissue** which secretes **protein**; *e.g.* **collagen** and elastic fibres.

■ **fibula** Leg bone, located below the knee and outside the **tibia** (shin bone) in the hind-limb of a four-legged (tetrapod) vertebrate. [2/3/a] [2/4/a] [2/5/a]

field *1.* In physics, region in which one object exerts a force on another object; *e.g.* **electric field, gravitational field, magnetic field**. *2.* In optics, area that is visible through an optical instrument. *3.* In computing, specific part of a **record**, or a group of characters that make up one piece of information.

■ **field coil** Coil used for producing a **magnetic field** in an **electromagnet** or other electric machine. [5/7/c]

field-effect transistor (FET) Type of **transistor** that consists of a conducting channel formed from a wafer of **semiconductor** material, the resistance of which is controlled by the voltage applied to one or more input **gates**.

field-emission microscope Microscope used for the observation of the positions of atoms in a surface.

field magnet Permanent **magnet** or **electromagnet**, the purpose of which is to provide a **magnetic field** in an electric machine.

filament *1.* In biology, the stalk of a **stamen**, which has an anther at its end. *2.* Also in biology, the **hypha** of a **fungus**. *3.* Also in biology, a string of cells that make up certain **algae**. *4.* In physics, a fine wire of high **resistance** which is heated by passing an electric current directly through it. Filaments are used in electric fires and incandescent lamps.

filler Inert substance added to paper, paints, plastics and rubbers during manufacture to increase their bulk or weight, or otherwise modify their properties.

film *1.* Thin layer of one substance on the surface of another substance, *e.g.* oil floating on water. Thin films can sometimes diffract light and produce rainbow colours (*see* **diffraction**). *2.* Plastic strip carrying a light-sensitive emulsion that is used in photography.

filoplume Small hair-like feathers which lack **vanes** and are scattered over a bird's body between the **contour** feathers. Birds use filoplumes for the fine control of flight direction.

filter *1.* In optics, light-absorbing semi-transparent material that passes only certain wavelengths (colours). *2.* In electronics, device that passes only certain a.c. frequencies.

filter feeding Acquisition of nutrients, characteristic of non-motile aquatic organisms, that involves straining out small particles of organic matter suspended in water.

filter pump Simple vacuum pump in which a jet of water draws air molecules from the system. It can produce only low pressures and is commonly used to increase the speed of filtration by drawing through the filtrate.

filtrate Liquid obtained after **filtration**.

filtration Method of separating solid particles from a mixture of the particles in a liquid by straining it through a porous material (such as filter paper or glass wool), through which only the liquid passes. The process can be accelerated using suction.

fin Locomotory appendage on a fish which provides a large surface area for steering and swimming.

fine chemical Chemical produced in pure form and in small quantities.

fine structure Splitting of certain lines in a **line spectrum** into a number of further discrete lines, which are observable only when high resolution is employed.

fireclay Clay that contains large amounts of **alumina** and **silica**, and is capable of withstanding high temperatures. It is used in the production of refractory brick, furnace linings, etc.

firedamp Explosive mixture of air and **methane** found in coal mines.

■ **fish** Aquatic vertebrate animals that are usually divided into three classes: the **Agnatha** (jawless fish, *e.g.* lampreys), the **Elasmobranchii** (cartilaginous fish, also called Chondrichthyes, *e.g.* sharks and rays) and the **Osteichthyes** (bony fish, *e.g.* cod, salmon, etc.). [2/3/b] [2/4/b]

■ **fission** Splitting. *1.* In atomic physics, disintegration of an atom into parts with similar masses, usually with the release of energy and one or more neutrons (*see* **nuclear fission**). [4/8/d] *2.* In biology, division of a cell or single-celled organism into two (*see* **meiosis; mitosis**).

■ **fission product Isotope** produced by **nuclear fission**, with a mass equal to roughly half that of the fissile material. [4/8/d]

fission spectrum Energy distribution of **neutrons** released during **nuclear fission**.

■ **fixation of nitrogen** Part of the **nitrogen cycle** that involves the conversion and eventual incorporation of atmospheric nitrogen into compounds that contain nitrogen. Nitrogen fixation in nature is carried out by nitrifying soil bacteria or blue-green algae (Cyanophyta) in the sea. Soil bacteria may exist symbiotically in the root nodules of leguminous plants or they may be free-living. Small amounts of nitrogen are also fixed, as **nitrogen monoxide** (nitric oxide), by the action of lightning. [2/8/d]

fixed point Standard temperature chosen to define a **temperature scale** or at which properties are measured, *e.g.* the **ice point** (0 °C) and the **steam point** (100 °C). Alternative name: fixed temperature.

fixing, photographic Process for removing unexposed silver halides after development of a photographic emulsion. It involves dissolving the silver salts by immersing the developed film in a fixing bath consisting of a solution of sodium or ammonium thiosulphate.

■ **flaccid** Describing tissue that has become soft and drooping, usually because of loss of water. [2/4/a] [2/5/a] [2/7/a]

flagellate Single-celled **protozoan** animal that moves by beating flagella (see **flagellum**).

flagellum Long hair-like **organelle** whose beating movement causes locomotion or the movement of fluid over a cell. Flagella are present in most motile **gametes** and unicellular plants or animals (*e.g.* protozoa), in which they occur singly or in small clusters. In some multicellular organisms (*e.g.* sponges and hydra) they are used for circulation of water containing food and respiratory gases. *See also* **cilium**.

flame cell Excretory cell possessed by some invertebrates. Waste products are drawn in and moved outside by **cilia**. Alternative name: solenocyte.

flame test Qualitative chemical test in which an element in a substance is identified by the characteristic colour it imparts to a Bunsen burner flame.

flash point Lowest temperature at which a substance or a mixture gives off sufficient vapour to produce a flash on the application of a flame.

flavin adenine dinucleotide (FAD) **Coenzyme** that functions in the oxidation-reduction reactions of **enzymes**, *e.g.* the oxidative degradation of **pyruvate**, **fatty acids** and **amino acids**, and in **electron transport**. Alternative name: flavine.

flavine Alternative term for **flavin adenine dinucleotide** (FAD).

flavonoid Aromatic, oxygen-containing **heterocyclic** organic compound. Many natural pigments are flavonoids.

flavoprotein Member of a group of conjugated **proteins** in which the **prosthetic group** constitutes a derivative of **riboflavin** (*e.g.* FAD or FMN). Flavoprotein dehydrogenases (enzymes) are involved in the **electron transport** chain of **aerobic respiration.**

F-layer Alternative name for the **Appleton layer** of the **ionosphere**.

Fleming's rules *See* **left-hand rule; right-hand rule**.

flint Natural crystalline form of **silica** that is used as an abrasive. It was important in the Stone Age as a material from which to make tools.

flip-flop Electronic component or circuit that can be switched from one of its two stable states to the other by an electric pulse; a bistable device. Flip-flops are frequently used in computers.

flocculation Coagulation of a finely divided precipitate into larger particles. *E.g.* in farming, flocculation of clay is deliberately brought about by the addition of lime, thus improving the drainage of clay soils.

floppy disk Flexible, portable magnetic disk that provides data and program storage for **microcomputers**. The disk may be enclosed in a flexible or a rigid casing. Alternative name: diskette. *See also* **hard disk**.

flotation, principle of A floating object displaces its own weight of fluid (liquid or gas). An object floats if its weight equals the **upthrust** on it. *See also* **Archimedes' principle**.

flotation process Method of concentrating ores by making the ore float on detergent-produced froth in a tank of liquid. Particles adhering to the bubbles are removed with the froth, and thus separated from the materials remaining in the slurry. Alternative name: froth flotation.

flower In flowering plants **(angiosperms)**, the organ of **sexual reproduction**, including the male **stamens** and female **carpels**. [2/1/a]

fluid Form of matter that can flow; thus both gases and liquids are fluids. Fluids can offer no permanent resistance to changes of shape. Resistance to flow is manifest as **viscosity**.

fluidics Study and use of fluid flow to control instruments or industrial processes.

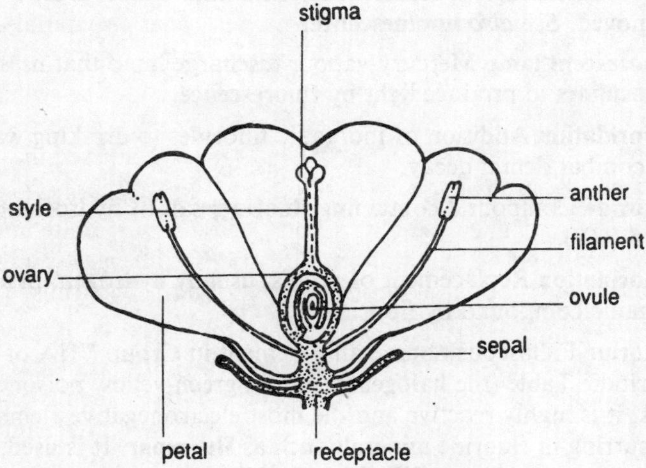

Structure of a flower

fluidity Property of flowing easily; the opposite of **viscosity**.

fluidization Technique whereby a finely divided solid acquires characteristics of fluid by upward passage of a gas through it.

fluke *1*. Any **endoparasitic** flatworm that belongs to the class Trematoda. See also **bilharzia**. *2*. Tail of a whale, dolphin or porpoise.

fluorescein $C_{20}H_{12}O_5$ Orange-red powder which dissolves in alkalis to give a green fluorescent solution. It is used as a chemical marker and for dyeing textiles. Alternative name: resorcinolphthalein.

fluorescence Emission of radiation (generally visible light) after **absorption** of radiation of another wavelength (usually ultraviolet or near-ultraviolet) or electrons; unlike

phosphorescence, it ceases when the stimulating source is removed. *See also* **luminescence**.

fluorescent lamp Mercury-vapour discharge lamp that uses **phosphors** to produce light by **fluorescence**.

fluoridation Addition of inorganic fluorides to drinking water to combat dental decay.

fluoride Compound containing **fluorine**; salt of hydrofluoric acid (HF).

fluorination Replacement of atoms, usually hydrogen, in an organic compound by fluorine.

■ **fluorine** F Gaseous nonmetallic element in Group VIIA of the Periodic Table (the **halogens**). A pale green-yellow poisonous gas, it is highly reactive and the most electronegative element, occurring in fluoride minerals such as **fluorspar**. It is used, as the gaseous uranium (VI) fluoride (UF_6), in the separation by diffusion of uranium **isotopes** and in making **fluorocarbons**. Inorganic **fluorides** are added to water supplies to combat tooth decay. At. no. 9; r.a.m. 18.9984. [4/8/a] [4/9/a]

fluorite Alternative name for **fluorspar**.

fluorocarbon Very stable organic compound in which some or all of the hydrogen atoms have been replaced by fluorine. Fluorocarbons are used as solvents, aerosol propellants and refrigerants. Their use is being limited because they have been implicated in damage to the ozone layer of the atmosphere. *See also* **Freons**.

fluoroscope Fluorescent screen that allows direct observation of X-ray images, often connected to a camera. It is used in medicine (radiography) and industrial X-ray applications.

fluorspar CaF_2 Naturally occurring calcium fluoride, used as a flux in glass and as a component of certain cements. Alternative name: fluorite.

flux *1*. Substance added to help **fusion**. *2*. Substance used in metallurgy to combine with unwanted materials and cause them to flow so that they can be removed from the metal as a slag. *3*. Rate of flow of mass, volume or energy per unit area normal to the direction of flow.

flux density Magnetic flux or luminous flux per unit of cross-sectional area.

fluxmeter Instrument for measuring **magnetic flux**.

FM Abbreviation of **frequency modulation**.

f-number Method of denoting the diameter of a lens aperture in a camera; for a simple lens it is the **focal length** divided by the diameter of the aperture. The smaller the *f*-number, the larger the aperture. In the usual sequence *f*22, *f*11, *f*8, *f*5.6, *f*4 etc., each aperture has twice the area (and hence admits twice the amount of light) as the preceding one in the series.

focal length Distance from the centre of a lens or mirror to its **focal point**. Alternative name: focal distance.

focal point In optics, the point at which light rays meet after refraction or reflection.

focusing Adjustment of an optical device so that it produces a sharp image. In the eye, for instance, ciliary muscles alter the shape of the lens to produce a clear image of an object on the retina.

foetus In mammals, an **embryo** after a certain stage of development, usually when it begins to resemble the developed animal (in human beings after about two or three months of pregnancy).

folic acid Water-soluble B-group **vitamin**. Its deficiency leads to anaemia and it is important in the formation of various coenzymes, which are in turn essential for growth and reproduction of cells.

■ **follicle** *1.* In botany, dry dehiscent fruit formed from a monocarpellary (with a single **carpel) ovary** which at maturity splits along one edge to release its seeds; *e.g.* columbine fruit. *2.*Cavity, sac or gland within an organ or tissue; *e.g.* **Graafian follicle**, hair follicle. [2/7/a]

■ **follicle-stimulating hormone** (FSH) Member of a group of **hormones** that are secreted by the anterior lobe of the **pituitary gland** in vertebrates. It stimulates the growth and maturation of **ovarian follicles** and the growth only of oöcytes, which are matured under the action of **luteinizing hormone**. In males FSH stimulates **sperm** formation in the **testes**. [2/7/a]

food additive *See* **additive**.

■ **food chain** Food relationship between organisms in an **ecosystem** in which energy is transferred from plants, the producers, through a series of organisms, the consumers. Each stage of the food chain is a trophic level. The first level is occupied by plants, which obtain their energy from the Sun. The second trophic level is occupied by **herbivores** (plant-eating animals), which are in turn eaten by the **carnivores** (meat-eating animals). Food chains in a **community** are interconnected because most organisms consume more than one type of food, thus forming a more complicated **food web** or food cycle. [2/7/d]

food consumer *See* **food chain**.

food producer *See* **food chain**.

food web Complex relationship formed between organisms due to mode of nutrition, involving a network of **food chains** in an environment.

force (*F*) Influence that can make a stationary object move, or a moving object change speed or direction, *i.e.* that changes the object's momentum. For a body of mass *m* and acceleration *a*, the force $F = ma$. Its SI unit is the **newton**.

force meter Instrument resembling a spring balance used for measuring force. *See also* **newton meter**.

force ratio Alternative name for **mechanical advantage**.

forebrain Largest and topmost portion of the vertebrate brain that comprises the **cerebral hemispheres** and the **basal ganglion**.

formaldehyde HCHO Colourless pungent organic gas, an **aldehyde**, which is readily soluble in water. It is used as a disinfectant and in the manufacture of plastics. Alternative name: methanal.

formalin 40% solution of **formaldehyde** (methanal) in water, which used to be employed for preserving biological specimens.

formic acid HCOOH Simplest **carboxylic acid**, made commercially by the catalytic combination of carbon monoxide and superheated steam. It occurs naturally in ant venom and nettles. Alternative name: methanoic acid.

formula Chemical composition of a substance indicated by the symbols of each element present in it and subscripts that show the number of each type of atom involved; *e.g.* the formula for water is H_2O, that for potassium dichromate is $K_2Cr_2O_7.2H_2O$. *See also* **empirical formula; molecular formula**. [4/9/b]

Fortin barometer Mercury **barometer**, used for the accurate measurement of atmospheric pressure. It was named after the French physicist Jean Fortin (1750–1831).

FORTRAN Acronym of FORmula TRANslation, a high-level computer programming language designed for mathematical and scientific use.

■ **fossil** Remains, impressions or casts of dead animals and plants preserved in rocks. Because organic matter rots away quickly, a fossil usually consists of skeletal material which is partly or wholly replaced by mineral deposits from circulating water. Burrows, footprints or faeces may also become fossilized. [2/3/b] [2/4/b]

■ **fossil fuel** Mineral fuel (*e.g.* **coal, natural gas, petroleum**) that is made from the remains of living organisms. [2/3/b]

Foucault pendulum Long pendulum with a heavy bob whose plane of swing appears slowly to turn because of the Earth's rotation. It was named after the French physicist Jean Foucault (1819–68).

fovea Part of the **retina** of the **eye** that has a concentration of **cones** (but no **rods**), which comes into play when acute vision is required. Alternative name: yellow spot.

f.p.s. units System of units based on the foot, pound and second.

fractional crystallization Separation of mixtures of substances by the repeated crystallization of a solution, each time at a lower temperature.

fractional distillation Separation of a number of liquids with different boiling points by **distillation** and collecting separately the liquids that come off at different temperatures. Alternative name: fractionation.

fractionating column Long vertical tube containing bubble-caps, sieve plates, or various irregular packing materials, used for industrial **fractional distillation**.

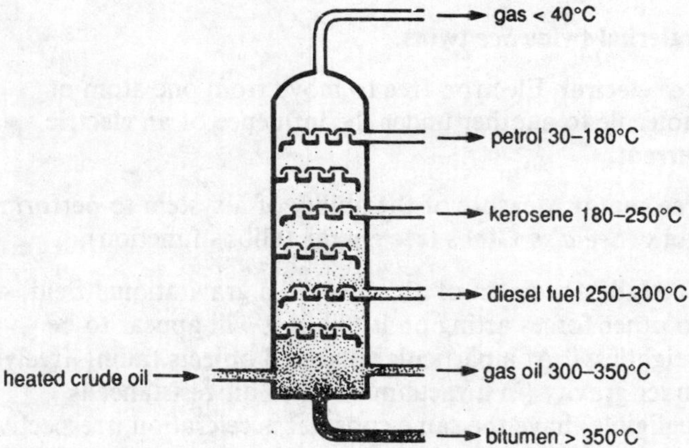

Fractionating column for crude oil

fractionation Alternative name for **fractional distillation.**

frame of reference Set of points, lines or planes for defining positions.

francium Fr Radioactive metallic element in Group IA of the Periodic Table (the **alkali metals**), made by proton bombardment of thorium. It has several **isotopes** with half-lives of up to 22 min. At. no. 87; r.a.m. (most stable isotope) 223.

■ **Frasch process** Process for the extraction of sulphur from underground deposits. It involves sinking three concentric tubes to the deposits. Superheated water is pumped down the outer tube to melt the sulphur, and when hot compressed air is injected down the central tube molten sulphur is forced up the remaining tube. It was named after the German chemist Herman Frasch (1851–1914). [4/4/b]

fraternal twins *See* **twins**.

free electron Electron free to move from one atom or molecule to another under the influence of an **electric current**.

free energy Measure of the ability of a system to perform work. *See also* **Gibbs free energy** (Gibbs function).

free fall Movement of an object in a gravitational field, with no other forces acting on it (when it will appear to be weightless). At a particular place all objects falling freely under gravity (in a vacuum or when air resistance is negligible) have the same constant acceleration irrespective of their masses. On Earth this is *g*, the **acceleration of free fall** (9.80665 ms^{-2}).

free radical Intermediate and highly reactive molecule that has an unpaired electron and so easily forms a chemical bond.

freeze drying Method of drying a heat-sensitive substance such as blood plasma or food by freezing it below 0C and then removing the frozen water by volatilization in a vacuum.

freezing Solidification of a liquid that occurs when it is cooled sufficiently (to below its freezing point).

freezing mixture Mixture of two substances (*e.g.* ice and salt) that absorbs heat and can be used to produce a temperature of below 0 °C. *See also* **eutectic mixture**.

freezing point Temperature at which a liquid solidifies. Alternative name: solidification point.

French chalk Powdered **talc**, used as a filler and lubricant.

Frenkel defect Crystal disorder that occurs when an ion occupies a vacant interstitial site, leaving its proper site empty.

Freon Trade name for certain **fluorocarbons** and **chlorofluorocarbons** derived from methane and ethane. They are used as refrigerants.

frequency (*f*) Rate of recurrence of wave, *i.e.* number of cycles, oscillations or vibrations in unit time, usually one second. The frequency of a wave is inversely proportional to the **wavelength** λ; *i.e.* $f = c/\lambda$, where c is the velocity of the wave. The SI unit of frequency is the **hertz** (which corresponds to 1 cycle per second).

frequency modulation (FM) Method of radio transmission, used in the short wavelengths, in which the information content is conveyed by means of variations in the **frequency** of the **carrier wave**.

Fresnel lens Lens that is formed by cutting a series of stepped concentric circles. It is flatter and lighter than a conventional lens of equivalent power, but the optical quality is not high. It was named after the French physicist Augustin Fresnel (1788–1827).

friction Resistance offered to sliding or rolling of one surface on another. It is a **force**, a type of **adhesion**.

Friedel-Crafts reaction Chemical reaction that introduces an **alkyl** or **acyl group** into a **benzene** ring, in the presence of a catalyst such as aluminium (III) chloride, boron trifluoride or hydrogen fluoride. It was named after two chemists, the Frenchman Charles Friedel (1832–99) and the American James Crafts (1839–1917).

froth Collection of fairly stable small bubbles in a liquid produced by shaking or aeration. Alternative name: foam.

froth flotation Process for the separation of finely divided materials in which a slurry is caused to froth by the addition of a foaming agent. Particles adhering to the bubbles are removed with the froth, and thus separated from the materials remaining in the slurry.

fructification Process of forming a **fruiting body**. The term also sometimes refers to the fruiting body itself.

fructose $C_6H_{12}O_6$ Fruit sugar, a **monosaccharide** carbohydrate (**hexose**) found in sweet fruits and honey. Alternative name: laevulose.

fruit Plant tissue that develops from the **ovary** of a flowering plant and forms around the maturing seeds, following **pollination** and subsequent **fertilization**. The term is also sometimes used to describe the mature seeds of some coniferous trees, *e.g.* juniper berries'. See also **berry**; **capsule**; **drupe**; **false fruit**; **follicle**.

fruiting body Structure that is developed by many **fungi** in which spores are formed.

FSH Abbreviation of **follicle-stimulating hormone**.

fucoxanthin Brown **carotenoid** plant pigment, present in brown **algae**.

fuel cell Device that uses the oxidation of a liquid or gaseous fuel to produce electricity. Thus the chemical energy of the fuel is converted directly to electrical energy, without the intermediate stages of a heat engine and a generator. *E.g.* hydrogen and oxygen flowing over porous platinum electrodes immersed in hot potassium hydroxide solution react to produce water and generate a voltage across the electrodes.

fuller's earth Green, blue or yellow-brown clay-like mineral, used for decolorizing solutions and oils. Alternative name: montmorillonite.

fuming sulphuric acid Alternative name for **oleum**.

functional group Atom or group of atoms that cause a chemical compound to behave in a particular way; *e.g.* the functional group in alcohols is the –OH (hydroxyl) group.

fundamental units Units of length, mass and time that form the basis of most systems of units.

fungi Plural of **fungus**.

■ **fungicide** Substance that kills **fungi**. [2/7/c] [2/10/e]

■ **fungus** Mainly terrestrial plant-like organism, different from other plants because of its lack of **chlorophyll**. Most fungi are **saprophytic** or **parasitic** organisms (*e.g.* moulds) whose walls consist of **chitin**, although a few produce cellulose as well. They are classified as the plant division (phylum) Mycota, although some authorities put them in a kingdom of their own. [2/3/b] [2/4/b] [2/7/c] [2/10/e]

funicle Stalk that connects the **ovule** to the **placenta** in the ovaries of **angiosperm** plants.

furan C_4H_4O Oxygen-containing **heterocyclic** liquid organic compound. Alternative name: furfuran.

furanose Any of a group of **monosaccharide** sugars (**pentose sugars**) whose molecules have a five-membered heterocyclic ring of four carbon atoms and one oxygen atom. *See also* **pyranose**.

furfuran Alternative name for **furan**.

fuse Device used for protecting against an excess **electric current** passing through a circuit. It consists of a piece of metal, connected into the circuit, that heats and melts (thereby breaking the circuit) when the current exceeds a certain value.

fusion Act of melting or joining together. *See also* **nuclear fusion.**

fusion reaction *See* **nuclear fusion; thermonuclear reaction.**

fusion reactor *See* **nuclear reactor.**

G

G-actin *See* **actin**.

gadolinium Gd Silvery-white metallic element in Group IIIB of the Periodic Table (one of the **lanthanides**), which becomes strongly magnetic at low temperatures. At. no. 64; r.a.m. 157.25.

gain Increase in power produced by an **amplifier**, the ratio of the amplitude of the output to that of the input.

galactose $C_6H_{12}O_6$ **Monosaccharide** sugar that occurs in milk and in certain gums and seaweeds as the **polysaccharide** galactan.

galena Widely distributed metallic-grey mineral that consists mainly of lead sulphide (PbS); it is the principal ore of **lead**.

Galilean telescope Optical refracting telescope that uses a converging (convex) lens as its objective and a diverging (concave) lens as its eyepiece. It produces an upright image, and is the arrangement used, *e.g.*, in opera glasses. It was named after the Italian astronomer and physicist Galileo Galilei (1564–1642).

■ **gall** *1.* In botany, swelling on a plant, usually caused by a parasite. *2.* In zoology, alternative name for **bile**. [2/4/a] [2/5/a] [2/7/a]

■ **gall bladder** Storage organ in some vertebrates that, stimulated by hormones, releases **bile** (along the bile duct) to the **duodenum** during **digestion**. [2/4/a] [2/5/a] [2/7/a]

gallium Ga Blue-grey metallic element in Group IIIa of the Periodic Table, used in low-melting-point alloys and

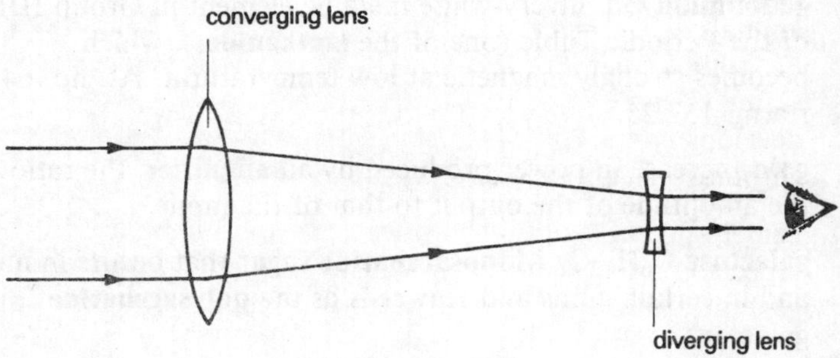

Galilean telescope is used in opera glasses

thermometers. Gallium arsenide is an important **semiconductor**, used also in lasers. At. no. 31; r.a.m. 69.72.

■ **gallstone** Accretion, usually of **cholesterol** or calcium salts, that occurs in the **gall bladder** or its ducts. Alternative name: biliary calculus. [2/4/a] [2/5/a] [2/7/a]

galvanic cell Alternative name for a **voltaic cell**.

■ **galvanized iron** Iron or, usually, steel coated with zinc (by dipping or electroplating) to prevent it going rusty. [4/7/b]

galvanizing Method of protecting a metal (*e.g.* iron or steel) from corrosion by covering it with a thin layer of **zinc** through dipping or electroplating.

galvanometer Device that detects or measures small electric currents passing through it.

gametangium In plants, an organ in which **gametes** are produced.

■ **gamete** Specialized **sex cell** (*e.g.* an ovum or a sperm in animals or an ovule or pollen in plants), which is **haploid** – *i.e.* containing half the normal number of **chromosomes**. Gametes combine at **fertilization** to form a **zygote** that develops into a new organism (with the normal, **diploid**, chromosome number). *See also* **parthenogenesis**.

gametophyte Stage in the life-cycle of certain plants, especially mosses and ferns, which show **alternation of generations**. It produces **gametes** and is **haploid** (as distinct from the following **sporophyte** stage, which is **diploid**).

gamma globulin Alternative term for **immunoglobulin**.

gamma iron Iron with a **face-centred cubic** structure. It is non-magnetic.

■ **gamma radiation** Penetrating form of **electromagnetic radiation** of shorter wavelength than **X-rays**, produced, *e.g.*, during the decay of certain **radio-isotopes**. [4/7/f]

■ **gamma rays** High-energy **photons** that make up **gamma radiation**.

ganglion Area of nervous tissue that contains a complex set of **synapses**. In vertebrates, ganglia constitute much of the **central nervous system** and occur in the **sympathetic** and **parasympathetic** nervous systems, but differ in their structure in each case.

■ **gas** Form (phase) of matter in which the atoms and molecules move randomly with high speeds, occupy all the space available, and are comparatively far apart; a vapour. A liquid heated above its boiling point changes into a gas.

gas chromatography Method of analysing mixtures of substances. The sample is volatilized and then introduced into a column containing the stationary phase (a solid or a non-volatile liquid on an inert support), and an inert carrier gas (*e.g.* argon) is passed through the column. Components of the mixture are removed from the column by the carrier gas at different rates. A detector measures the conductivity of the gas leaving the column, which is recorded on a chart as a series of peaks corresponding to each of the components. The chart is calibrated by passing samples of known composition through the machine.

gas constant (R) Constant in the **gas equation,** value 8.31 J mol^{-1} K^{-1}. Alternative name: universal molar gas constant.

gas-cooled reactor Nuclear reactor in which the cooling medium is a gas, usually carbon dioxide.

gas equation For n moles of a gas, $pV = n\mathrm{R}T$, where p = pressure, V = volume, n = number of moles, R = the **gas constant** and T = absolute temperature.

■ **gas exchange** Part of **respiration** in which organisms exchange gases (carbon dioxide and oxygen) with their environment: air for terrestrial plants and animals, water for aquatic ones. It may involve the use of **lungs** (mammals, birds, adult amphibians and reptiles), **gills** (larval amphibians, fish and other aquatic animals), **spiracles** (insects and other terrestrial arthropods) or **stomata** (green plants). In other organisms (*e.g.* aquatic plants, fungi) gas exchange takes place directly between cells and the environment. [2/5/a]

gas laws Relationships between pressure, volume and temperature of a gas. The combination of **Boyle's, Charles'** and **Gay-Lussac's laws** is the **gas equation**.

gas-liquid chromatography (GLC) Type of **gas chromatography** in which the column contains a non-volatile liquid on an inert support.

gas oil One of the main fractions of crude oil, used as fuel oil. Alternative name: heavy oil.

gas thermometer Thermometer based on the variation in pressure or volume of a gas. Alternative name: constant-volume gas thermometer.

■ **gastric** Relating to the stomach or digestion. [2/5/a]

■ **gastric juice** Fluid secreted by glands in the stomach wall during **digestion**; it contains two principal **enzymes, pepsin** and **rennin**, and hydrochloric acid. [2/5/a]

gastropod Member of the mollusc class **Gastropoda**.

Gastropoda Class of **molluscs** characterized by a single flattened foot, a head bearing eyes on stalks and usually a coiled shell. The gastropods include slugs (which have no external shells), snails and whelks.

gastrula Cup-shaped structure in early embryonic development, the stage that succeeds the **blastula**.

gas turbine Form of internal combustion engine in which the expansion of the hot gases resulting from the combustion of the fuel is used to drive a turbine (coupled to the compressor). Alternative name: jet engine.

gate In computing, an electronic circuit (switch) that produces a single output signal from two or more input signals. Alternative name: logic element.

gauss (G) Unit of magnetic **flux density** equal to 10^{-4} **tesla**, by which the gauss has been replaced in the SI system.

Gay-Lussac's law of volume When gases react their volumes are in a simple ratio to each other and to the volume of products, at the same temperature and pressure. It was named after the French chemist and physicist Joseph Gay-Lussac (1778–1850).

gear Toothed wheel that engages with another toothed wheel to transfer rotation from one shaft to another. If the two gears have different numbers of teeth (the ratio of these numbers is the gear ratio), the second shaft rotates at a different speed from the drive shaft.

■ **Geiger counter** Instrument for detecting atomic and subatomic particles (*e.g.* **alpha** and **beta particles**), used for radioactivity measurements. It was named after the German physicist Hans Geiger (1882–1945). Alternative name: Geiger-Müller counter. *See also* **counter**.

Geiger-Nuttal law Empirical law for calculating the distance that an **alpha particle** can travel once it is emitted from a radioactive substance.

Geissler tube Electric discharge tube, containing traces of gas at very low pressure, that glows when a high-voltage current flows between metal electrodes sealed into it. It was named after the German apparatus-maker Heinrich Geissler (1814–79).

gel Jelly-like colloidal solution. *See* **colloid**.

gelatin Protein extracted from animal hides, skins and bones, which forms a stiff jelly when dissolved in water. It is used as a clarifying agent, in foodstuffs, and in the manufacture of adhesives. Alternative name: gelatine.

gel filtration Type of **chromatography** in which compounds are separated according to their molecular size. Molecules of the components of a mixture penetrate the surface of an inert porous material in proportion to their size. Alternative name: gel permeation.

gelignite Explosive made from a mixture of nitroglycerine (glyceryl trinitrate), gun-cotton (nitrocellulose, cellulose trinitrate), sodium nitrate and wood pulp. Alternative name: blasting glycerin.

gemmation *See* **budding**.

■ **gene** Segment of **DNA** that specifies a complete **polypeptide** chain, which can be regarded as an independent inheritable unit. Genes specify particular traits and are passed from parent to offspring. Some genes may be **dominant** to others that may be **recessive**. Different forms of a gene are **alleles**. They may undergo **mutation** to give different characteristics. Every cell of an organism has a set of genes on its **chromosomes** that are identical to those in every other cell, having ultimately arisen from mitotic division of a **zygote**. [2/5/b] [2/10/b]

gene pool Total of all the different **genes** in a population.

genera Plural of **genus**.

general formula Expression representing the common chemical **formula** of a group of compounds. *E.g.* C_nH_{2n+2} is the general formula for an **alkane**. A series of compounds of the same general formula constitute a **homologous series**.

General Theory of Relativity Part of the theory of **relativity** (along with the Special Theory).

generation time Time within a population of cells (*e.g.* micro-organisms) that it takes them to undergo division to form pairs of **daughter cells**; *i.e.* the doubling time.

generator Machine for converting mechanical energy into electrical energy. Alternative name: electric generator. *See also* **alternator; dynamo.**

gene therapy Treatment of disorders by altering the genetic material in a cell; *e.g.* by microinjection of favourable **genes** into a **germ cell.**

■ **genetic code** Code possessed by **DNA** which contains instructions for **protein synthesis** in a **cell**. Each **amino acid** is specified by a triplet of **bases** located in the DNA, and each protein is in turn specified by a particular sequence of **amino acids**. The code thus contains instructions for all **enzymes** produced in the cell, and consequently the characteristics of the organism. [2/5/b]

■ **genetic engineering** Manipulation of genetic material such as **DNA** for practical use, *e.g.* the introduction of foreign **genes** into micro-organisms for the production of a useful **protein** (such as human insulin). The technique is also used in the study of genetic material. [2/10/c]

■ **genetic variation** Differences between members of the same species resulting from slight changes in the make-up of the genes they inherit from their parents. Such differences may in turn be inherited and are significant in adaptation (and therefore species survival) in a changing environment. *See also* **variation**. [2/6/b]

genetics Study of inheritable characteristics of organisms – *i.e.* **heredity**.

genome Entire genetic material present in a cell of an organism.

genotype Genetic constitution of an organism, *i.e.* the characteristics specified by its **alleles**. It is the genetic make-up of the organism, as opposed to the way its genes are expressed in its appearance (which is the **phenotype**).

■ **genus** In biological **classification**, one of the groups into which a family is divided, and which is itself divided into species; *e.g. Vulpes* (foxes). *See also* **binomial nomenclature.** [2/4/b]

geodesic Describing structures, *e.g.* domes, that are made from large numbers of identical components, and that have the load distributed evenly throughout the structure.

geomagnetism Earth's magnetic field.

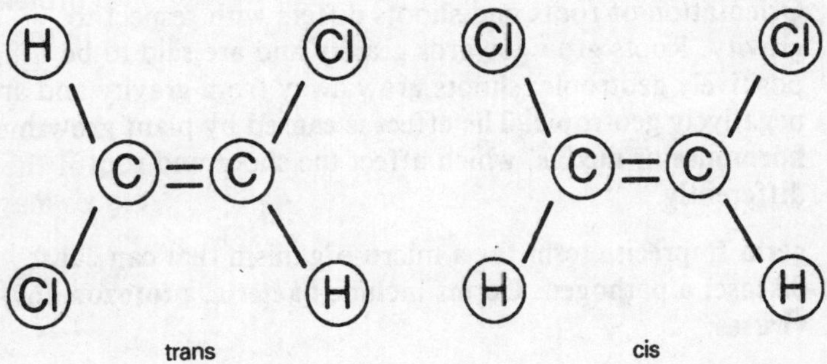

Geometric isomerism

geometric isomerism Form of **stereoisomerism** that results from there being no free rotation about a bond between two

atoms. Groups attached to each atom may be on the same side of the bond (the *cis*-isomer) or on opposite sides (the *trans*-isomer). Alternative name: cis-trans isomerism.

geophysics Study of the physics of the Earth.

geostationary orbit Path of an artificial Earth satellite at such a height (35,900 km) that it takes 24 hours to make one complete **orbit**, during which time the Earth makes one revolution on its axis and so the satellite remains above the same place on the Earth's surface. It is used mainly for communications satellites. Alternative names: stationary orbit, synchronous orbit.

geotaxis Change in the direction of movement of a mobile organism that is made in response to gravity. *See also* **geotropism**.

geotropism Growth in response to **gravity** in plants. Orientation of roots and shoots differs with respect to gravity. Roots grow towards gravity and are said to be positively geotropic; shoots grow away from gravity and are negatively geotropic. The effect is caused by plant growth **hormones** or **auxins**, which affect the shoot and root differently.

germ Imprecise term for a micro-organism that can cause disease; a pathogen. Germs include **bacteria**, **protozoa** and **viruses**.

germanium Ge Grey-white semimetallic element in Group IVA of the Periodic Table, which occurs in some silver ores. It is an important **semiconductor**, used for making solid-state diodes and transistors. Its oxide is used in optical instruments and infra-red cameras. At. no. 32; r.a.m. 72.59.

German silver Alloy of copper, nickel and zinc, used in

coinage and as a base metal for silver-plated articles. Alternative name: nickel silver.

germ cell Sex cell, or **gamete**.

■ **germination** Beginning of the development of an embryo in a **seed** into a plant, which commences with the growth of the **radicle** and **plumule**, and is initiated by the right conditions of moisture, oxygen supply and temperature. *See also* **epigeal**; **hypogeal**.

germ layer Layer of cells present in an **embryo**. In **triploblastic** organisms the layers consist of **ectoderm**, **mesoderm** and **endoderm**, which each give rise to particular **tissues**.

gestation Time of development of an **embryo** from fertilization to birth in **viviparous** animals; the duration of **pregnancy**.

GeV Abbreviation of giga-electron-volt, which is 10^9 electron volts.

gibberellin Any one of a group of plant **hormones** that stimulate rapid growth.

Gibbs free energy (G) Measure of the energy that would be liberated or absorbed during a **reversible process**. $G = H - TS$, where H is heat content, T thermodynamic temperature and S entropy. It was named after the American chemist and physicist Josiah Gibbs (1839–1903). Alternative name: Gibbs function.

■ **gill** *1*. In zoology, respiratory organ in fish and certain other aquatic animals that extracts dissolved oxygen from water. *2*. In botany, thin vertical structure under the fruiting body of a fungus that carries the spore-bearing hymenium. [2/5/a]

■ **gizzard** Part of the gut of animals that eat indigestible food. It has thick muscular walls and a tough grinding inner surface specialized for breaking up food. Gizzards of birds and earthworms contain small stones or grit; those of the **arthropods** have chitin patches and spines. Alternative name: gastric mill. [2/4/a] [2/5/a] [2/7/a]

glacial Describing a compound of ice-like crystalline form, especially that of the solid form of a liquid; *e.g.* **glacial acetic** (ethanoic) **acid**.

glacial acetic acid Crystalline form of **acetic** (ethanoic) **acid** below its freezing point.

■ **gland** *1.* In animals, organ that synthesizes and secretes specific chemicals, either directly into the bloodstream (**endocrine gland**) or through a duct (**exocrine gland**) into tubular organs or onto the body surface. *2.* In plants, specialized unicellular or multicellular structure involved in the secretion of various substances formed as by-products of plant metabolism. [2/7/a]

glass Hard brittle amorphous mixture of the silicates of sodium and calcium, or of potassium and calcium. In some particularly strong or heat-resistant forms of glass, boron replaces some of the atoms of silicon. Glass is usually transparent or translucent.

glass ceramic Material that consists of lithium and magnesium aluminium silicates. It is thermally stable, and used for making ovenware. *See also* **ceramics**.

glass electrode Glass membrane electrode used to measure hydrogen ion concentration or pH.

glass wool Material that consists of fine glass fibres, used in filters, as a thermal insulator and for making fibre-glass.

Glauber's salt Alternative name for **sodium sulphate**, named after the German physicist Johann Glauber (1603–68).

globulin Water-insoluble protein that is soluble in aqueous solutions of certain salts. Globulins generally contain **glycine** and are coagulated by heat; *e.g.* immunoglobulin.

■ **glomerulus** Ball of **capillaries** located in the **Bowman's capsule** of the kidney. [2/5/a] [2/7/a]

■ **glottis** Opening of the **larynx,** through which air passes from the pharynx to the **trachea** (windpipe) of vertebrates.[2/4/a] [2/5/a] [2/7/a]

glove box Closed box that has gloves fixed into holes in the walls, and in which operations involving hazardous substances, such as radioactive materials or toxic chemicals, may be carried out safely. Alternative name: dry box.

glucagon In animals, a polypeptide hormone that (like **insulin**) is synthesized and secreted by the **islets of Langerhans** in the **pancreas**. Produced in response to low blood pressure, it stimulates **glycogen** breakdown in the liver, with release of **glucose** into the bloodstream. Its action is therefore opposite to that of insulin (which reduces blood glucose levels).

gluconic acid $CH_2OH(CHOH)_4COOH$ Soluble crystalline organic acid, an oxidation product of **glucose**, used in paint strippers. Alternative name: dextronic acid.

■ **glucose** $C_6H_{12}O_6$ **Monosaccharide** carbohydrate, a soluble colourless crystalline **sugar** (hexose) which occurs abundantly in plants. It is the principal product of photosynthesis and source of energy in animals (it is a product of carbohydrate digestion and is the sugar in blood), used in the manufacture of confectionery and the production of beer. Its natural polymers include **cellulose** and **starch**. Alternative names: dextrose, grape sugar. [2/3/d]

glucoside *See* **glycoside**.

glue *1*. Adhesive obtained by extracting bones. *2*. Any adhesive made by dissolving a substance such as rubber or plastic in a volatile **solvent**.

gluon Subatomic particle of the type believed to hold **quarks** together.

gluten Protein that occurs in cereals, particularly wheat flour. People with coeliac disease cannot tolerate gluten, and have to eat a gluten-free diet.

glyceride Ester of **glycerol** with an organic acid. The most important glycerides are **fats and oils**.

glycerol $HOCH_2CH(OH)CH_2OH$ Colourless sweet syrupy liquid, a trihydric **alcohol** that occurs as a constituent of **fats and oils** (from which it is obtained). It is used in foodstuffs, medicines and in the preparation of alkyd **resins** and nitroglycerine (glyceryl trinitrate). Alternative names: glycerin, glycerine, propan-1,2,3-triol.

glyceryl trinitrate Alternative name for **nitroglycerine**.

glycine $CH_2(NH_2)CO_2H$ Simplest **amino acid**, found in many **proteins** and certain animal excretions. It is a precursor in the biological synthesis of **purines, porphyrins** and **creatine**. It is also a component of **glutathione** and the bile salt **glycocholate**. It acts as a **neurotransmitter** at inhibitory nerve **synapses** in vertebrates. Alternative names: aminoacetic acid, aminoethanoic acid.

glycogen Polysaccharide carbohydrate, the main energy store in the liver and muscles of vertebrates ('animal starch'), also found in some algae and fungi. **Amylase** enzymes convert it to **glucose**, for use in metabolism.

glycol Alternative name for any **diol** or, specifically, ethylene glycol **(ethanediol)**.

glycolipid Member of a family of compounds that contain a **sugar** linked to **fatty acids**. Glycolipids are present in higher plants and neural tissue in animals. Alternative names: glycosylacylglycerols, glycosyldiacylglycerols.

glycolysis Conversion of **glucose** to lactic or pyruvic acid with the release of energy in the form of **adenosine triphosphate** (ATP). In animals it may occur during short bursts of muscular activity.

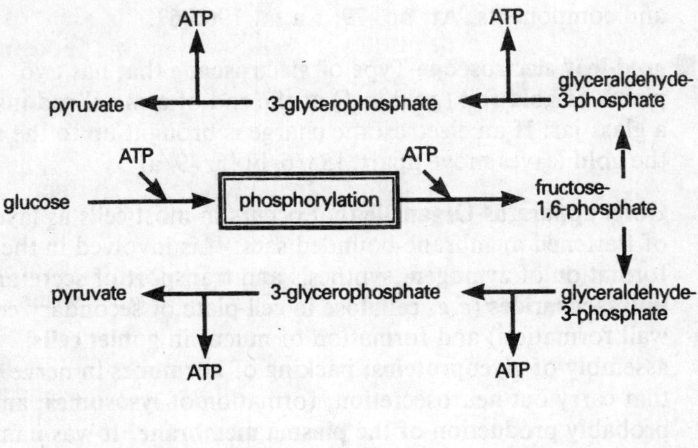

Production of ATP by glycolysis

glycoprotein Member of a group of compounds that contain **proteins** attached to **carbohydrate** groups. They include blood glycoproteins, some **hormones** and **enzymes**.

glycoside Compound formed from a **monosaccharide** in which an alcoholic or phenolic group replaces the first hydroxyl group. If the monosaccharide is **glucose**, it is termed a glucoside.

Gnathostomata Superclass containing all vertebrates that possess jaws.

goblet cell Mucus-secreting cell in **mucous membranes**.

gold Au Yellow metallic element in Group IB of the Periodic Table. It occurs as the free metal (native) in lodes and placer (alluvial) deposits. Most gold is held in currency reserve stocks, although some is used (usually as an alloy) in coinage, jewellery, dentistry and in electroplating electronic circuits and components. At. no. 79; r.a.m. 196.967.

■ **gold-leaf electroscope** Type of **electroscope** that has two pieces of gold foil (gold leaf) at the end of a metal rod inside a glass jar. If an electrostatic charge is brought up to the rod, the gold leaves move apart. [5//6/b] [5//9/a]

Golgi apparatus Organelle that occurs in most cells as layers of flattened membrane-bounded sacs. It is involved in the formation of **zymogen**, synthesis and transport of secretory **polysaccharides** (*e.g.* **cellulose** in cell plate or secondary cell-wall formation) and formation of mucus in **goblet cells**; assembly of **glycoproteins**; packing of **hormones** in nerve cells that carry out neurosecretion; formation of **lysosomes**; and probably production of the **plasma membrane**. It was named after the Italian histologist Camillo Golgi (1843–1926). Alternative names: Golgi body, Golgi complex.

■ **gonad** Reproductive organ of an animal, *e.g.* **ovary** or **testis**, in which **ova** (eggs) and **sperm** are formed respectively. Gonads may also function as **endocrine glands**, secreting sex **hormones**.

■ **gonadotrophic hormone Hormone** that acts on the **gonads**, controlling the initiation of puberty, the menstrual cycle and lactation in females and sperm-formation in males. It is produced by the pituitary gland. It is also used for the treatment of infertility in women. Alternative name: gonadotrophin.[2/4/a] [2/5/a] [2/7/a]

■ **Graafian follicle** Fluid-filled ball of cells in the mammalian **ovary** inside which an **oöcyte** develops. It matures periodically and then bursts at the surface of the ovary (at **ovulation**) to release an **ovum** (egg). The follicle then temporarily becomes a solid body, the **corpus luteum**. It was named after the Dutch anatomist Regnier de Graaf (1641–73). Alternative name: ovarian follicle. [2/7/a]

gradient Rate of rise or fall of a variable quantity such as temperature or pressure.

graft Transplantation of an **organ** or **tissue** from one organism into the body of another, although, apart from cornea and skin grafts, the term is applied mostly to plants (other animal tissue grafts are usually called transplants).

graft hybrid Plant that is an unusual mixture of two genetically different tissues; *e.g.* grafting purple broom with laburnum produces a tree that has the shape of laburnum but with purple flowers. A graft hybrid is a form of **chimaera**.

grafting Method of propagating plants using a **graft** (a stem or bud from the scion inserted into the tissues of the stock).

Graham's law Velocity of diffusion of a gas is inversely proportional to the square root of its density. It was named after the British chemist Thomas Graham (1805–69).

gram Unit of mass in the metric system. Alternative name: gramme.

■ **Graminae** Large family of flowering plants that consists of the grasses. There are more than 90,000 species world-wide, a very few of which (wheat, rice, maize and millet) provide the staple food of most of the world's population. The bamboos are the largest, and have woody stems. [2/3/b] [2/4/b]

gramme Alternative name for **gram**.

gram molecule Molecular weight of a substance in grams; mole.

Gram's stain Staining technique that differentiates between and classifies two major groups of bacteria, known as Gram-positive and Gram-negative bacteria, which stain deep purple or red respectively. The different uptake of stain is due to differences in cell-wall structure. It was named after the Danish physician Hans Christian Gram (1853–1938).

grape sugar Alternative name for **glucose**.

■ **graphite** Soft black natural **allotrope** of **carbon**. It is used as a **moderator** in nuclear reactors, as a lubricant, in paints, in pencil 'leads' and as a coating for foundry moulds. [4/8/d] [4/8/e]

gravimeter Instrument that measures variations in the Earth's gravitational field by detecting changes in the force (of gravity) acting on a suspended weight. It is used in prospecting for deposits of minerals (including petroleum), which affect the gravitational field near them.

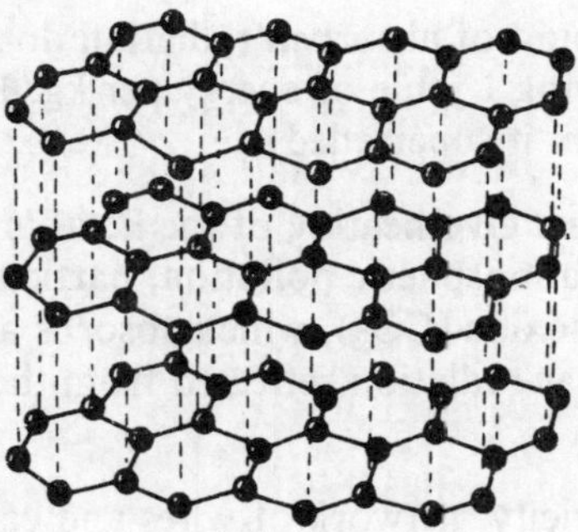

Atomic structure of graphite

gravimetric analysis Quantitative chemical analysis made ultimately by weighing substances.

gravitational constant (G) Constant in **Newton's law of gravitation**. Alternative name: universal gravitational constant.

gravitational field Region in which an object with mass exerts a force of attraction on another such object because of **gravity**.

gravitation, law of Alternative name for **Newton's law of gravitation**.

graviton Hypothetical particle that is believed to cause the gravitational effect between subatomic particles.

■ **gravity** Force of attraction that one object has on another because of their masses. The strength of the force is described by **Newton's law of gravitation**. The term is often used to refer to the force of gravity of the Earth.

gray (Gy) Amount of absorbed radiation dose in SI units, equal to supplying 1 joule of energy per kg. It is equivalent to 100 rad (the unit it superseded).

■ **greenhouse effect** Overheating of the Earth's atmosphere resulting from atmospheric pollution, particularly the build-up of carbon dioxide (CO_2), which absorbs and thus traps some of the solar radiation reflected from the Earth's surface.

grid *1.* In electricity, network of wires and cables that distribute electricty from power stations to where it is used. *2.* In electronics, an electrode (other than the anode and cathode) that controls the flow of current through a **thermionic valve** (*e.g.* triode, tetrode).

Grignard reagent Member of a class of **organometallic compounds** that have the general formula RMgX, where R = an **alkyl group** and X = Br or I. The reagents are very reactive, giving rise to the highly nucleophilic radical R^-. They are important in organic synthesis. They were named after the French chemist Victor Grignard (1871–1935).

ground state Lowest energy state of an atom or molecule, from which it can be raised to a higher energy state by **excitation**. *See also* **energy level**.

■ **group** Column, or vertical row, of elements in the Periodic Table (horizontal rows are periods). The group number (I to VIII and 0) indicates the number of **electrons** in the atom's outermost shell. [4/7/a]

growth Increase in size, dry mass or numbers that is a characteristic of living organisms.

■ **growth hormone** Any of a group of **polypeptides** that control growth and differentiation in animals and plants. In animals it is secreted by the anterior **pituitary gland** and acts directly on the cells of the body, particularly those of bones. Alternative names: (animals) somatotrophin; (plants) auxin; growth substance. [2/7/a]

growth ring Alternative name for **annual ring**.

growth substance Any of the substances (sometimes called plant hormones) that regulate the growth and development of plants. Growth substances include auxins and gibberellins.

guanidine $HN=C(NH_2)_2$ Strongly basic organic compound, used in the manufacture of explosives. Alternative name: imido-urea.

guanine $C_5H_5N_5O$ Colourless crystalline organic base (a **purine** derivative) that occurs in **DNA**.

■ **guard cell** One of a pair of cells that control the opening (and closing) of **stomata**, in turn controlling the **gas exchange** and **transpiration** in a leaf. [2/3/d]

gum camphor Alternative name for **camphor**.

■ **gut** Alternative name for **alimentary canal**.[2/4/a] [2/5/a] [2/7/a]

gymnosperm Member of the **Gymnospermae**.

Gymnospermae Subdivision of the plant division (phylum) **Spermatophyta**. Gymnosperms are cone-bearing plants (typically cycads and conifers) whose seeds have an **ovule** that is not enclosed by the **carpel**.

gynaecium Female reproductive part of a flower. Alternative name: carpel.

gypsum $CaSO_4.2H_2O$ Very soft calcium mineral, used in making cement and plasters. Alternative name: calcium sulphate dihydrate.

gyromagnetic ratio Ratio of the **magnetic moment** of an atom or nucleus to its **angular momentum**.

gyroscope Device in which a flywheel rotates at a high speed and because of this resists any change in direction of its axis of rotation. Gyroscopes are used in stabilizers for ships and guns and in guidance systems.

H

■ **Haber process** Industrial process for making **ammonia** from hydrogen and atmospheric nitrogen at high temperature and pressure in the presence of a **catalyst**. The hydrogen is obtained by the **Bosch process** or from **synthetic gas**. It was named after the German chemist Fritz Haber (1868–1934). Alternative name: Haber-Bosch process. [4/7/c,d]

■ **habitat** Part of the **environment** in which an organism or a community of plants and animals lives; *e.g.* a lake or a forest. [2/2/b] [2/2/c]

habituation Process by which the nervous system becomes accustomed to a particular **stimulus** and after a time is no longer irritated by it; *e.g.* the feel of one's clothes or the background noise of machinery.

hadron Any **elementary particle** that interacts strongly with other particles, including **baryons** and **mesons**.

■ **haem** Iron-containing group of atoms attached to a **polypeptide** chain; *e.g.* in **haemoglobin** and **myoglobin**.

■ **haematite** Mineral containing iron (III) oxide, one of the principal ores of iron. One form, kidney ore, occurs as distinctive brown crystals. [3/2/b]

haemocyanin Blue **blood** pigment that contains copper as its prosthetic group for the transport of oxygen. It is confined to lower animals, *e.g.* molluscs.

■ **haemoglobin** Red **blood** pigment that contains four iron **haem** prosthetic groups for the transport of oxygen. It occurs in red blood cells (erythrocytes) of vertebrates and some invertebrates. [2/4/c] [2/7/a]

haemolysis Breakdown of red blood cells (erythrocytes) which results in the release of **haemoglobin**.

haemophilia Inherited disease that affects human males in which there is a deficiency in the blood-clotting process. Sufferers are at constant risk of bleeding excessively from even minor injuries.

hafnium Hf Silvery metallic element in Group IVB of the Periodic Table (a **transition element**), used to make **control rods** for nuclear reactors. At. no. 72; r.a.m. 178.49.

hahnium Ha Element no. 105 (a **post-actinide**). It is a radioactive metal with short-lived **isotopes**, made in very small quantities by bombardment of an **actinide** with atoms of an element such as carbon or oxygen.

■ **hair** *1.* Derivative of **ectoderm** composed of insoluble **proteins** or **keratins**. Its role in mammals includes assisting regulation of body temperature. *2.* In plants, any of various outgrowths from the epidermis, *e.g.* root hairs, which absorb water.

half-cell Half of an **electrolytic cell**, consisting of an electrode immersed in an **electrolyte**.

■ **half-life** *1.* Time taken for something whose decay is exponential to reduce to half its value. *2.* More specifically, time taken for half the nuclei of a **radioactive** substance to decay spontaneously. The half-life of some unstable substances is only a few seconds or less, whereas for other substances it may be thousands of years; *e.g.* lawrencium has a half-life of 8 seconds, and the **isotope** plutonium-239 has a half-life of 24,400 years. [4/9/d]

halide Binary compound containing a **halogen**: a fluoride, chloride, bromide or iodide.

halite Naturally occurring form of **sodium chloride**. Alternative names: common salt, rock salt.

halo White or rainbow-coloured ring sometimes observed around the Sun or Moon, caused by light **refraction** through minute ice crystals in the Earth's atmosphere.

haloalkane Alternative name for **halogenoalkane**.

haloform Organic compound of the type CHX_3, where X is a **halogen** (chlorine, bromine or iodine); *e.g.* chloroform (trichloromethane), iodoform (tri-iodomethane). The compounds are prepared by the action of the halogen on heating with **ethanol** in the presence of sodium hydroxide.

■ **halogen** Element in Group VIIA of the Periodic Table: **fluorine**, **chlorine**, **bromine**, **iodine** or **astatine**. [4/9/a]

halogenation Chemical reaction that involves the addition of a **halogen** to a substance; *e.g.* the chlorination of benzene (C_6H_6) to form chlorobenzene (C_6H_5Cl).

halogenoalkane **Halogen** derivative of an **alkane**, general formula $C_nH_{2n+1}X$, where X is a **halogen** (fluorine, chlorine, bromine or iodine); *e.g.* chloromethane (methyl chloride), CH_3Cl. Alternative names: haloalkane, alkyl halide; monohalogenoalkane.

halophilic Exhibiting a preference for an environment containing salt (*e.g.* sea-water). The term is usually applied to **bacteria**.

■ **halophyte** Plant that lives in a salty environment. [2/2/c]

■ **haploid** Having half the number of **chromosomes** of the organism in the cell **nucleus**. The haploid state is found in **gametes** and in the **gametophytes** of plants that exhibit **alternation of generations**. *See also* **diploid**. [2/6/b] [2/7/b]

hard copy In computing, document that people can read (*e.g.* a print-out in plain language).

hard disk Rigid magnetic disk that provides data and program storage for computers, including microcomputers. Hard disks can hold a high density of data. When used in a microcomputer, they are usually not portable. Alternative name: diskette. *See also* **floppy disk**; **winchester**.

hardness of water Property of water that prevents it forming a lather with soap because of the presence of dissolved compounds of calcium or magnesium. Such compounds form an insoluble scum with soap, and fur up kettles and hot-water pipes. Water with little or no hardness is termed soft.

hardware Electronic, electrical, magnetic, and mechanical parts that make up a **computer** system. *See also* **software**.

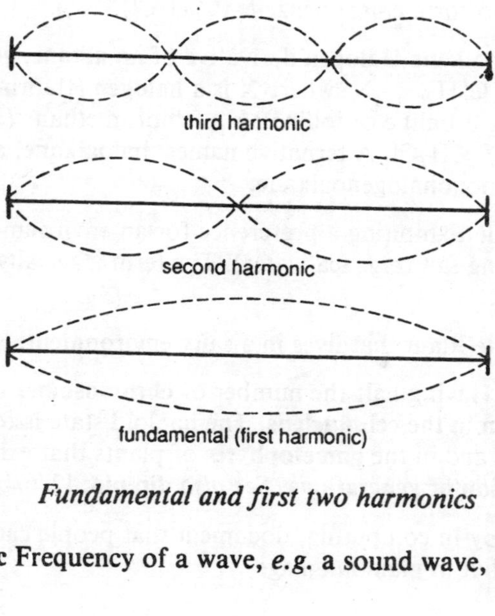

Fundamental and first two harmonics

■ **harmonic** Frequency of a wave, *e.g.* a sound wave, that is a whole-number multiple of the frequency of another wave (known as the fundamental). Alternative name: overtone. [5/8/c]

harmonic motion Alternative name for **simple harmonic motion**.

harmonic series Series of *e.g.* sounds in which each is a **harmonic** of a fundamental sound.

Haversian canal Any of the numerous channels that occur in **bone** tissue, containing blood vessels and nerves. An organic matrix is laid down in layers encircling each Haversian canal. It was named after the English physician Clopton Havers (?–1702).

■ **hay fever** Excessive secretion of mucus from the nasal passages caused by the inhalation of an **antigen** such as pollen or fungal spores. It is a type of **allergy**. [2/7/a]

head In a tape recorder, video recorder, record player or computer input/output device, an electromagnetic component that can read, erase or write signals off or onto tapes and disks.

■ **heart** Muscular organ in vertebrates that pumps **blood** into a system of **arteries**. Blood returns to the heart from the tissues via **veins**, and is passed to the **lungs** to become reoxygenated. The mammalian heart is divided into four chambers: the right and left atria (also known as auricles) and right and left, thick-walled ventricles, so that deoxygenated and oxygenated blood remain separate.[2/4/a] [2/5/a] [2/7/a]

heartbeat Alternate contraction and relaxation of the **heart**, corresponding to **diastole** and **systole**.

■ **heat** Form of **energy**, the energy of motion (kinetic energy) possessed by the atoms or molecules of all substances at temperatures above **absolute zero**. [3/6/c]

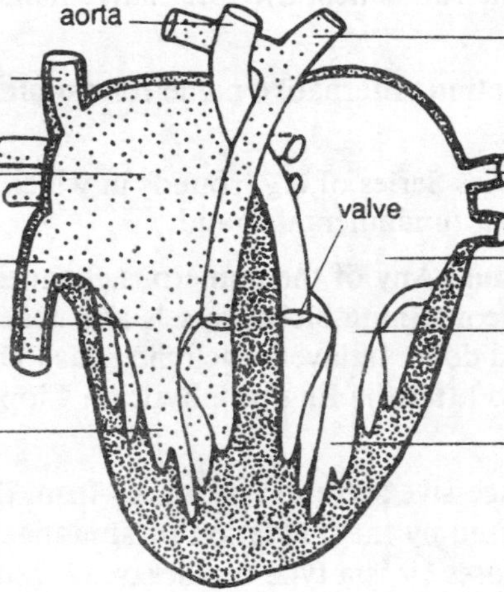

The human heart

heat balance Statement of all sources of heat and all of its uses for an industrial process or piece of equipment.

heat capacity Quantity of heat required to produce unit rise of temperature in an object. *See also* **specific heat capacity**.

heat engine Machine that converts heat into mechanical energy (for doing useful work).

heat exchanger Device that employs two separate streams of fluid (gas or liquid) for heating or cooling one of them; *e.g.* a car radiator uses air flow to cool water.

heat of activation Difference between the values of the thermodynamic functions for the activated complex and the **reactants** for a **chemical reaction** (all the substances being in their **standard states**).

heat of atomization Amount of heat that is required to convert 1 mole of an element into the gaseous state.

heat of combustion Heat change that accompanies the complete combustion of 1 mole of a substance.

heat of formation Heat change that occurs when 1 mole of a compound is formed from its elements, in their normal states.

heat of neutralization Heat change that occurs when 1 mole of aqueous **hydrogen ions** are neutralized by a **base** in dilute solution.

heat of reaction Heat change that occurs when the molar quantities (lowest possible multiples of 1 mole) of reactants as stated in a chemical equation react together.

heat of solution Heat change that occurs when 1 mole of a substance is dissolved in so much water that further dilution with water produces no further heat change.

Heaviside layer Region of the **ionosphere** at an altitude of 110–120 km. It is the most regular of the ionized layers that reflect radio waves, and has its greatest effect during the hours of daylight. It was named after the British scientist Oliver Heaviside (1850–1935). Alternative name: E-layer. *See also* **F- layer**.

heavy oil Alternative name for **gas oil**.

heavy water D_2O Compound of oxygen and **deuterium** (heavy hydrogen), analogous to **water**. It differs from ordinary water in most of its properties; *e.g.* it will not support life. Alternative name: deuterium oxide.

hecto- Metric prefix for a multiple of 10^2; denotes 100 times.

Heisenberg uncertainty principle The precise position and momentum of an electron cannot be determined simultaneously. It was named after the German physicist Werner Heisenberg (1901–76).

■ **helium** He Gaseous element in Group 0 of the Periodic Table (the **rare gases**), which occurs in some natural gas deposits. It is inert and non-inflammable, used to fill airships and balloons (in preference to inflammable hydrogen) and in helium-oxygen 'air' mixtures for divers (in preference to the nitrogen-oxygen mixture of real air, which can cause the bends). It is also used in gas lasers. Liquid helium is employed as a coolant in cryogenics. At. no. 2; r.a.m. 4.0026. [4/9/a]

Helmholtz free energy (*F*) Thermodynamic quantity equal to $U - TS$, where U is the **internal energy**, T the thermodynamic temperature and S the **entropy**. It was named after the German physicist Hermann Helmholtz (1821–94).

Hemichordata Subdivision of the **Chordata** (consisting of the acorn worms). Hemichordates lack a true **notochord** and have a very primitive nervous system.

hemichordate Member of the **Hemichordata**.

hemihydrate Compound containing one water molecule for every two molecules of the compound; *e.g.* $CaSO_4.\frac{1}{2}H_2O$.

Hemiptera Order of winged sap-sucking insects, often referred to as the true bugs. Hemipterans include aphids and leaf-hoppers.

henry (H) Unit of electrical **inductance**, defined as the inductance that produces an induced electromotive force of 1 volt for a current change of 1 ampere per second. It was named after the American physicist Joseph Henry (1797–1878).

Henry's law Weight of gas dissolved by a liquid is proportional to the gas pressure.

■ **heparin Polysaccharide** substance that prevents the clotting of blood by inhibiting the conversion of prothrombin to thrombin; used medicinally as an anticoagulant. [2/7/a]

■ **hepatic** Relating to the **liver**.[2/4/a] [2/5/a] [2/7/a]

■ **hepatic portal system** In vertebrates, system of blood **capillaries** into which dissolved foods (except for **fatty acids** and **glycerol**, which enter the **lymphatic system**) pass from the intestine lining for transport to the liver. [2/4/a] [2/5/a] [2/7/a]

heptane C_7H_{16} Liquid **alkane** hydrocarbon, the seventh member of the methane series, present in petrol.

heptavalent With a valency of seven. Alternative name: septivalent.

■ **herbicide** Chemical that kills plants; a weed-killer. Herbicides include defoliants, contact herbicides (which kill only the parts they touch) and selective herbicides (which kill only certain plants). [2/10/a,e]

■ **herbivore** Animal that feeds on plants, with teeth and a digestive system adapted for that purpose; *e.g.* cattle, deer, goats, rabbits, sheep. *See also* **carnivore; omnivore**. [2/3/b] [2/4/b]

■ **heredity** Mechanism by which offspring inherit certain characteristics from their parents. *See* **genetics**. [2/10/a]

hermaphrodite Plant or animal that possesses both male and female organs. Alternative name: bisexual (particularly of plants). [2/6/a]

herpes virus Any of a group of animal DNA **viruses**

responsible for various diseases; *e.g.* chickenpox (variola) and herpes simplex (coldsore), and herpes zoster (shingles).

■ **hertz** (Hz) Unit of frequency in the SI system. 1 Hz = 1 cycle per second. It was named after the German physicist Heinrich Hertz (1858–94).

Hess's law Total energy change resulting from a chemical reaction is dependent only on the initial and final states, and is independent of the reaction route. It was named after the Austrian physicist Victor Hess (1883–1964).

hetero- Prefix denoting other or different. *See also* **homo-**.

heterocyclic compound Cyclic organic compound that contains atoms other than carbon in the ring, *e.g.* **pyridine**. *See also* **homocyclic compound**.

■ **heterogametic** Describing an organism that produces two kinds of **gametes**, each possessing a different **sex chromosome**. These are usually produced by the male; *e.g.* in human males half the **sperms** contain an **X-chromosome** and half a **Y-chromosome**. [2/6/a]

heterogeneous Relating to more than one **phase** (*e.g.* describing a **chemical reaction** that involves one or more **solids**, in addition to a **gas** or a **liquid** phase); describing a system of non-uniform composition. *See also* **homogeneous**.

heterolytic fission Breaking of a two-electron **covalent bond** to give two fragments, with one fragment retaining both electrons. Alternative name: heterolytic cleavage. *See also* **homolytic fission**.

■ **heterotrophism** Mode of nutrition exhibited by most animals, fungi, bacteria and some flowering plants. It involves the intake of organic substances from the **environment** due to the inability of the organism to synthesize them from inorganic materials. *See also* **autotrophism**; **holophytic**. [2/2/c]

■ **heterozygous** Describing an organism that possesses two dissimilar **alleles** in a pair of **chromosomes**. A **dominant** allele can be expressed in the heterozygous or **homozygous** state, but a **recessive** allele can be expressed only in the homozygous state. [2/6/b]

hexamine $C_6H_{12}N_4$ Organic compound made by condensing **formaldehyde** (methanal) with ammonia, used as a camping fuel and antiseptic drug; It can be nitrated to make the high explosive **cyclonite**. Alternative name: hexamethylenetetramine.

hexane C_6H_{14} Colourless liquid **alkane**, used as a **solvent**.

hexanedioic acid Alternative name for **adipic acid**.

hexose **Monosaccharide** carbohydrate (sugar) that contains six carbon atoms and has the general formula $C_6H_{12}O_6$; *e.g.* **glucose** and **fructose**.

■ **hibernation** Winter dormancy that occurs in some animals. It involves the slowing of **metabolism** and a drop in body temperature, and provides a means of avoiding the necessity to maintain a high body temperature during winter. *See also* **aestivation**.

high frequency (HF) *1.* Radio frequency between 30,000 and 3,000 KHz, used in radio communications. *2.* Rapidly alternating electric current or wave.

high-vacuum distillation Alternative name for **molecular distillation**.

■ **hinge joint** Simple form of articulation between bones that allows for movement in a single plane (*e.g.* elbow joint, knee joint). [2/3/a]

hippocampus *1.* Often refers to *Hippocampus*, the sea horse, a genus of fish. *2.* Structure of nervous tissue within the vertebrate **brain**, which in human beings is believed to be involved in the process of memory.

Hirudinea Class of **annelids** consisting of the leeches, some of which are blood-sucking.

Hirundinidae Family of birds with pointed wings and forked tails, which includes swallows and martins.

histamine Organic compound that is released from cells in connective tissue during an allergic reaction. It causes dilation of capillaries and constriction of bronchi.

histidine $(C_3H_3N_2)CH_2CH(NH_2)COOH$ Crystalline soluble solid, an optically active basic **essential amino acid**. Alternative name: 2-amino-3-imidazolylpropanoic acid.

histocompatibility lymphocyte-A system (HL-A system) In animals, genetically determined **glycoprotein antigens** that are secreted by the **macrophages** and **lymphocytes** in the cell membrane. They show a strong immunological response, leading to rejection of transplanted tissue or organs. Alternative name: histocompatibility antigens.

histogram Type of **bar chart** in which the area of a bar or block represents a quantity.

histology Scientific study of the structure of **tissues**.

histone One of a group of small **proteins** with a large proportion of basic **amino acids**; *e.g.* arginine or lysine. Histones are found in combination with nucleic acid in the chromatin of eukaryotic cells.

HL-A system Abbreviation of **histocompatibility lymphocyte-A system**.

holmium Ho Silvery metallic element in Group IIIB of the Periodic Table (one of the **lanthanides**). At. no. 67; r.a.m. 164.930.

holoenzyme Enzyme that forms from the combination of a coenzyme and an apoenzyme. The former determines the nature and the latter the specificity of a reaction.

holography Method of reproducing a three-dimensional image that employs **interference** and **diffraction** of **laser** light to store and subsequently release light waves.

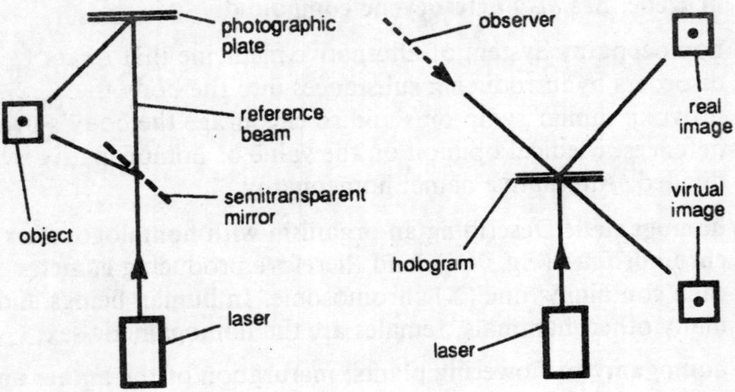

Principle of holography

■ **holophytic** Describing an organism that lives by **photosynthesis**; *e.g.* most green plants. *See also* **autotrophism**. [2/2/c]

Holothuroidea Class of **Echinodermata**, elongated marine invertebrates that may be free-living or attached to a surface; the sea-cucumbers.

■ **holozoic** Describing organisms that feed on solid organic matter or other organisms; *e.g.* most animals and insect-eating plants. *See also* **heterotrophism**. [2/2/c]

■ **homeostasis** Maintenance of a constant chemical and physical state in the internal environment of a cell, organ or organism, *e.g.* maintenance of body temperature or the balance of salts in the blood. [2/7/a] [2/9/a]

homo- Prefix denoting the same or similar. *See also* **hetero-**.

homocyclic compound Chemical compound that contains one or more closed rings comprising carbon atoms only, *e.g.* **benzene**. *See also* **heterocyclic compound**.

■ **homoeopathy** System of alternative medicine that treats disorders by introducing substances into the body that provoke similar symptoms and so encourage the body's own defences. Medical opinion on the value of homoeopathy is divided. Alternative name: homeopathy.

homogametic Describing an organism with homologous **sex chromosomes** (*e.g.* XX), and therefore producing **gametes** each containing one (X) chromosome. In human beings and many other mammals, females are the homogametic sex.

homogamy In flowering plants, maturation of the **anther** and **stigma** at the same time.

homogeneous *1.* In biology, describing similar structures found in different species that are thought to have originated from a common ancestor. *2.* In chemistry, relating to a single **phase** (*e.g.* describing a chemical reaction in which all the reactants are solids, or liquids or gases); describing a system of uniform composition.

■ **homoiothermic** Describing animals that maintain a more or less constant body temperature (*e.g.* mammals, birds). Alternative name: warm-blooded. [2/7/a]

homologous Describing things with common origin, but not necessarily the same appearance or function (*e.g.* the arms of a human and the wings of a bat). *See also* **analogous**.

■ **homologous chromosome Chromosome** that undergoes pairing during **meiosis**. Each one of the pair carries the same **genes** but not always the same **alleles** for a given character as the other member of the pair. Thus they have the same **gene** loci, **centromeres** at the same points and are very similar in length and shape. At **fertilization** each parent contributes one homologue of each pair. [2/5/b]

homologous series Family of organic chemical compounds with the same general formula; *e.g.* **alkanes, alkenes** and **alkynes**.

homologue *1*. In biology, a **homologous chromosome**. *2*. In chemistry, member of a **homologous series**.

homolytic fission Breaking of a two-electron **covalent bond** in such a way that each fragment retains one electron of the bond. Alternative name: homolytic cleavage. *See also* **heterolytic fission**.

homopolymer Polymer that is formed by polymerization of a single substance (monomer); *e.g.* polyethylene, polypropylene. *See also* **copolymer**.

■ **homozygous** Having the same two **alleles** for a given **gene**. Two organisms that are homozygous for the same alleles breed true for the character in question, thus producing progeny which are homozygous and identical to the parent with respect to that gene. *See also* **heterozygous**. [2/5/b]

Hooke's law For a material being stretched within its **elastic limit**, the extension is proportional to the force producing it. It was named after the British scientist Robert Hooke (1635–1703). [2/5/b]

■ **hormone** *1.* In animals, chemical 'messenger' of the body that is secreted directly into the bloodstream by an **endocrine gland**. Each gland secretes hormones of different composition and purpose which exert specific effects on certain target tissues. *2.* In plants, organic substance that at very low concentrations affects growth and development; *e.g.* auxin, gibberellin, abscisin, florigen. *See* **growth substance**. [2/7/a]

horn Hard structure generally made of **keratin** usually borne in pairs on the heads of certain animals (*e.g.* most antelopes, cattle and deer). Antlers (horns of deer) are shed and regrown every year.

hornblende Dark-coloured mineral, mainly composed of silicates of iron, calcium and magnesium, which is a major component of granite and many other metamorphic and igneous rocks.

horsepower (hp) Power needed to lift 33,000 pounds one foot in one minute. 1 horsepower = 745.7 watts.

horticulture Commercial growing of plants for gardens and parks. *See also* **agriculture**.

■ **host** *1.* Organism on which a parasite lives (*see* **parasitism**). *2.* Tree on which an **epiphyte** grows. [2/4/d] [2/6/a]

■ **humerus** Upper bone in the forelimb of a tetrapod vertebrate; in human beings, the upper arm bone. [2/3/a] [2/4/a] [2/5/a]

humidity Measure of the amount of water vapour in a gas, *e.g.* the air, usually expressed as a percentage. Relative humidity is the amount of vapour divided by the maximum amount of vapour the gas will hold (at a particular temperature).

humus *1.* Dark brown organic material produced in soil by the action of **decomposers** on dead plant and animal matter. *2.* Topmost layer of soil that contains a high proportion of decomposed organic material, known also as leaf litter.

■ **hybrid** Progeny of plants or animals produced from the cross of two genetically dissimilar parents. The term is usually limited in use to offspring of parents from two different (but closely related) species; *e.g.* a hybrid of a mare (female horse) and a male donkey is a mule.

hybridization Crossing of animals or plants to produce a **hybrid**.

hybrid vigour General improvement of physical characteristics of a **hybrid** in comparison to its parents resulting from an increase in genetical variation.

hydra Small freshwater animal of the phylum **Cnidaria**.

hydrate Chemical compound that contains **water of crystallization**.

hydrated *1.* Describing a substance after treatment with water. *2.* Describing a compound that contains chemically bonded water, a **hydrate**.

hydrated lime Alternative name for **calcium hydroxide**.

hydration Attachment of water to the particles (particularly **ions**) of a **solute** during the dissolving process.

hydraulics Branch of engineering concerned with the flow of liquids in pipes and channels, the pumping of liquids, and their use for the transmission and generation of power.

hydrazine NH_2NH_2 Colourless liquid, a powerful **reducing agent** used in organic synthesis and as a rocket fuel.

hydrazone Member of a family of organic compounds that contain the group $-C=NNH_2$. Hydrazones are formed by the action of **hydrazine** on an **aldehyde** or **ketone**, and are used in identifying them.

hydride Compound formed between **hydrogen** and another element (*e.g.* calcium hydride, CaH_2).

hydro- Prefix denoting water.

hydrobromic acid HBr Colourless acidic aqueous solution of **hydrogen bromide**; its salts are **bromides**.

■ **hydrocarbon** Organic compound that contains only carbon and hydrogen. The chief naturally occurring hydrocarbons are bitumen, coal, methane, natural gas and petroleum. Most of these, and hydrocarbons derived from them, are used as fuels. The **aliphatic** hydrocarbons form three homologous series: **alkanes, alkenes** and **alkynes**. Aromatic hydrocarbons are **cyclic** compounds.

hydrocephalus Commonly known as water on the brain, an excess of **cerebrospinal fluid** within the **ventricles** of the brain, eventually leading to malformation of brain tissue.

■ **hydrochloric acid** HCl Colourless acidic aqueous solution of **hydrogen chloride**, a strong acid that dissolves most metals with the release of hydrogen; its salts are **chlorides**. It is used to make **chlorine** and other chemicals, and for cleaning metals before electroplating or galvanizing.

hydrocyanic acid HCN Very poisonous solution of **hydrogen cyanide** in water; its salts are **cyanides**. Alternative name: prussic acid.

hydrofluoric acid HF Colourless corrosive aqueous solution of **hydrogen fluoride**; its salts are **fluorides**. It is used for etching glass (and must be stored in plastic bottles).

■ **hydrogen** H Gaseous element usually given its own place at the beginning of the Periodic Table, but sometimes assigned to Group IA. Colourless, odourless and highly inflammable, it is the lightest gas known and occurs abundantly in combination in water (H_2O), coal and petroleum (mainly as hydrocarbons) and living things (mainly as carbohydrates). It is also the major constituent of the Sun and other stars, and is the most abundant element in the Universe. Hydrogen is made commercially by **electrolysis** of aqueous solutions, by **cracking** of petroleum, by the **Bosch process** or as **synthetic gas**. In the laboratory hydrogen is generally prepared by the action of a dilute acid on zinc. It has many uses: for hydrogenating (solidifying) oils, to make ammonia by the **Haber process**, as a fuel (particularly in rocketry), in organic synthesis, and as a **moderator** for nuclear reactors. In addition to the common form (sometimes called protium, r.a.m. 1.00797) there are two other isotopes: **deuterium** or heavy hydrogen (r.a.m. 2.01410) and the radioactive **tritium** (r.a.m. 3.0221). At. no. 1; r.a.m. (of the naturally occurring mixture of isotopes) 1.0080. [4/9/a]

hydrogenation Method of chemical synthesis by adding hydrogen to a substance. It forms the basis of many important industrial processes, such as the conversion of liquid oils to solid fats.

hydrogenation of coal Industrial synthesis of mineral oil from coal by **hydrogenation**.

hydrogenation of oil Manufacture of margarine by the **hydrogenation** of liquid vegetable oils to edible fats.

hydrogen bomb Powerful explosive device that uses the sudden release of energy from **nuclear fusion** (of **deuterium** and **tritium** atoms). Alternative name: thermonuclear bomb.

■ **hydrogen bond** Strong chemical **bond** that holds together some **molecules** that contain hydrogen, *e.g.* water molecules, which become associated as a result. A hydrogen atom bonded to an electronegative atom interacts with a (non-bonding) **lone pair of electrons** on another electronegative atom.

hydrogen bromide HBr Pale yellow gas which dissolves in water to form **hydrobromic acid**.

hydrogencarbonate Acidic salt containing the ion HCO_3^-. Alternative name: bicarbonate.

hydrogen chloride HCl Colourless gas which dissolves readily in water to form **hydrochloric acid**. It is made by treating a **chloride** with concentrated sulphuric acid or produced as a by-product of electrolytic processes involving chlorides.

hydrogen cyanide HCN Colourless poisonous gas, which dissolves in water to form **hydrocyanic acid** (prussic acid). It has a characteristic smell of bitter almonds.

hydrogen electrode Half-cell that consists of hydrogen gas bubbling around a platinum electrode, covered in platinum black (very finely divided platinum). It is immersed in a molar acid solution and used for determining standard **electrode potentials**. Alternative name: hydrogen half-cell.

hydrogen fluoride HF Colourless fuming liquid, which is extremely corrosive and dissolves in water to form **hydrofluoric acid**.

hydrogen half-cell Alternative name for **hydrogen electrode**.

hydrogen halide Compound of hydrogen and a **halogen**; *e.g.* hydrogen fluoride, HF, hydrogen iodide, HI.

■ **hydrogen ion** H^+ Positively charged hydrogen atom; a **proton**. A characteristic of an **acid** is the production of hydrogen ions, which in aqueous solution are hydrated to hydroxonium ions, H_3O^+. Hydrogen ion concentration is a measure of acidity, usually expressed on the **pH** scale. A hydrogen ion concentration of 10^{-7} mol dm^{-3} (corresponding to pH 7) is neutral.

hydrogen peroxide H_2O_2 Colourless syrupy liquid with strong oxidizing powers, soluble in water in all proportions. Dilute solutions are used as an **oxidizing agent**, disinfectant and bleach; in concentrated form it is employed as a rocket fuel.

hydrogen spectrum Spectrum produced when an electric discharge is passed through hydrogen gas. The hydrogen molecules dissociate and the atoms emit light at a series of characteristic frequencies. *See also* **Balmer series**.

hydrogensulphate Acidic salt containing the ion HSO_4^-. Alternative name: bisulphate.

hydrogen sulphide H_2S Colourless poisonous gas with a characteristic smell (when impure) of bad eggs. It is formed by rotting sulphur-containing organic matter and the action of acids on sulphides.

hydrogensulphite Acidic salt containing the ion HSO_3^-. Alternative name: bisulphite.

hydrological cycle Alternative name for **water cycle**.

hydrolysis Chemical decomposition of a substance by water, with a hydroxyl group (–OH) from the water taking part in the reaction; *e.g.* esters hydrolyse to form alcohols and acids.

■ **hydrometer** Instrument for measuring the density of a liquid. It consists of a weighted glass bulb with a long graduated stem, which floats in the liquid being tested.

hydrophilic Possessing an affinity for water.

hydrophobia Popular name for **rabies**, although in fact it refers to only one of the symptoms, the fear of drinking water.

hydrophobic Water-repellent; having no attraction for water.

■ **hydrophyte** Plant that lives in water. [2/2/c]

hydrosol Aqueous solution of a **colloid**.

hydrosphere Portion of the Earth's crust that consists of water: the oceans, seas, lakes, rivers, etc.

hydrostatics Branch of science that deals with fluids at rest.

hydrostatic skeleton Supporting structure, consisting of fluid under pressure, possessed by some invertebrate animals (*e.g.* sponges).

hydrotropism Growth in plants in response to the presence of water; *e.g.* roots are hydrotropic.

hydrous Containing water.

hydroxide Compound of a metal that contains the **hydroxyl group** (–OH) or the hydroxide ion (OH^-). Many metal hydroxides are **bases**.

hydroxonium ion Hydrated **hydrogen ion**, H_3O^+.

hydroxybenzene Alternative name for **phenol**.

hydroxybenzoic acid Alternative name for **salicylic acid**.

hydroxyl group (–OH) Group containing oxygen and hydrogen, characteristics of **alcohols** and some **hydroxides**.

hydroxypropionic acid Alternative name for **lactic acid**.

■ **hygiene** Practice of contributing to good health through cleanliness, particularly of one's body and food.

hygrometer Instrument for measuring the **humidity** of air, the amount of water vapour in the atmosphere.

hygroscopic Having the tendency to absorb moisture from the atmosphere.

hyperfine structure Fine structure that is observable only when very high resolution is employed.

hypermetropia Long-sightedness, a visual defect in which the eyeball is too short (front to back) so that light rays entering the eye from nearby objects would be brought to a focus at a point behind the retina. It can be corrected by spectacles or contact lenses made from converging (convex) lenses. Alternative name: hyperopia. *See also* **myopia**.

hyperon Member of a group of short-lived **elementary particles** which are greater in mass than the **neutron**.

■ **hypha** Microscopic hollow filament characteristic of **fungi**. Hyphae form a network called a mycelium. [2/3/b]

hyphae Plural of **hypha**.

hypo Popular name for **sodium thiosulphate**.

hypochlorite Salt of **hypochlorous acid** (containing the ion ClO^-). Hypochlorites are used as bleaches and disinfectants.

hypochlorous acid ($HOCl$) Weak liquid **acid** stable only in solution, used as an oxidizing agent and bleach. Its salts are hypochlorites.

hypogeal Describing plant germination in which the **cotyledons** remain underground. *See also* **epigeal**.

hypophysis Alternative name for **pituitary**.

hyposulphuric acid Alternative name for **dithionic acid**.

hypothalamus Floor and sides of the vertebrate forebrain, which is concerned with physiological coordination of the body; *e.g.* regulation of body temperature, heart rate, breathing rate, blood pressure, sleep pattern as well as drinking, eating, water excretion and other metabolic functions.

■ **hypothermia** Condition in which body temperature is much lower than normal, as a result of which metabolic processes slow down. It may be induced deliberately to treat certain disorders (such as high fever), or may arise accidentally through exposure to freezing temperatures.

hypotonic Having a lower than normal **osmotic pressure**.

hysteresis In physics, the lag or delay between a cause and its effect, as in the magnetization of a magnetic substance.

I

ice Water in its solid state (*i.e.* below its **freezing point**, 0 °C). It is less dense than water, because **hydrogen bonds** give its crystals an open structure, and it therefore floats on water. This also means that water expands on freezing.

Iceland spar Very pure transparent form of **calcite** (calcium carbonate), and noted for the property of double refraction.

ice point Freezing point of water, 0 °C, used as a fixed point on temperature scales.

Ichneumonoidea Large family of parasitic wasps, often misleadingly referred to as ichneumon flies. A female ichneumonoid lays her eggs inside the body of a host (a spider or an insect) that is specific to her sub-family. When the eggs hatch, the larvae devour the host from within.

id Term used by some psychologists to denote the part of the human mind that is the source of primitive **instincts** and urges, and drives the **unconscious**. *See also* **ego**.

ideal crystal Crystal structure that is considered as perfect, containing no defects.

ideal gas Hypothetical gas with molecules of negligible size that experience no intermolecular forces. Such a gas would in theory obey the **gas laws** exactly. Alternative name: perfect gas.

ideal solution Hypothetical solution that obeys **Raoult's law** exactly.

identical twins Two offspring that arise from a single fertilized ovum (**zygote**) by mitotic division of the zygote or during the early embryo stage. Alternative name: maternal twins. *See also* **twins**.

ignis fatuus Light seen over marshy ground caused by the ignition of **methane** (generated in rotting vegetation) by traces of phosphorus compounds, *e.g.* phosgene. Alternative name: will-o'- the-wisp.

ignition Initial combustion of a substance, particularly of an explosive mixture (*e.g.* of petrol vapour and air) in an internal combustion engine.

ignition temperature Temperature to which a substance must be heated before it will burn in air (or in some other specified oxidant).

■ **ileum** Last section of the small **intestine** in mammals, where both digestion and absorption take place. *See also* **duodenum**; **jejunum**. [2/4/a] [2/5/a] [2/7/a]

illumination (*E*) Brightness of light on a surface, given as the intensity per unit area, expressed in lux (lumens per square metre). Alternative name: illuminance.

■ **image** In optics, point from which rays of light entering the **eye** appear to have originated. A real image, *e.g.* one formed by a converging lens, can be focused on a screen; a virtual image, *e.g.* one formed in a plane mirror, can be seen only by the eyes of the observer. [5/3/d]

image converter Electron tube for converting infra-red or other invisible images into visible images.

imago Sexually mature adult form of an insect that develops after previous stages of **metamorphosis**.

imbricate Describing any structure made up of overlapping parts, *e.g.* fish scales.

imide Organic compound derived from an acid anhydride, general formula R-CONHCO-R′, where R and R′ are organic **radicals**. Alternative name: imido compound.

imido compound Alternative name for **imide**.

imido-urea Alternative name for **guanidine**.

imine Secondary **amine**, an organic compound derived from **ammonia**, general formula RNHR′, where R and R′ are organic **radicals**; *e.g.* dimethylamine, $(CH_3)NH$. Alternative name: imino compound.

imino compound Alternative name for **imine**.

imino group The group =NH, characteristic of an **imine**.

immiscible Describing two or more liquids that will not mix (when shaken together they separate into layers).

■ **immune response** Response of vertebrates to invasion by a foreign substance (**antigen**). It involves the production of specific **antibody** molecules, which combine with the antigen to form an antigen-antibody complex. Antibodies may be present in body fluids or carried by **lymphocytes**.

immune system Organs, substances and mechanisms involved in **immunity** in a particular organism.

■ **immunity** Protection by an organism against infection. Defence may be divided into passive and active mechanisms. Passive processes prevent the entry of foreign invasion, *e.g.* skin, mucous membranes. Active mechanisms include **phagocytosis** by **leucocytes** and the **immune response** in animals. Plants can have immunity, *e.g.* by means of **phytoalexins**.

immunization Stimulation of active artificial **immunity** by injection of small amounts of **antigen** (**a vaccine**). A specific

antibody or immunoglobulin is formed in response and may persist in the body, preventing further infection by the same organism.

immunoglobulin *See* **antibody**; **immunization**.

immunology Scientific study of **immunity**.

immunosuppressive Describing a drug that suppresses the **immune response**, given to recipients of transplanted organs to minimize that chance of rejection.

impedance (Z) Property of an electric circuit or circuit component that opposes the passage of current. For **direct current** (d.c.) it is equal to the **resistance** (R). For **alternating current** (a.c.) the **reactance** (X) also has an effect, such that $Z^2 = R^2 + X^2$, or $Z = R + iX$, where $i^2 = -1$.

imperial system Comprehensive system of weights and measures (feet and inches, **avoirdupois** weights, pints and gallons, etc.) that was formerly used throughout the British Empire. Many units are still in widespread use in Britain, but **SI units** have replaced the imperial system for scientific measurement.

implantation Process in which a fertilized ovum (egg) or **embryo** becomes attached to the lining of the uterus (womb) of a mammal. It is the beginning of pregnancy.

implosion Opposite of explosion; an inward burst, *e.g.* as when a vacuum-filled vessel crumples in air.

imprinting Behavioural attachment of an animal to a parent figure for protection. It occurs during the early stages of the life of many animals.

impulse *1*. In biology, transmission of a message along a **nerve fibre**. The nerve impulse is an electrical phenomenon which results in depolarization of the nerve membrane. This **action potential** lasts for a millisecond before the **resting potential** is restored. *See also* **all-or-none response**. *2*. In physics, when two objects collide, over the period of impact there is a large reactionary force between them whose time integral is the impulse of the force (equal to either object's change of momentum).

inbreeding Reproduction between closely related organisms of a species. *See also* **outbreeding**.

incandescence Light emission that results from the high temperature of a substance; *e.g.* the filament in an electric lamp is incandescent. *See also* **luminescence**.

incisor Chisel-shaped cutting tooth of mammals located at the front of the upper and lower jaws; human beings have eight incisors. They grow continually in rodents, which use them for gnawing.

incompatibility *1*. Failure of a group of plants, algae or fungi to achieve fertilization over a period of time. *2*. Mismatching of biochemical components, *e.g.* between **blood groups**, or between a transplanted donor organ and its recipient, or between a **scion** and the plant on which it is grafted. With an organ transplant it can lead to immunological rejection (counteracted by **immunosuppressive drugs**).

■ **incomplete dominance** Condition in which neither of a pair of **alleles** (genes) is **dominant**; their effects merge to produce an intermediate characteristic. [2/7/b]

incubation *1*. Period of time between laying and hatching in which an embryo (of a reptile or bird) develops in its egg. *2*. Period of time during which bacteria develop in a host's body before symptoms of an illness appear.

incus One of the **ear ossicles**. Alternative name: anvil.

indehiscent Describing a **fruit** that does not spontaneously open to release its seeds (*i.e.* the seed or seeds remain inside the fruit when it falls from the plant).

■ **independent assortment** The second of **Mendel's laws**, which states that **genes** are transmitted independently from parents to offspring and assort freely. Thus there is an equal chance of any particular gene being transmitted to the **gametes**. It does not apply to genes that exhibit **linkage**. [2/7/b]

■ **indicator** *1*. In biology, organism that survives only in certain environments; its presence gives information about the environment. *E.g.* the presence in water of certain bacteria that normally live in faeces indicates that the water is polluted with sewage. *2*. In chemistry, substance that changes colour to indicate end of a chemical reaction or the **pH** of a solution; *e.g.* litmus, methyl orange, phenolphthalein. Indicators are commonly used in titrations in **volumetric analysis**. [4/5/a]

indigo $C_{16}H_{10}N_2O_2$ Blue organic dye, a derivative of **indole**, that occurs as a glucoside in plants of the genus *Indigofera*.

indium In Silvery-white metallic element of Group IIIA of the Periodic Table, used in making mirrors and **semiconductors**. At. no. 49; r.a.m. 114.82.

indole C_8H_7N Colourless solid, consisting of fused **benzene** and **pyrrole** rings, which occurs in coal-tar and certain plants. It is the basis of the **indigo** molecule. Alternative name: benzpyrrole.

induced magnetism Creation of a magnet by aligning the magnetic **domains** in a ferromagnetic substance by placing it in the **magnetic field** of a permanent magnet or electromagnet.

induced radioactivity Alternative name for **artificial radioactivity**.

■ **inductance** *1*. Property of a current-carrying electric circuit or circuit component that causes it to form a magnetic field and store magnetic energy. *2*. Measurement of electromagnetic **induction**. [5/7/c]

■ **induction** Magnetization or electrification produced in an object. *1*. Electromagnetic induction is the production of an electric current in a conductor by means of a varying magnetic field near it. *2*. Magnetic induction is the production of a magnetic field in an unmagnetized metal by a nearby magnetic field. *3*. Electrostatic induction is the production of an electric charge on an object by a charged object brought near it. [5/7/c,d]

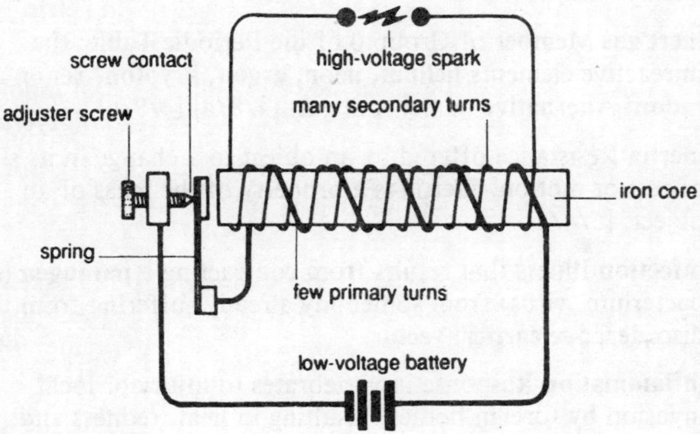

Simple induction coil

induction coil Type of **transformer** for producing high-voltage alternating current from a low-voltage source. Induction coils can be used to produce a high voltage pulse, *e.g.* for firing spark plugs in a petrol engine.

induction heating Heating effect that arises from the electric current induced in a conducting material by an alternating **magnetic field**.

induction motor Electric motor that consists of two coils, one stationary (the stator) and one moving (the rotor). An electric current is fed to the stator, which creates a magnetic field and induces currents in the rotor. The interaction between the magnetic field and the induced currents causes the rotor to turn.

inductor Any component of an electrical circuit that possesses significant **inductance**. Alternative names: choke, coil.

inert Chemically nonreactive, *e.g.* gold is inert in air at normal temperatures.

■ **inert gas** Member of Group 0 of the Periodic Table; the unreactive elements **helium, neon, argon, krypton, xenon** and **radon**. Alternative name: rare gas. [4/8/a] [4/9/a]

■ **inertia** Resistance offered by an object to a change in its state of rest or motion. Inertia is a property of the **mass** of an object. [5/4/a]

infection Illness that results from contracting a **pathogen** (*e.g.* bacterium, virus) from somebody already suffering from the disorder. *See* **carrier**; **vector**.

■ **inflammation** Response in vertebrates to injury or local invasion by foreign bodies, resulting in heat, redness and swelling. It involves dilation of **blood** vessels, migration of **leucocytes** to the site of injury, and movement of fluid and **plasma proteins** into the inflamed tissue. [2/7/a]

inflorescence Any of the various types of arrangement of flowers on a plant's (single) main stem.

■ **infra-red radiation Electromagnetic radiation** in the wavelength range from 0.75 μm to 1 mm approximately; between the visible and microwave regions of the **electromagnetic spectrum**. It is emitted by all objects at temperatures above absolute zero, as heat (thermal) radiation. [5/9/a,c]

infrasound Sound waves with a frequency below the threshold of human hearing, *i.e.* less than about 20 Hz.

inheritance Process by which characteristics are passed on from parents to offspring, *i.e.* from one generation to another.

■ **inhibitor** Substance that prevents an action from occurring; *e.g.* some germinating seeds produce an inhibitor in the soil that stops other seeds from germinating. [2/9/c]

■ **inner ear** Fluid-filled part of the **ear** that contains both the organs that convert sound to nerve impulses and the organs of balance. [2/4/a] [2/5/a] [2/7/a]

inoculation *See* **vaccine**.

■ **inorganic chemistry** Study of chemical elements and their **inorganic compounds**. *See also* **organic chemistry**.

inorganic compound Compound that does not contain **carbon**, with the exception of carbon's oxides, metallic carbides, carbonates and hydrogencarbonates. *See also* **organic compounds**.

input device Part of a **computer** that feeds in with **data** and **program** instructions. The many types of input devices include a keyboard, punched card reader, paper tape reader, optical character recognition, light pen (with a VDU) and various types of devices equipped with a **read head** to input magnetically recorded data (*e.g.* on magnetic disk, tape or drum).

■ **insect** Arthropod animal of the class **Insecta**.

■ **Insecta** Large class of arthropods. Most adult insects possess a three-part segmented body consisting of a **head**, **thorax** and **abdomen**, with an **exoskeleton** composed of **chitin**, one pair of **antennae**, and three pairs of walking legs; many also possess wings, typically two pairs. Respiration is by means of **tracheae**. Many insects have life-cycles that involve **metamorphosis**. [2/3/b] [2/4/b]

■ **insecticide** Substance used to kill insects. There are two main types: those that are eaten by insects (after being applied to their food) or inhaled by insects, and those that kill by contact. Non-biodegradable insecticides (such as DDT) may persist for a long time and become concentrated in **food chains**, where they have a damaging ecological effect. [2/5/c] [2/9/d]

■ **Insectivora** Order of insect-eating mammals. Insectivores include shrews, hedgehogs and moles. They resemble the ancestral mammals that co-existed with the dinosaurs. [2/4/d]

insectivore Strictly an animal of the order **Insectivora**, although often extended to include any insect-eating animal (such as ant-eaters and some bats).

insemination Placing of **sperm** within the body of a female so that fertilization of an **ovum** (egg) is probable.

insoluble Describing a substance that does not dissolve in a given **solvent**; not capable of forming a **solution**.

instar The form assumed by an insect during any of the stages of its development prior to its final moult (*see* **metamorphosis**).

instinct General term for any animal behaviour, or fixed response to a particular stimulus, that has not been learned but inherited; it is therefore common to all the members of a species. Instinct can be modified by learning; *e.g.* a bird instinctively knows how to sing but may have to learn a particular song.

insulation Layer of material (an **insulator**) used to prevent the flow of **electricity** or **heat**.

■ **insulator** Substance that is a poor conductor of **electricity** or **heat**; a non-conductor. Most non-metallic elements (except carbon) and polymers are good insulators. [5/2/b]

■ **insulin Hormone** secreted in vertebrates that controls the level of **glucose** in the blood. It is produced by groups of cells, called islets of Langerhans, in the **pancreas**. A deficiency of insulin results in the disorder diabetes mellitus, whose symptoms include excessive thirst and high levels of glucose in the blood and urine. [2/4/a] [2/5/a] [2/7/a]

integrated circuit Very small solid-state circuit consisting of interconnected **semiconductor** devices such as diodes, transistors, capacitors and resistors printed into a single silicon chip.

integument *1*. Animal's outer protective covering; *e.g.* cuticle, skin. *2*. Part of the outer coating of a seed (*see* **testa**).

■ **intensity** Power of sound, light or other wave-form (*e.g.* the loudness of a sound or the brightness of light), determined by the **amplitude** of the wave. [2/4/c]

interaction In atomic physics, exchange of energy between a particle and a second one or an **electromagnetic wave**.

intercellular Between cells; *e.g.* intercellular fluid surrounds cells, maintaining a constant internal environment. *See also* **intracellular**.

■ **intercostal muscle** Any of the muscles between a mammal's ribs, which are important in breathing movements. [2/5/a]

interface Boundary of contact (the common surface) of two adjacent **phases**, either or both of which may be solid, liquid or gaseous.

■ **interference** Interaction between two or more waves of the same frequency emitted from coherent sources. The waves may reinforce each other or tend to cancel each other; the resultant wave is the algebraic sum of the component waves. The phenomenon occurs with **electromagnetic waves** and **sound**. [5/10/c]

interferon Substance that is produced by animal cells, which prevents the multiplication of **viruses**. It is a protein, and its action is not specific to any particular group of viruses. There is some debate as to whether it may be useful in the treatment of certain forms of cancer.

intermediate *1*. In industrial chemistry, compound to be subjected to further chemical treatment to produce finished products such as dyes and pharmaceuticals. *2*. Short-lived species in a complex chemical reaction.

intermediate compound Compound of two or more metals that are present in definite proportion although they frequently do not follow normal **valence** rules.

intermediate frequency Frequency to which radio signal is converted during heterodyne reception (which involves the superimposition of two radio waves of different frequencies).

■ **intermediate host** Organism that acts as host to a maturing

parasite (*e.g.* to the larval stage of an insect), but which is not a host to the mature parasite. [2/2/c]

intermediate neutron Neutron that has energy between that of a **fast neutron** and a **thermal neutron**.

intermolecular force Force that binds one molecule to another. Intermolecular forces are much weaker than the bonding forces holding together the atoms of a molecule. *See also* **van der Waals' force**.

internal combustion engine Engine that produces power through the combustion of a fuel/air mixture inside an enclosed space (in petrol and diesel engines, a cylinder fitted with a piston).

internal conversion Effect on the nucleus of an atom produced by a gamma-ray **photon** emerging from it and giving up its energy on meeting an electron of the same atom.

internal energy Total quantity of energy in a substance, the sum of its **kinetic energy** and **potential energy**.

internal friction *See* **viscosity**.

internal resistance Electrical **resistance** in a circuit of the source of current, *e.g.* a cell.

international candle Former unit of **luminous intensity**.

■ **interphase** State of cells when not undergoing division. Preparation for division **(mitosis)** is carried out during this phase, including replication of **DNA** and cell constituents. [2/8/b]

interstitial atom Atom that is in a position other than a normal **lattice** place.

interstitial cells In mammals, cells present in the male and female **gonads**. In males they are found between the **testis**

tubules and in females in the **ovarian follicle**. When stimulated by **luteinizing hormone**, they produce **androgens** in males and **oestrogen** in females.

■ **intestine** In vertebrates, the part of the gut between the **stomach** and **anus**. It usually consists of two sections: the small and the large intestine. The small intestine is concerned with the further digestion of food and the absorption into the bloodstream of already digested **amino acids** and **monosaccharides**. The large intestine is mainly concerned with the absorption of water from semi-solid indigestible remains, which form the **faeces**. [2/4/a] [2/5/a] [2/7/a]

intracellular Occurring within the boundary of a cell or cells. *See also* **intercellular**.

■ **inverse square law** Law that quantifies the falling off of an effect (*e.g.* electrostatic, gravitational and magnetic fields, light radiation) with the square of the distance to the source. For instance in optics, the quantity of light from a given source on a surface of definite area is inversely proportional to the square of the distance between the source and the surface.

inversion In organic chemistry, splitting of **dextrorotatory** higher sugars (*e.g.* sucrose) into equivalent amounts of **laevorotatory** lower sugars (*e.g.* fructose and glucose).

invertase Plant **enzyme** that brings about the hydrolysis of **sucrose** (cane-sugar) to form **invert sugar**.

■ **invertebrate** Animal that does not possess a backbone. *See also* **vertebrate**. [2/3/b] [2/4/a] [2/4/b]

invert sugar Natural **disaccharide** sugar that consists of a mixture of **fructose** and **glucose**, found in many fruits. It is also formed by the hydrolysis of **sucrose** (cane-sugar), which can be brought about by the enzyme **invertase**.

in vitro Literally 'in glass'. It describes techniques or processes that are carried on outside a living organism, *e.g.* in a test-tube.

in vivo Literally 'in life'. It describes biological experiments or processes that are carried on inside living organisms.

■ **involuntary muscle** Muscle not under conscious control, located in internal organs and tissues, *e.g.* in the alimentary canal and blood vessels. Alternative name: smooth muscle. *See also* **voluntary muscle**. [2/5/a]

iodide Compound of **iodine** and another element; salt of hydriodic acid (HI).

■ **iodine** I Non-metallic element in Group VIIA of the Periodic Table (the **halogens**), extracted from Chile saltpetre (in which it occurs as an iodate impurity of sodium nitrate) and certain seaweeds. It forms purple-black crystals that sublime on heating to produce a violet vapour. It is essential for the correct functioning of the thyroid gland. Iodine and its organic compounds are used in medicine; silver iodide is used in photography. At. no. 53; r.a.m. 126.9044. [4/8/a] [4/9/a]

iodine number Number that indicates the amount of iodine taken up by a substance, *e.g.* by fats or oils; it gives a measure of the number of **unsaturated** bonds present. Alternative names: iodine value, iodine absorption.

iodoform Alternative name for **triiodomethane**.

■ **ion** Atom or molecule that has positive or negative electric charge because of the loss or gain of one or more electrons. Many inorganic compounds dissociate into ions when they dissolve in water; *e.g.* sodium chloride (NaCl) forms sodium ions (Na^+) and chloride ions (Cl^-) in solution. Ions are the electric current-carriers in **electrolysis** and in **discharge tubes**. [4/7/e]

ion engine Theoretical motor for propulsion in outer space that would use a high-velocity 'jet' of **ions** or **electrons,** accelerated by an electromagnetic field. Such an engine would provide drive in much the same way as the exhaust in a conventional rocket.

ion exchange Reaction in which **ions** of a solution are retained by oppositely-charged groups covalently bonded to a solid support, such as **zeolite** or a synthetic **resin**. The process is used in water softeners, **desalination** plants and for **isotope** separation.

■ **ionic bond** Chemical bond that results from the attractive electrostatic force between oppositely charged **ions** in molecules or ionic crystals. Alternative name: electrovalent bond. *See also* **covalent bond**. [4/8/d,e]

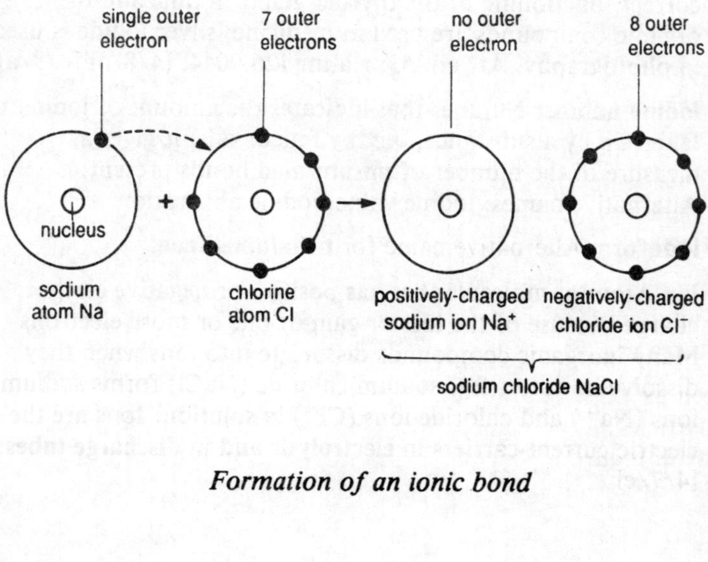

Formation of an ionic bond

ionic crystal Crystal composed of **ions**. Alternative names: electrovalent crystal, polar crystal.

ionic product Product (in moles per litre) of the concentrations of the **ions** in a liquid or solution, *e.g.* in sodium chloride solution, the ionic product of sodium chloride is given by $[Na^+][Cl^-]$. In a pure liquid, it results from the **dissociation** of molecules in the liquid.

ionic radius Radius of an **ion** in a crystal.

ionization Formation of **ions**. It is generally achieved by chemical or electrical processes, or by dissociation of ionic compounds in solution, although at extremely high temperatures (such as those in stars) heat can cause ionization.

ionization chamber Apparatus consisting of a gas-filled container with a pair of high-voltage electrodes. It is used to study the **ionization** of gases or ionizing radiation.

ionization potential Electron bonding energy, the energy required to remove an **electron** from a neutral atom.

■ **ionizing radiation** Any **radiation** that causes **ionization** by producing **ion pairs** in the medium it passes through. [4/7/f]

ionosphere Layer of the Earth's upper atmosphere that is characterized by the presence of many **ions** and free **electrons**. It is important in radio transmission.

ion pair Two charged fragments that result from simultaneous ionization of two uncharged ones; a positive and a negative **ion**.

ion propulsion Type of rocket drive that uses charged particles (*e.g.* lithium **ions**) accelerated by an **electrostatic field**, used in an **ion engine**. Only experimental models have as yet been built.

ion pump High-vacuum pump for removing a gas from a system by ionizing its atoms or molecules and adsorbing the resulting **ions** on a surface.

iridium Ir Steel-grey metallic element in Group VIII of the Periodic Table (a **transition element**). It is used (with platinum or osmium) in hard alloys for bearings, surgical tools and crucibles. At. no. 77; r.a.m. 192.2.

■ **iris** Pigmented part of the human **eye** that controls the amount of light entering the eye. *See also* **iris diaphragm**.

■ **iron** Fe Silver-grey magnetic metallic element in Group VIII of the Periodic Table (a **transition element**). It is the fourth most abundant element in the Earth's crust and probably forms much of the core. It is also the most widely used metal, particularly (alloyed with carbon and other elements) in **steel**. It occurs in various ores, chief of which are **haematite**, **limonite** and **magnesite**, which are refined in a **blast furnace** to produce **pig iron**. Inorganic iron compounds are used as pigments; the blood pigment **haemoglobin** is an organic iron compound. At. no. 26; r.a.m. 55.847.

iron (II) Alternative name for ferrous in iron compounds.

iron (III) Alternative name for ferric in iron compounds.

iron (III) chloride $FeCl_3.6H_2O$ Brown crystalline compound, used as a catalyst, a **mordant** and for etching copper in the manufacture of printed circuits. Alternative name: ferric chloride.

iron (III) oxide Fe_2O_3. Red insoluble compound, the principal constituent of **haematite**. It is used as a pigment, catalyst and polishing compound. Alternative name: ferric oxide.

iron (II) sulphate $FeSO_4.7H_2O$. Green crystalline compound, used in making inks, in printing and as a wood preservative. Alternative name: ferrous sulphate.

irradiation *1*. Radiant energy per unit of intercepting area. *2*. Exposure of an object to radiant energy or **ionizing radiation**.

irreversible reaction Chemical reaction that takes place in one direction only, therefore proceeding to completion.

irritability Ability to respond to a stimulus, evident in all living material.

■ **islets of Langerhans** In vertebrates, group of specialized secretory cells in the **pancreas**. They control the level of **glucose** in the **blood** by secreting **insulin** and **glucagon**. These hormones in turn either stimulate the liver to convert glucose to glycogen (function of insulin) or cause glycogen to be broken down to release glucose (function of glucagon). [2/7/a]

isobar *1*. Curve that relates to qualities measured at the same pressure. *2*. Line drawn on a map through places having the same atmospheric pressure at a given time. *3*. One of a set of atomic nuclei having the same total of **protons** and **neutrons** (*i.e.* the same nucleon number or **mass number**) but different numbers of protons and therefore different identities.

isocyanide Organic compound of general formula RNC, where R is an organic **radical**; *e.g.* methyl isocyanide, CH_3NC. Alternative names: isonitrile, carbylamine.

isoelectric point Hydrogen ion concentration **(pH)** at which a system is electrically neutral.

isogamy Sexual fusion of **gametes** that are similar in structure and size. It occurs in some protozoa, fungi and algae. *See also* **anisogamy**.

isoleucine Crystalline **amino acid**; it is a constituent of **proteins**, and essential in the diet of human beings.

isomer Substance that exhibits chemical **isomerism**.

isomerism Existence of substances that have the same molecular composition, but different structures. *See* **optical isomerism**.

isomorphism *1*. In plants, the existence of morphologically identical male and female individuals or generations. *2*. In mineralogy, the phenomenon in which two or more minerals that are closely similar in chemical composition crystallize in the same forms.

isonitrile Alternative name for **isocyanide**.

isopod Crustacean of the order **Isopoda**.

Isopoda Order of small **crustaceans**. Most of the 4,000 species of isopods are marine, but some, notably woodlice, have adapted to life on land.

isoprene $CH_2{=}CH(CH_3)CH{=}CH_2$ Colourless unsaturated liquid hydrocarbon, used in making synthetic rubber. Alternative name: 2-methyl-1,3-butadiene.

isopropanol $(CH_3)_2CHOH$ One of the two **isomers** of **propanol**. Alternative name: isopropyl alcohol.

isopropyl alcohol Alternative name for **isopropanol**.

Isoptera Order of social insects related to cockroaches, which includes termites.

isothermal process Process that occurs at a constant or uniform temperature; *e.g.* the compression of a gas under constant temperature conditions. *See also* **adiabatic process**.

isotonic Having the same **osmotic pressure** as blood, or the same as the cell sap in a particular plant.

isotope One form of an atom that has the same **atomic number** but a different **atomic mass** to other forms of that atom. This results from there being different numbers of **neutrons** in the nuclei of the atoms, *e.g.* uranium-238 (also written as U-238 or $_{238}$U) and uranium-235 are two isotopes of uranium with atomic weights of 238 and 235 respectively. The isotopes of an atom are chemically identical, although with very light isotopes the relative difference in masses may make them react at different rates. The existence of isotopes explains why most elements have non-integral atomic masses. A few elements have no naturally occurring isotopes, including fluorine, gold, iodine and phosphorus.

isotopic number Difference between the number of **neutrons** in an **isotope** and the number of **protons**. Alternative name: neutron excess.

isotopic weight **Atomic mass** of an **isotope**. Alternative name: isotopic mass.

J

■ **jaws** *1.* In vertebrates, bony structure enclosing the mouth, often furnished with teeth for grasping prey and/or chewing food, consisting of an upper **maxilla** and lower **mandible**. *2.* In invertebrates, grasping structure surrounding the mouth. [2/4/a] [2/5/a] [2/7/a]

jejunum Part of the **intestine** of mammals, located between the **duodenum** and the **ileum**, the main function of which is absorption.

jet *1.* Stream of liquid or gas issuing from an orifice or nozzle, or the nozzle itself (*e.g.* the jet of a carburettor). *2.* **Gas turbine** engine, particularly one used in aircraft.

jet engine Alternative name for a **gas turbine** engine used in aircraft.

■ **joint** In anatomy, point of articulation of limbs or **bones**. The bones are connected to each other by connective tissue **ligaments**, and are well lubricated. Common types of joints include ball-and-socket joints (*e.g.* the hip and shoulder joints), hinge joints (*e.g.* the elbow and knee joints) and sliding joints (*e.g.* between vertebrae). [2/3/a] [2/5/a]

joule (J) SI unit of **work** and **energy**. 1 joule is the work done by a force of 1 **newton** moving 1 metre in the direction of the force; 4.2 J = 1 calorie; 4.2 kilojoules = 1 kilocalorie, or 1 Calorie (with a capital C). The unit is named after the British physicist James Joule (1818–89).

Joule's law *1.* **Internal energy** of a given mass of gas is dependent only on its temperature and is independent of its pressure and volume. *2.* If an electric current *I* flows through a resistance R for a time t, the heat produced Q, in joules, is given by $Q = I^2Rt$.

Joule-Thompson effect When a gas is allowed to undergo adiabatic expansion through a porous plug, the temperature of the gas usually drops. This results from the work done in breaking the intermolecular forces in the gas, and is a deviation from **Joule's law**. The effect is important in the liquefaction of gases by cooling. Alternative name: Joule-Kelvin effect.

J-particle Alternative name for **psi particle**.

■ **jugular vein** In vertebrates, main **vein** carrying blood from the head (including the brain) to the heart. [2/5/a]

junction diode Type of **diode** consisting of a layer of *n*-type semiconductor and *p*-type semiconductor. An applied votage flows in one direction only across the junction because of the diffusion of electrons from the *n*-type to the *p*-type.

■ **jungle** Form of dense tropical vegetation, a tangle of creepers (lianas), bamboo, shrubs and palms. It occurs in areas of rainforest that have not yet reached the status of a **climax community**, typically on the site of a former clearing. [2/2/b,c]

juvenile hormone In insects, a **hormone** that is secreted at certain stages of development. The hormone inhibits **metamorphosis** to the adult form of the insect, but promotes larval growth and development. *See also* **neoteny**.

K

kainite Hydrated **magnesium sulphate** mineral containing **potassium chloride**; it crystallizes in the **monoclinic** system.

Kalanite Proprietary hard electrical insulating material not affected by oils.

kaolin China clay, a mineral consisting of **kaolinite**.

kaolinite Hydrated aluminium silicate mineral, consisting of minute crystals derived from **feldspars**.

karyokinesis *See* **mitosis**.

karyotype Visible trait on the **chromosome** of a typical **cell**. It provides information about the species or strain, because a karyotype is characteristic to the cell of a particular organism.

Kekulé structure Structural forms of **benzene** with alternate single and double bonds, proposed by the German chemist Friedrich August Kekulé (1829–96).

■ **kelvin** (K) SI unit of thermodynamic temperature, named after the British physicist Lord Kelvin (William Thomson) (1824–1907). It is equal in magnitude to a degree Celsius (°C). *See also* **absolute temperature**; **Kelvin temperature**.

Kelvin effect Alternative name for the **Thomson effect**.

■ **Kelvin temperature** Scale of temperature that originates at **absolute zero**, with the **triple point** of water defined as 273.16K. The freezing point of water (on which the Celsius scale is based) is 273.16K. Alternative name: Kelvin thermodynamic scale of temperature.

keratin Tough, non-living outermost layer of skin which forms a protective covering in vertebrates. Formed from **epidermis**, it may be modified to make *e.g.* hair, feathers, horns or claws.

keratinization Replacement of the **cytoplasm** of cells in the **epidermis** by **keratin**, thus resulting in hardening of skin. Alternative name: cornification.

kerosene Oily mixture of **hydrocarbons**, obtained mainly from **petroleum** and **oil shale**. It has a boiling range from about 150 °C to 300 °C, and is used as a fuel and as a solvent. Alternative names: kerosine, paraffin oil.

Kerr cell Chamber of liquid between two crossed polaroids that darkens or lightens in an electric field (applied between two electrodes). It can be used as a shutter or to modulate a light beam. It was named after the British physicist John Kerr (1824–1907).

ketene Member of a group of unstable organic compounds of general formula $R_2C=CO$, where R is hydrogen or an organic **radical**; *e.g.* ketene, $CH_2=CO$. Ketenes react with other **unsaturated** compounds to form 4-membered rings.

keto-enol tautomerism Existence of a chemical compound in two double-bonded structural forms, keto and enol, which are in equilibrium. The keto form (containing the group $-CH_2-C=O-$) changes to the enol (containing $-CH=C(OH)-$) by the migration of a hydrogen atom to form a **hydroxyl group** with the oxygen of the **carbonyl group**; the position of the double bond also changes.

ketone Member of a family of organic compounds of general formula RCOR′, where R and R′ are organic **radicals** and $=CO$ is a **carbonyl group**. Ketones may be made in various ways, such as the oxidation of a secondary **alcohol**; *e.g.* oxidation of **isopropanol**, $(CH_3)_2OH$, gives **acetone** (propanone), $(CH_3)_2CO$.

■ **ketonuria** Presence in the urine of compounds containing **ketones**, usually associated with the disorder diabetes mellitus. A newborn baby's urine is tested for ketonuria soon after birth. [2/7/a]

keV Abbreviation of kilo-electron-volt, a unit of particle energy equivalent to 10^3 **electron-volts**.

■ **key** Set of paired descriptions used to identify an organism. [2/4/b]

keyboard Computer **input device** which a human operator uses to type in **data** as **alphanumeric** characters. It consists of a standard qwerty keyboard, usually with additional function keys.

■ **kidney** In vertebrates, one of a pair of excretory organs that filter waste products (particularly nitrogenous waste) from the blood and concentrate them in **urine**. Kidneys also have an important function in the regulation of the balance of water and salts in the body. The main processes of the kidney occur in a large number of tubular structures called nephrons. Water and waste products pass from the kidneys to the bladder via the ureters. [2/4/a] [2/7/a]

kieselguhr **Silica**-containing mineral, a whitish powder that consists mainly of diatom skeletons. It is used in fireproof cements and in the manufacture of **dynamite**. Alternative name: diatomite.

kilo- Metric prefix meaning a thousand times ($x10^3$).

kilocalorie (C) Unit of energy equal to 1,000 **calories**. 1 C = 4.2 kilojoules. Alternative name: Calorie (capital C).

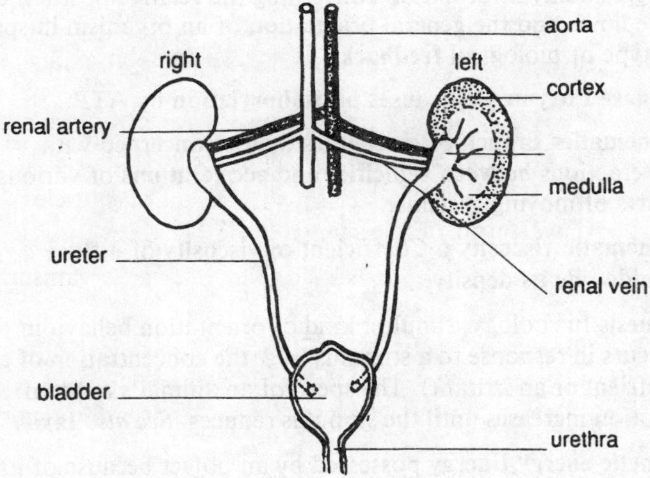

The human kidneys

kilogram (kg) Unit of mass in the SI system, equal to 1,000 grams. 1 kg = 2.2046 lb.

kilohertz (kHz) Unit of frequency equal to 1,000 **hertz**.

kilojoule (kJ) Unit of energy equal to 1,000 **joules**.

kilometre (km) Unit of length equal to 1,000 metres. 1 km = 0.62137 miles.

kiloton Unit for the power of a nuclear explosion or warhead, equivalent to 1,000 tons of TNT.

■ **kilowatt-hour** (kWh) Unit of electrical energy equal to a rate of consumption of 1,000 watts per hour. Alternative name: unit. [3/7/c]

kinaesthesis Process by which sensory cells in muscles and organs relay information concerning the relative position of the limbs and the general orientation of an organism in space; a type of biological **feedback**.

kinase **Enzyme** that causes **phosphorylation** by ATP.

kinematics Branch of **mechanics** that is concerned with interactions between **velocities** and **accelerations** of various parts of moving systems.

kinematic viscosity υ Coefficient of **viscosity** of a fluid divided by its density.

kinesis In zoology, simplest kind of orientation behaviour that occurs in response to a stimulus (*e.g.* the concentration of a nutrient or an irritant). The speed of an animal's random motion increases until the stimulus reduces. *See also* **taxis**.

■ **kinetic energy** Energy possessed by an object because of its motion, equal to $\frac{1}{2}mv^2$, where m = mass and v = velocity. The kinetic energy of the particles that make up any sample of matter (*see* **kinetic theory**) determines its **heat** energy and therefore its **temperature** (except at **absolute zero**, when both are equal to zero). [5/9/a]

kinetics Study of the rates at which chemical reactions take place.

■ **kinetic theory** Theory that accounts for the properties of substances in terms of the movement of their component particles (atoms or molecules). The theory is most important in describing the behaviour of gases (when it is referred to as the kinetic theory of gases). An ideal gas is assumed to be made of perfectly elastic particles that collide only occasionally with each other. Thus, *e.g.*, the pressure exerted by a gas on its container is then the result of gas particles colliding with the walls of the container. [4/6/e]

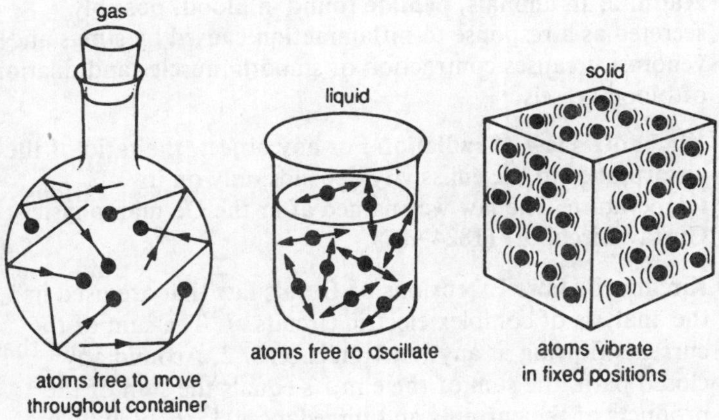

Kinetic theory accounts for how substances behave

kinetochore Point of attachment of the **spindle** in a **cell**.

king crab Large member of a class of marine **arthropods**, most of which are extinct, that are related to the scorpions. Alternative name: horseshoe crab.

■ **kingdom** Highest rank in the **classification** of living organisms, which encompasses phyla (for animals) and divisions (for plants). Criteria determining members of a kingdom are broad, and consequently members are very diverse. Traditionally, there were two kingdoms: the plants and the animals. However, **fungi**, **protists** and **prokaryotes** are now often placed in kingdoms of their own. [2/3/b] [2/4/b]

kinin *1.* In plants, growth substance that stimulates cell division; if naturally occurring it is termed a cytokinin; *e.g.* zeatin. *2.* In animals, **peptide** found in blood, possibly secreted as a response to inflammation caused by stings and venoms; it causes contraction of smooth muscles and dilation of blood vessels.

Kirchhoff's law of radiation For any object, the ratio of the absorptance to the emissivity depends only on its temperature. The law was named after the German physicist Gustav Kirchhoff (1824–87).

■ **Kirchhoff's laws** Extensions of **Ohm's law** that are used in the analysis of complex electric circuits. *1.* The sum of the currents flowing at any junction is zero. *2.* Around any closed path, the sum of the e.m.f.s equals the sum of the products of the currents and impedances. [5/6/b] [5/9/b]

klino-taxis Movement of an animal in respose to light.

knock Pre-ignition, the instantaneous violent explosion of a compressed mixture of fuel vapour and air in an internal combustion engine. It reduces the available mechanical energy, but can be prevented by adding an anti-knock agent such as **tetraethyllead** to the fuel. Alternative names: knocking, pinking.

■ **Krebs cycle** Energy-generating series of biochemical reactions that occur in living cells. It forms the second stage of **aerobic respiration**, in which **pyruvate** or **lactic acid** produced by **glycolysis** is oxidized to carbon dioxide and water, thus producing a large amount of energy in the form of **ATP** molecules. The Krebs cycle takes place in **mitochondria**. It was named after the German-born British biochemist Hans Krebs (1900–82). Alternative names: citric acid cycle, tricarboxylic acid cycle. [2/3/b]

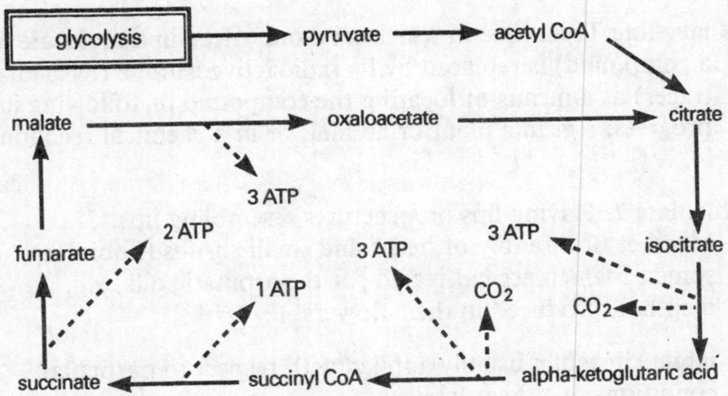

The Krebs cycle

krypton Kr Gaseous nonmetallic element of Group 0 of the Periodic Table (the **rare gases**), which occurs in trace quantities in air (from which it is extracted). It is used in gas-filled lamps and discharge tubes. At. no. 36; r.a.m. 83.80.

k-selection Survival strategy based on maximum competitiveness that is one of the processes in **natural selection**. K-selection species are characterized by low birth rates, prolonged development of young, and high survival rates for offspring. *See also:* **r-selection**.

kurchatovium Alternative name for the **post-actinide** element **rutherfordium**.

L

■ **labelling** Technique in which an atom (often in a molecule of a compound) is replaced by its radioactive **isotope** (termed a tracer) as a means of locating the compound or following its progress; *e.g.* in a plant or animal, or in a chemical reaction. [4/7/f]

■ **labiate** *1.* Having lips or structures resembling lips. *2.* Member of a family of herbs and small shrubs (Labiatae), some of which are cultivated for the aromatic oils, *e.g.* menthol, produced in their flowers. [2/4/b]

labile Unstable; usually applied with respect to particular conditions, *e.g.* heat-labile.

labium *1.* In insects, the lower lip, used for manipulating food during feeding. *2.* In mammals, one of the folds of flesh at the entrance to the **vagina**.

labrum *1.* In crustaceans and insects, a ridge of **cuticle** that forms the upper lip. *2.* In gastropod molluscs, the outer margin of the shell.

■ **labyrinth** Part of the inner **ear** of vertebrates concerned with balance, consisting of three semi-circular canals (each in a different plane) containing fluid. A chalky body (otolith) in each canal moves against sensitive receptors, so detecting movements of the head. [2/4/a] [2/5/a] [2/7/a]

lacrimal gland Gland that produces tears. Fluid is continuously secreted to protect and moisten the **cornea**; it also contains the bactericidal enzyme **lysozyme**. Alternative name: lachrimal gland.

β-lactam Member of a group of **antibiotics** that include the penicillins.

lactate *1.* Salt or ester of **lactic acid** (2-hydroxypropanoic acid). *2.* To produce milk.

■ **lactation** Milk-production; it occurs in mammals for feeding the young. [2/4/c]

lacteal **Lymph** vessel of the **villi** in the **intestine** of vertebrates. Fat passes into the lacteals as an emulsion of globules to be circulated in the **lymphatic system**.

■ **lactic acid** $CH_3CH(OH)COOH$ Colourless liquid organic acid. A mixture of (+)-lactic acid (**dextrorotatory**) and (−)-lactic acid (**laevorotatory**) is produced by bacterial action on the sugar **lactose** in milk during souring. The (+)-form is produced in animals when **anaerobic respiration** takes place in muscles because of an insufficient oxygen supply during vigorous activity. Lactic acid is used in the chemical and textile industries. Alternative name: 2-hydroxypropanoic acid.

lactoflavin Alternative name for **riboflavin**.

lactone Unstable internal ester, a **cyclic** compound containing the group −CO.O− in the ring, formed by heating γ- and δ-hydroxycarboxylic acids.

lactose $C_{12}H_{22}O_{11}$ White crystalline **disaccharide** sugar that occurs in milk, formed from the union of **glucose** and **galactose**. It is a reducing sugar. Alternative name: milk-sugar.

laevorotatory Describing a compound with **optical activity** that causes the plane of polarized light to rotate in an anti-clockwise direction. Indicated by the prefix (−)- or *l*-.

laevulose Alternative name for **fructose**.

■ **Lamarckism** Theory proposed by the French naturalist Jean-Baptiste Lamarck (1744–1829) that evolutionary change could be achieved by the transmission of acquired characteristics from parents to offspring. The theory was superseded by **Darwinism**, and there has yet to be any firm evidence that the inheritance of acquired characteristics ever occurs at all. *See also* **Lysenkoism**. [2/9/c]

lambda particle Type of **elementary particle** with no electric charge.

lambda point Temperature at which liquid helium (helium I) becomes the **superfluid** known as helium II.

Lambert's law Equal fractions of incident light radiation are absorbed by successive layers of equal thickness of the light-absorbing substance. It was named after the German mathematician and physicist Johann Lambert (1728–77).

■ **lamella** Thin plate-like structure, *e.g.* the membrane in a **chloroplast** that forms folded structures containing chlorophyll. [2/3/d] [2/7/a]

lamellibranch Alternative name for a bivalve **mollusc**.

lamina Thin flat sheet of a material of uniform thickness.

laminar flow Streamlined, or non-turbulent, flow in a gas or liquid.

lamp black Type of **carbon black**.

lamprey Jawless parasitic eel-like fish of the class Agnatha.

■ **lanolin** Yellowish sticky substance obtained from the grease that occurs naturally in wool. It is used in cosmetics, as an ointment and in treating leather. Alternative names: lanoline, wool fat.

lanthanide Member of the Group IIIB elements of atomic number 57 to 71. The properties of these metals are very similar, and consequently they are difficult to separate. Alternative names: lanthanoid, rare-earth element.

lanthanum La Silver-white metallic element in Group IIB of the Periodic Table, the parent element of the **lanthanide** series. It is used in making lighter-flints. At. no. 57; r.a.m. 138.91.

lapis lazuli Semi-precious deep blue gem, a sodium aluminium silicate that contains sulphur.

Laplace law Alternative name for **Ampère's law**.

■ **large intestine** *See* **colon**; **intestine**. [2/4/a] [2/5/a] [2/7/a]

Larmor precession Orbital motion of an **electron** about the **nucleus** of an atom when it is subjected to a small magnetic field. The electron precesses about the direction of the magnetic field. It was named after the British physicist Joseph Larmor (1857–1942).

■ **larva** Juvenile form of some animals which, while radically different from the adult in form, is capable of independent existence; *e.g.* a caterpillar, grub, maggot or tadpole. The larva changes to the adult by **metamorphosis**, *e.g.* in some insects and amphibians. *See also* **imago**; **pupa**.

■ **larynx** Region of **trachea** (windpipe) that usually houses the **vocal cords** (composed of membrane folds that vibrate to produce sounds). [2/5/a]

laser Acronym of 'light amplification by stimulated emission of radiation'. It is a device that produces a powerful monochromatic and coherent beam of light by stimulation of emission from excited atoms. Lasers are used in **holography**, for measuring and surveying, as industrial and medical cutting tools and for many other purposes. *See also* **maser**.

latch Electric circuit designed to stay functioning even when the original signal that triggered it is switched off (*e.g.* in an alarm circuit or counter read-out).

■ **latent heat** Heat energy that is needed to produce a change of state during the melting (solid-to-liquid change) or vaporization (liquid-to-vapour/gas change) of a substance; it causes no rise in temperature. This heat energy is released when the substance reverts to its former state (by freezing/solidifying or condensing/liquefying). [5/3/b]

■ **lateral inversion** Apparent reversal of an image left to right when viewed in a plane mirror, resulting from the actual front- to-back reversal as required by the laws of reflection. [5/5/c]

■ **lateral line** Line of receptor cells, sensitive to water pressure and vibration, that is found along the sides of the body and head of all fish and some amphibians. [2/5/a]

laterite Fine-grained clay produced by the weathering of igneous rocks in a tropical climate. The presence of iron (III) hydroxide gives it a distinctive red colour.

latex Milky fluid produced in some plants after damage, containing sugars, proteins and alkaloids. It is used in manufacture, *e.g.* of rubber. A suspension of synthetic rubber is also called latex.

■ **latitude** Imaginary line drawn parallel to the equator around the Earth. The equator is 0° latitude, and the distance between the equator and the poles is divided into 90° of latitude. Using latitude in combination with **longitude**, any position on the Earth's surface can be denoted. Alternative name: parallel.

lattice Regular network of atoms, ions or molecules in a **crystal**.

lattice energy Strength of an ionic bond; the energy required for the separation of the ions in 1 mole of a crystal to an infinite distance from each other. Alternative name: lattice enthalpy.

laughing gas Alternative name for **dinitrogen oxide**.

launch window Period of time during which a rocket may be launched on a particular **trajectory**. The term was previously applied only to spacecraft, but is now used by the military to describe opportunities for much smaller projectiles such as missiles.

lauric acid $CH_3(CH_2)_{10}COOH$ White crystalline **carboxylic acid**, used in making **soaps** and **detergents**. Alternative names: dodecanoic acid, dodecylic acid.

law In science, simple statement or mathematical expression for the generalization of results relating to a particular phenomenon or known facts. There are articles on many scientific laws listed under their individual names in this dictionary (*e.g.* **conservation of mass, law of**).

lawrencium Lr Radioactive element in Group IIIB of the Periodic Table (one of the **actinides**), the heaviest element definitely identified. At. no. 103; r.a.m. 257 (most stable isotope).

LD–50 (lethal dose 50) Toxicity test in which the end-point is the quantity of a substance that causes death in 50%of the organisms tested.

LDR Abbreviation of **light-dependent resistor**.

Le Chatelier's principle If a change occurs in one of the factors (such as temperature or pressure) under which a system is in equilibrium, the system will tend to adjust itself so as to counteract the effect of that change. It was named after the French physicist Henri le Chatelier (1850–1936). Alternative name: Le Chatelier-Braun principle.

leaching Washing out of a soluble material from a solid by a suitable liquid.

lead Pb Silver-blue poisonous metallic element in Group IVA of the Periodic Table, obtained mainly from its sulphide ore **galena**. Various **isotopes** of lead are final elements in **radioactive decay** series. The metal is used in building, as shielding against ionizing **radiation**, as electrodes in **lead-acid accumulators** and in various alloys (such as **solder**, metals for bearings and type metal). Its inorganic compounds are used as pigments; tetraethyllead is employed as an anti-knock agent in petrol. At. no. 82; r.a.m. 207.19.

lead (II) Alternative name for plumbous in lead compounds.

lead (IV) Alternative name for plumbic in lead compounds.

lead-acid accumulator Rechargable **electrolytic cell** (battery) that has positive electrodes of lead (IV) oxide (PbO_2), negative electrodes of lead, and a solution of sulphuric acid as the electrolyte.

lead-chamber process Obsolete process for the manufacture of **sulphuric acid** that used sulphur dioxide, air and nitrogen oxides as a catalyst. Alternative name: chamber process. *See also* **contact process**.

lead dioxide Alternative name for **lead (IV) oxide**.

lead equivalent Factor that compares any form of shielding against radioactivity to the thickness of lead that would provide the same measure of protection.

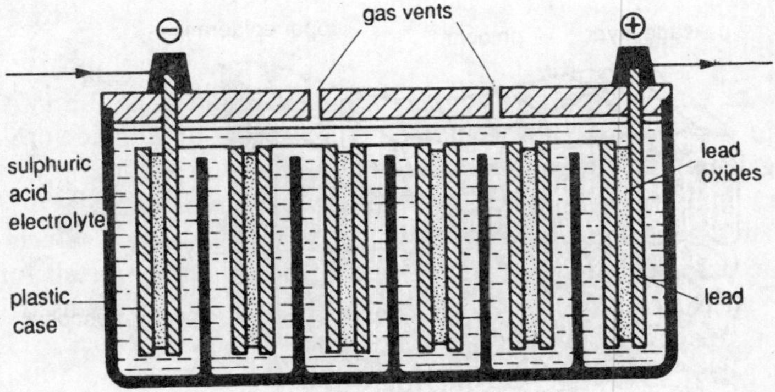

Construction of a lead-acid accumulator

lead-free petrol Petrol that does not contain the anti-knocking agent **tetraethyllead** (which causes atmospheric pollution).

lead monoxide Alternative name for **lead (II) oxide**.

lead (II) oxide PbO Yellow crystalline substance, used in the manufacture of glass. Alternative names: lead monoxide, litharge.

lead (IV) oxide PbO_2 Brown amorphous solid, a strong oxidizing agent, used in lead-acid accumulators. Alternative names: lead dioxide, lead peroxide.

lead tetraethyl (IV) Alternative name for **tetraethyllead**.

■ **leaf** Part of a plant, usually flat and green, that grows from the stem. In green plants leaves are the sites of **photosynthesis** and **transpiration**. [2/3/d]

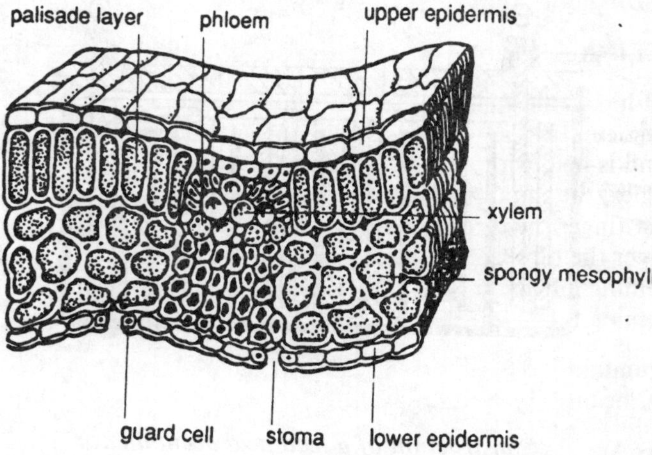

Section of a leaf

leaf litter Layer of decaying plant material on the surface of the soil. *See also* **humus**.

lecithin Type of **phospholipid**, a **glyceride** in which one organic acid residue is replaced by a group containing **phosphoric acid** and the base **choline**. It is a major component of a cell **membrane**. Alternative name: phosphatidyl choline.

Leclanché cell **Primary cell** that has a zinc cathode and carbon anode dipping into an electrolyte of ammonium chloride solution. A porous pot of crushed carbon and manganese (IV) oxide (managanese dioxide) surrounds the anode to prevent **polarization**. It is the basis of the **dry cell** used in most batteries. It was named after the French chemist Georges Leclanché (1839–82).

LED Abbreviation of **light-emitting diode**.

LEED Abbreviation of **low-energy electron diffraction**.

■ **left-hand rule** Rule that relates the directions of current, magnetic field and movement in an electric motor. If the left hand is held with the thumb, first and second fingers at right angles, the thumb indicates the direction of movement, the first finger the direction of the magnetic field and the second finger the direction of current flow. Alternative name: Fleming left-hand rule, after the British physicist Ambrose Fleming (1849–1945).

■ **leguminous** Describing any plant of the pea family (legumes), *e.g.* beans, clover, lucerne (alfalfa), peas. [2/8/d]

■ **lens** Any device for focusing or modifying the direction of a beam of rays (usually light) passing through it. By analogy also a current-carrying coil that focuses a beam of electrons (as in an **electron microscope**).

lenticel Raised pore on the surface of a woody stem that allows entry and exit of gases.

■ **Lenz's law** When a wire moves in a **magnetic field**, the electric current induced in the wire generates a magnetic field that tends to oppose the motion. It was named after the Russian physicist Heinrich Lenz (1804–65). [3/6/d]

■ **Lepidoptera** Large order of insects, consisting of moths and butterflies, characterized by wings made up of overlapping scales. [2/3/b] [2/4/b]

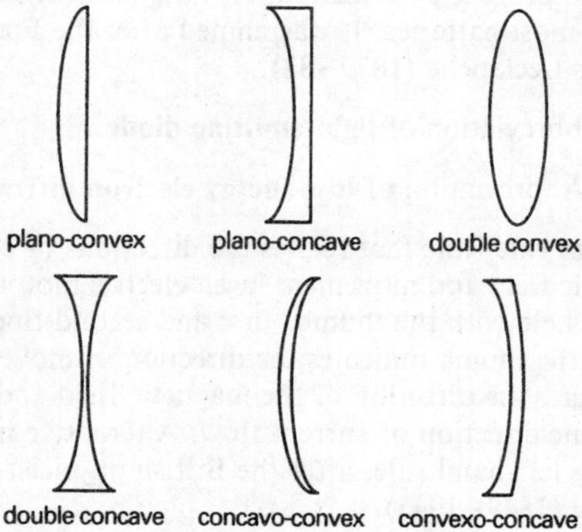

Various types of lenses

lepton Subatomic particle that does not interact strongly with other particles, *e.g.* an electron. *See also* **hadron**.

■ **leptotene** Stage of **prophase** in the first cell division in **meiosis**. At this stage, **chromosomes** can be seen to carry chromomeres. [2/6/b]

leucine $(CH_3)_2CHCH_2CH(NH_2)COOH$ Colourless crystalline **amino acid**; a constituent of many **proteins**. Alternative name: 2-amino-4-methylpentanoic acid.

leuco- Prefix denoting white; *e.g.* leucocyte. (The prefix is spelled *leuk-* in *leukaemia*.)

leucocyte White blood cell. Leucocytes are generally amoeboid, and play a major role in defending the body against disease.

leucoplast Colourless **plastid**, often containing reserves of starch.

■ **leukaemia** Disorder characterized by overproduction of white blood cells, often called blood cancer. There are many forms, some of which can now be successfully treated. [2/7/a]

■ **lever** Rigid bar supported or pivoted at some point (called the fulcrum) along its length. An effort applied at one point on the bar can move a load at another point; a lever is thus a simple machine. Every lever belongs to one of three classes or orders, depending on whether the fulcrum is between the load and effort (class 1), or the load and effort are on the same side of the fulcrum, with the load nearer the fulcrum (class 2) or the effort nearer the fulcrum (class 3).

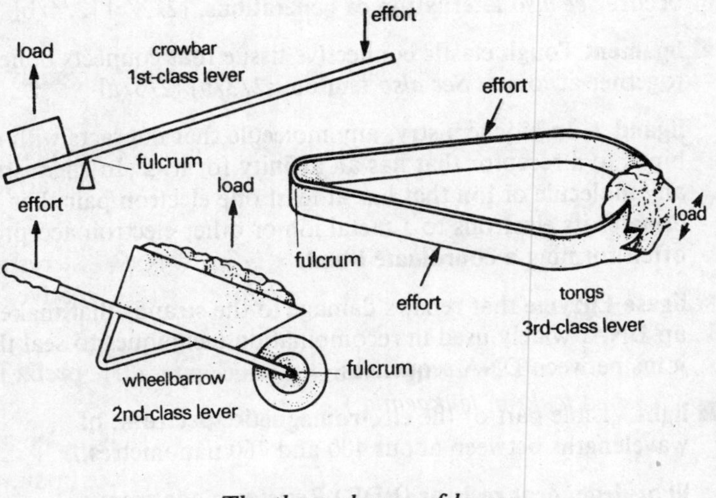

The three classes of levers

Lewis acid and base Concept of **acids** and **bases** in which an acid is defined as a substance capable of accepting a pair of electrons, whereas a base is able to donate a pair of electrons to a bond. It was named after the American chemist Gilbert Lewis (1875–1946).

Leyden jar Original and obsolete form of **capacitor**, named after the Dutch town (now Leiden).

■ **lichen** Organism formed from a symbiotic relationship (*see* **symbiosis**) between a **fungus** and an **alga**, classified in the plant division Lichenes. [2/2/b] [2/2/c]

Lichenes Division of the plant kingdom that contains **lichens**.

■ **life cycle** Progressive sequence of changes that an organism undergoes from fertilization to death. In the course of the cycle a new generation is usually produced. Reproduction may be sexual or asexual; both **meiosis** and **mitosis** may occur. *See also* **alternation of generations**. [2/2/c] [2/4/b]

■ **ligament** Tough elastic connective tissue that connects bones together at joints. *See also* **tendon**. [2/3/a] [2/5/a]

ligand *1*. In biochemistry, any molecule that interacts with or binds to a receptor that has an affinity for it. *2*. In chemistry, any molecule or **ion** that has at least one electron pair that donates its electrons to a metal ion or other electron acceptor, often forming a **coordinate bond**.

ligase Enzyme that repairs damage to the strands that make up **DNA**, widely used in recombination techniques to seal the joins between DNA sequences.

■ **light** Visible part of the **electromagnetic spectrum**, of wavelengths between about 400 and 760 nanometres.

light-dependent resistor (LDR) Resistor made from a

semiconductor (*e.g.* cadmium sulphide, selenium) whose resistance decreases when light falls on it.

light-emitting diode (LED) Semiconducting **diode** that gives off light, used for displaying numerals and letters in calculators, watches and other equipment.

lightning Gigantic electric spark within a cloud, between clouds or between a cloud and the earth. The spark occurs when clouds become highly electrically charged, different areas of one cloud often having an opposite charge.

lightning conductor Metal rod with its upper end projecting to attract lightning and its lower end buried in the ground. It is used to protect buildings against lightning strikes.

light pen Computer **input device**, used in association with special software, that enables an operator to write or draw on a VDU screen more or less as if using an ordinary pen.

light petroleum ether Mixture of **pentane** and **hexane** hydrocarbons derived from crude petroleum; boiling range 30 to 60 °C. It is used as a solvent.

■ **light, speed of** *See* **speed of light**. [5/7/e]

light-year Unit of distance in astronomy, the distance light travels in one year (about 9.4650×10^{12} km, or 5.88×10^{12} miles). The nearest star to the Sun, Proxima Centauri, is approximately 4 light-years away.

■ **lignin** Complex **polymer** found in many plant cell walls, that glues together fibres of **cellulose** and provides additional support for the cell wall.

lignite Soft brown form of **coal**, between peat and bituminous coals in quality. It generally has a high moisture content and when burnt gives only about half as much heat as good-quality coal. Alternative name: brown coal.

ligroin Mixture of hydrocarbons derived from crude petroleum, boiling range 80 to 120 °C, used as a general solvent.

Liliopsida Class of **angiosperm** plants that comprises the **monocotyledons**.

lime *1.* General term for quicklime (**calcium oxide**, CaO), slaked lime and hydrated lime (both **calcium hydroxide**, $Ca(OH)_2$). They are obtained from **limestone**. *2.* Ground limestone used as a fertilizer and in iron smelting.

limestone Sedimentary rock of marine origin. Its main constituent is **calcium carbonate**, and it is used as a building stone, in iron smelting and in the manufacture of **lime**.

lime water Solution of **calcium hydroxide** ($Ca(OH)_2$) in water, used as a test for carbon dioxide (which turns lime water milky when bubbled through it due to the precipitation of calcium carbonate; after prolonged bubbling the solution goes clear again due to the formation of soluble calcium hydrogencarbonate).

limiting friction Maximum value of a frictional force.

limonite Major iron ore that consists of oxides and hydroxide of **iron**.

Linde process Technique for liquefying air, and extracting liquid oxygen and liquid nitrogen. It was named after the German scientist Carl von Linde (1842–1934).

linear absorption coefficient Measure of a medium's ability to absorb a beam of radiation passing through it, but not to scatter or diffuse it.

linear accelerator Apparatus for accelerating charged particles.

linear attenuation coefficient Measure of a medium's ability to diffuse and absorb a beam of radiation passing through it.

linear energy transfer (LET) Linear rate of energy dispersion of separate particles of radiation when they penetrate an absorbing medium.

linear molecule Molecule whose atoms are arranged in a line.

■ **linear momentum** Product of the mass (*m*) of a moving object and its velocity (*v*); *mv*. [5/10/a]

linear motor Type of **induction motor** in which the 'rotor' travels along a rail that acts as the stator.

line printer Comparatively fast computer **output device** that prints out **data** one whole line at a time.

line spectrum **Spectrum** that consists of separate lines of definite wavelengths. The spectral lines are produced by the excited electrons of atoms falling back to lower **energy levels** with the emission of photons.

line transect In an ecological survey, a method of systematically sampling along a line or narrow band.

■ **linkage** Occurrence of two **genes** on the same **chromosome**. Genes that are close together are likely to be inherited together; genes that are farther apart may become separated during **crossing over**. [2/10/b]

■ **Linnaean system** System that classifies and names all organisms according to scientific principles. Each species has two names; the first indicates the organism's general type (genus), the second gives the unique species, *e.g. Canis* (dog) *domesticus* (household). The system was called after the Swedish botanist Carl Linné (Linnaeus) (1707–78). Alternative name: binomial classification. [2/3/b] [2/4/b]

linoleic acid $C_{17}H_{31}COOH$ **Unsaturated fatty acid**, used in paints, which occurs in **linseed oil** and other oils derived from plants. Alternative name: linolic acid.

linseed oil Oil extracted from the seeds of flax. Because it contains **linoleic acid**, it is a drying oil (hardening on exposure to air), used in paints, putty, varnishes and enamels.

lipase In vertebrates, an **enzyme** in intestinal juice and pancreatic juice that catalyses the **hydrolysis** of **fats** to **glycerol** and **fatty acids**.

lipid Member of a group of naturally occurring fatty or oily compounds that share the property of being soluble in organic solvents, but sparingly soluble in water. Also, all lipids yield **monocarboxylic acids** on **hydrolysis**.

lipochrome Yellow pigment in butterfat.

lipolyte Lipid-containing **cell**. Alternative name: fat cell.

liposome Droplet of **fat** in the **cytoplasm** of a cell, particularly that of an **egg**.

■ **liquefaction of gases** All gases can be liquefied by a combination of cooling and compression. The greater the pressure, the less the gas needs to be cooled, but there is for each gas a certain **critical temperature** below which it must be cooled before it can be liquefied.

liquefied natural gas (LNG) Liquid **methane**. *See* **natural gas**.

liquefied petroleum gas (LPG) Mixture of hydrocarbons derived from crude petroleum which has **propane** as its major constituent. It is used as a fuel.

■ **liquid** Fluid that, without changing its volume, takes the shape of all or the lower part of its container. According to the **kinetic theory**, the molecules in a liquid are not bound together as rigidly as those in a solid but neither are they as free to move as those of a gas. It is therefore a **phase** that is intermediate between a solid and a gas.

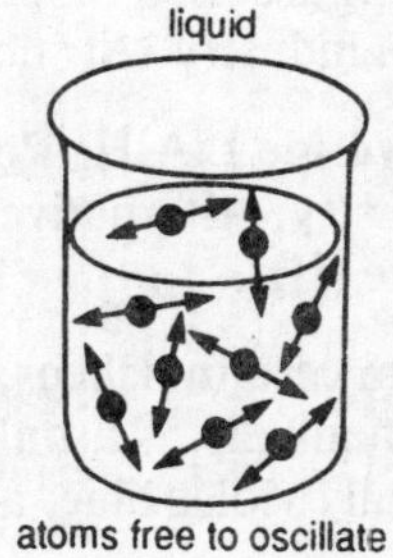

Molecules in a liquid are free to move around

liquid crystal Compound that is liquid at room temperature and atmospheric pressure but shows characteristics normally expected only from solid crystalline substances. Large groups of its molecules maintain their mobility but nevertheless also retain a form of structural relationship. Some liquid crystals change colour according to the temperature.

liquid-crystal display (LCD) **Digital display** based on liquid crystal cells used in calculators, watches and other equipment.

liquid-liquid extraction Alternative name for **solvent extraction**.

Lissajous' figure Plane curve formed by the combination of two or more simple periodic motions. It was named after the French physicist Jules Lissajous (1822–80). Alternative name: Lissajous' circle.

litharge Alternative name for **lead (II) oxide**.

lithium Li Silver-white metallic element in Group IA of the Periodic Table (the **alkali metals**), the solid with the least density. Its compounds are used in lubricants, ceramics, drugs and the plastics industry. At. no. 3; r.a.m. 6.939.

lithium aluminium hydride $LiAlH_4$ Powerful reducing agent, used in organic chemistry. Alternative name: lithium tetrahydridoaluminate (III).

■ **litmus** Dye made from certain lichens, used as an **indicator** to distinguish acids from alkalis. Neutral litmus solution or litmus paper is naturally violet-blue; acids turn it red, alkalis turn it blue. [4/5/a]

litre (l) Unit of volume in the metric system, defined as 1 dm^3, *i.e.* 1,000 cm^3 (formerly defined as the volume of 1 kg of water at 4 °C). 1 litre = 1.7598 pints.

■ **liver** In vertebrates, a large organ in the abdomen, the main function of which is to regulate the chemical composition of the blood by removing surplus **carbohydrates** and **amino acids**, converting the former into **glycogen** for storage and the latter into **urea** for excretion. Other functions include storage of **iron**, metabolism and storage of **fats**, secretion of **bile** and production of the blood-clotting factors prothrombin and **fibrinogen**. [2/4/a] [2/5/a] [2/7/a]

Lloyd's mirror Device for producing **interference** bands of contrasting brightness or darkness. A plane glass plate (acting as a mirror) is illuminated by monochromatic light from a slit parallel to the plate. Interference occurs between direct light from the slit and that reflected from the plate. It was named after the British physicist Humphrey Lloyd (1800–81).

LNG Abbreviation of **liquefied natural gas**.

local oscillator Oscillator that supplies the **radio frequency** oscillation with which the received wave is combined in a **superheterodyne** radio receiver.

locomotion Movement from place to place, one of the distinguishing characteristics of animals (as opposed to plants).

■ **locus** In biology, position of a **gene** on a **chromosome**. [2/6/d]

lodestone Fe_3O_4 Naturally occurring magnetic oxide of iron, a piece of which was reputedly used (by being suspended on a thread) as a primitive compass. Alternative names: loadstone, magnetite.

logarithmic scale Non-linear scale of measurement. For common logarithms (to the base 10), an increase of one unit represents a tenfold increase in the quantity measured.

logic In electronic data-processing systems, the principles that define the interactions of data in the form of physical entities.

logic gate Electronic switching circuit that gives an output only when specified input conditions are met. It is part of a **computer** that performs a particular logical operation (*e.g.* 'AND', 'OR', 'NOT', etc.). Alternative names: logic circuit, logic element.

lone pair of electrons Pair of unshared electrons of opposite spin (in the same **orbital**) that under suitable conditions can form a **coordinate bond**. *E.g.* the nitrogen atom in ammonia (NH_3) has a lone pair of electrons, which form bonds in various co-ordination compounds.

longitude Imaginary line that passes round the Earth through both poles. The angular distance around the globe is 360°, which is measured as 180° east of Greenwich (designated 0°) and 180° west of Greenwich. Using longitude in combination with **latitude**, any position on the Earth's surface can be denoted. Alternative name: meridian.

longitudinal wave Elastic wave in which the particles of a medium vibrate in the direction of propagation.

long sight Alternative name for **hypermetropia**.

long-sightedness Visual defect in which the eyeball is too short (front to back) so that light rays entering the **eye** from near objects would be brought to a focus at a point behind the retina. It can be corrected by spectacles or contact lenses made from converging (convex) lenses. Alternative name: hypermetropia.

Lorentz-Fitzgerald contraction Contraction in the length of a moving object in its direction of motion at near the velocity of light and relative to the frame of reference from which measurements are made, proposed independently by the Dutch physicist Hendrik Lorentz (1853–1928) and the Irish physicist George Fitzgerald (1851–1901). Alternative names: Lorentz contraction; Fitzgerald-Lorentz contraction.

Lorentz relation Alternative name for **Wiedmann-Franz law**.

Lorentz transformation Set of equations for correlating space and time co-ordinates in two frames of reference.

■ **loudness** Property of sound determined by the **amplitude** of the sound waves, usually expressed in **decibels**. [5/6/c]

loudspeaker Device for converting electrical impulses into audible sound, usually by making a **solenoid** vibrate a conical diaphragm.

low-energy electron diffraction Electron diffraction that uses low-energy electrons, which are strongly diffracted by surface layers of atoms.

low frequency (LF) Radio frequency of between 30 and 300 kHz.

Lowry-Brönsted theory *See* **Brönsted-Lowry theory**.

LPG Abbreviation of **liquefied petroleum gases**.

LSD Abbreviation of **lysergic acid diethylamide**.

lubricant Any substance used to reduce friction between surfaces in contact; *e.g.* oil, graphite, molybdenum disulphide, silicone grease.

luciferase Enzyme that initiates the **oxidation** of **luciferin**.

luciferin Substance that occurs in the light-producing organ of some animals, *e.g.* firefly. When oxidized (through the action of the enzyme luciferase) it produces **bioluminescence**.

lumen *1*. In biology, the space enclosed by a duct, vessel or tubular organ. *2*. In physics, the SI unit of luminous flux, equal to the amount of light emitted by source of 1 candela through unit solid angle.

luminance Measure of surface brightness expressed as the luminous flux per unit solid angle per unit projected area.

luminescence Emission of light by a substance without any appreciable rise in temperature. *See* **fluorescence**; **incandescence**; **phosphorescence**.

luminosity Brightness, *e.g.* of an image or a star.

luminous intensity Amount of light emitted in a given direction per second in unit solid angle by a point source.

■ **lung** In air-breathing vertebrates, paired or single respiratory organ located in the **thorax.** Its surface contains a large area of thinly folded, moist **epithelium** membrane so that it occupies little volume. This membrane is richly supplied by blood **capillaries** which allow for efficient and easy gaseous exchange. Air enters and leaves lungs through the **bronchus,** which in mammals branches into **bronchioles** ending in clusters of **alveoli,** where the main gaseous exchange takes place. [2/4/a] [2/5/a] [2/7/a]

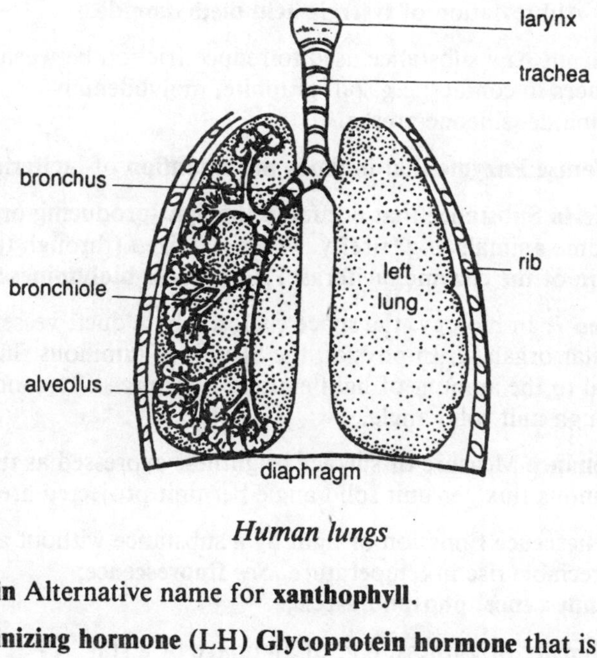

Human lungs

lutein Alternative name for **xanthophyll.**

luteinizing hormone (LH) **Glycoprotein hormone** that is secreted by the **pituitary** under regulation by the **hypothalamus.** In female mammals, it stimulates the formation of the **corpus luteum** and **Graafian follicle**, which in turn stimulates **oestrogen** production. LH is also essential for **ovulation.** In males, it stimulates interstitial cells of the **testis** to produce **androgens.**

luteotrophin Alternative name for **prolactin.**

lutetium Metallic element in Group IIIB of the Periodic Table (one of the **lanthanides**). The irradiated metal is a beta-particle emitter, used in catalytic processes. At. no. 71; r.a.m. 174.97.

lux SI unit of illumination, equal to one lumen per square metre. Alternative name: metre-candle.

Lycopsida One of the most primitive classes of vascular plants, the club mosses.

lyddite Explosive consisting of **picric acid.**

Lyman series Series of lines in the ultraviolet **spectrum** of hydrogen. It was named after the American physicist Theodore Lyman (1874–1954).

■ **lymph** Colourless alkaline tissue fluid that drains into the lymphatic system from intercellular spaces. It is similar in salt concentration to **plasma**, but possesses a lower **protein** concentration. It contains cells that are chiefly **lymphocytes.**

■ **lymphatic system** System of vessels and nodes (glands) that circulates **lymph** throughout the body, transporting the products from the digestion of fats and producing **antibodies** and **leucocytes.**

lymph node Flat oval structure made of lymphoid tissue that lies in the lymphatic vessels and occurs in clusters in the neck, armpit or groin. Its main function is the manufacture of **antibodies** and **leucocytes**. Lymph nodes also act as a defence barrier against the spread of infection by filtering out foreign bodies and bacteria, thus preventing their entry into the bloodstream. Alternative name: lymph gland.

lymphocyte Type of agranular **leucocyte** with a large nucleus, rich in **DNA**. Lymphocytes form 25% of all leucocytes and produce **antibodies**, which are important in defence against disease. They are produced in **myeloid tissue** of red bone marrow, **spleen, tonsils, lymph nodes** and **thymus.**

lymphoid tissue Tissue found in dense aggregations in **lymph nodes, tonsils, thymus** and **spleen**. It produces **lymphocytes** and macrophagocytic cells, which ingest **bacteria** and other foreign bodies. Alternative name: lymphatic tissue.

lyophilic Possessing an affinity for liquids.

lyophobic Liquid-repellent, having no attraction for liquids.

Lysenkoism Doctrine adopted in the Soviet Union from the 1940s to the 1960s that held all variation in organisms to be the direct result of environmental influence, denied the existence of genes and revived the tenets of **Lamarckism**. The doctrine was adopted for purely political reasons and named after its main advocate, Trofim Lysenko (1898–1976).

lysergic acid diethylamide (LSD) Synthetic substance, similar to some fungus **alkaloids**, which provokes hallucinations and extreme mental disturbance if taken, even in extremely small quantities.

lysine $H_2N(CH_2)_4CH(NH_2)COOH$. **Essential amino acid** that occurs in **proteins** and is responsible for their basic properties because of its two $-NH_2$ groups. Alternative name: diaminocaproic acid.

lysis Degeneration and subsequent breakdown of a cell. Under normal conditions, cell lysis is carried out by **phagocytes**, which also lyse invading cells. Under rare conditions, lysis may occur from within a cell.

lysosome Membrane-bound organelle of cells that contains a range of digestive **enzymes**, such as **proteases**, phosphatases, **lipases** and **nucleases**. The functions of lysosomes include contributing enzymes to white **blood cells** during **phagocytosis** and the destruction of cells and tissue during normal development, *e.g.* the loss of a tadpole's tail. Lysosomes may be produced directly from the **endoplasmic reticulum** or by budding of the **Golgi apparatus**.

lysozyme Enzyme in saliva, egg white, tears and mucus. It catalyses the destruction of bacterial cell walls by **hydrolysis**, and thus has a bactericidal effect.

M

machine Device for doing **work** in which a small **force** (the effort) overcomes a larger force (the load); *e.g.*, in the simplest case, levers and pulleys, the inclined plane and the wheel-and-axle. This mechanical definition can be extended to include also electrical devices, such as a transformer. *See also* **efficiency; mechanical advantage; velocity ratio.**

machine code Code in which instructions are given to a **computer**. Many computer languages (used for **programs**) have to be translated into machine code before they can be 'understood' by a computer.

Mach number Indication of the speed of an object in relation to the speed of sound in a particular medium (1,229 km/h in air at sea level), which is given the value Mach 1. An aircraft travelling at Mach 2 is flying at twice the speed of sound.

macro- Prefix meaning large or long in size or duration.

macrocyclic Describing a chemical compound whose molecules have a large ring structure.

macromolecular Describing the structure of a chemical compound that consists of **macromolecules**.

macromolecule Very large **molecule** containing hundreds or thousands of **atoms**; *e.g.* natural **polymers** such as cellulose, rubber and starch, and synthetic ones, including **plastics**.

macronutrient Food substance needed in fairly large amounts by living organisms, which may be an inorganic element (*e.g.* phosphorus or potassium in plants) or an organic compound (*e.g.* amino acids and carbohydrates in animals). *See also* **trace element; vitamin.**

macrophage Large **phagocyte** found in vertebrate tissue, particularly in liver, spleen and lymph glands. It removes foreign particles at the site of infection.

macula Any small blemish, coloured patch, growth or shallow depression that is visible on plant or animal tissue, especially on external surfaces. *E.g.* the macula lutea is the yellow spot (**fovea**) at the centre of the retina.

magnesia **Magnesium oxide**, MgO, particularly a form that has been processed and purified. It is used as an antacid.

magnesite Naturally occurring **magnesium carbonate**, $MgCO_3$, used as a refractory for furnace linings and to make various grades of **magnesium oxide** (magnesia).

magnesium Mg Reactive silver-white metallic element in Group IIA of the Periodic Table (the **alkaline earths**). It burns in air with a brilliant white light, and is used in flares and lightweight alloys. It is the metal atom in chorophyll and an important trace element in plants and animals. At. no. 12; r.a.m. 24.305.

magnesium carbonate $MgCO_3$ White crystalline compound, soluble in acids and insoluble in water and alcohol, which occurs naturally as **magnesite** and **dolomite**. It is used, often as the basic carbonate, as an antacid.

magnesium chloride $MgCl_2$ White crystalline compound obtained from sea-water and the mineral **carnallite**, used as a source of magnesium. The hexahydrate is hygroscopic and used as a moisturizer for cotton in spinning.

magnesium hydroxide $Mg(OH)_2$ White crystalline compound, used as an antacid.

magnesium oxide MgO White crystalline compound, insoluble in water and alcohol, made by heating **magnesium**

carbonate. It is used as a refractory and antacid. Alternative name: magnesia.

magnesium silicate *See* **talc**.

magnesium sulphate $MgSO_4.7H_2O$ White crystalline salt, used in mineral waters, as a laxative and in the leather industry. Alternative name: Epsom salt.

■ **magnet** Object possessing the property of **magnetism**, either permanently (a permanent magnet, made of a ferromagnetic material) or temporarily under the influence of another magnet or the magnetic field associated with an electric current (an **electromagnet**). *See also* **paramagnetism**.

magnetic amplifier **Transducer** so arranged that a small controlling direct current input can produce large changes in coupled alternating current circuits.

magnetic circuit Completely closed path described by a given set of lines of **magnetic flux**.

magnetic core Computer storage device consisting of a ferromagnetic ring wound with wires; a current flowing in the wires polarizes the core, which can therefore adopt one of two states (making it a bistable device).

magnetic declination Angle by which north as shown by a magnetic compass (magnetic north) deviates from **true north**. The position of magnetic north is subject to significant variation with time, and for this reason magnetic declination is not a constant factor.

magnetic dip *See* **dip**.

magnetic disk Device for direct-access storage and retrieval of data, used in computers and similar systems. It consists of a rotatable flexible or rigid plastic disc (*i.e.* a floppy or hard disk) coated on one or both surfaces with magnetic material, such as iron oxide. Data is stored or retrieved through one or more **read-write heads**. Alternative name: magnetic disc.

magnetic domain Group of atoms with aligned magnetic moments that occur in a **ferromagnetic** material. There are many randomly oriented domains in a permanent **magnet**.

magnetic drum Computer storage device consisting of a rotatable drum coated with magnetic material, such as iron oxide. Data is stored or retrieved through one or more **read-write heads**.

magnetic element *See* **geomagnetism**.

magnetic field Field of force in the space around the magnetic poles of a **magnet**.

magnetic field strength Alternative name for **magnetic intensity**.

magnetic flux Measure of the total size of a **magnetic field**, defined as the scalar product of the flux density and the area. Its SI unit is the weber.

magnetic flux density Product of **magnetic intensity** and **permeability**. Its SI unit is the tesla (formerly weber m^{-2}).

■ **magnetic induction** In a magnetic material, magnetization induced in it, *e.g.* by placing it in the electromagnetic field of a current-carrying coil or by stroking it with a permanent **magnet**. [4/10/e]

magnetic intensity Magnitude of a magnetic field. Its SI unit is the ampere m^{-2}. Alternative names: magnetic field strength, magnetizing force.

magnetic lens Arrangement of electromagnets used to focus a beam of charged particles (*e.g.* electrons in an **electron microscope**).

magnetic mirror Region of high magnetic field strength that reflects particles from a **plasma** back into a magnetic bottle. *See also* **fusion reactor.**

magnetic moment Property possessed by an individual atom or molecule, a moving charge, a permanent magnet, or a current-carrying coil, used as a measure of its magnetic strength. Multiplied by the magnetic induction it equals the torque (turning force) on the magnet.

magnetic monopole Theoretically possible, but so far unobserved, single **magnetic pole**.

magnetic north Point on the Earth's surface towards which a magnetic compass points. The exact location of magnetic north varies seasonally, and from year to year. *See also* **magnetic declination**; **true north**.

magnetic pole *1.* One of the two points, called north and south, to and from which a **magnetic field** appears to radiate. Unless deliberately made otherwise, *e.g.* in a horseshoe magnet, the magnetic poles tend to occur at opposite sides of any object exhibiting magnetism. *2.* Position on Earth to or from which a magnetized needle (a compass) points. The positions of the magnetic poles vary over a period of time and do not coincide with the geographic poles. The angular difference between the directions of magnetic north and true north is the **magnetic declination**. *See also* **dip**.

magnetic storm Local disturbances in the Earth's magnetic field, which can disrupt **telecommunications**, and which are probably caused by gusts of charged particles emanating from the Sun.

magnetic susceptibility Difference between **relative permeability** and unity; equal to the intensity of magnetization divided by the applied field.

magnetic tape Medium for the storage of electronic signals by magnetizing particles (of *e.g.* iron oxide) in a coating on plastic tape. It is used in (audio) tape recorders, video recorders and computers.

magnetism Presence of magnetic properties in materials. **Diamagnetism** is a weak effect common to all substances and results from the orbital motion of electrons. In certain substances this is masked by a stronger effect, **paramagnetism,** due to electron spin. Some paramagnetic materials such as iron also display **ferromagnetism**, and are permanently magnetic.

magnetite Minéral form of a black iron oxide (Fe_3O_4), used as a source of iron, as a flux and in ceramics. It is strongly magnetic (*see* **lodestone**).

magnetization Difference between the ratio of the **magnetic induction** to the **permeability** and the **magnetic intensity**; its SI unit is the ampere m^{-1}. It represents departure from randomness of **magnetic domains**.

magnetohydrodynamics Branch of physics that deals with the behaviour of a conducting fluid under the influence of a **magnetic flux**.

magneton Unit for the **magnetic moment** of an electron. Alternative name: Bohr-magneton.

magnetron Electronic valve that is used to provide pulsed emissions in **microwave** telecommunications, **radar** transmitters, and microwave cookers.

magnification In optics, the ratio y'/y, where y is the height of an object perpendicular to the optical axis and y' is the corresponding height of its magnified image. For a single lens this is equivalent to the ratio of the image distance to the object distance.

Magnoliophyta Subdivision of the plant division **Spermatophyta**, consisting of the **angiosperms**.

Magnoliopsida Class of **angiosperm** plants that comprises the **dicotyledons**.

malachite Bright green mineral consisting of a mixture of copper (II) carbonate and copper (II) hydroxide, $CuCO_3.CU(OH)_2$.

Malacostraca Subclass of crustaceans that includes lobsters, crabs and shrimps.

malic acid $COOHCH_2CH(OH)COOH$ Colourless crystalline **dicarboxylic acid** with an agreeable sour taste resembling that of apples, found in unripe fruit. It is used as a flavouring agent.

malignant Describing a (medical) condition that tends to spread or get worse; opposite of benign.

malleable Describing a metal that can easily be beaten into a sheet (*e.g.* lead, gold).

malleus One of the **ear** ossicles.

Mallory cell Alternative name for **mercury cell**.

Malpighian body Part of the mammalian **kidney**, encompassing the **Bowman's capsule** and **glomerulus**. Its function is to filter blood. It was named after the Italian biologist Marcello Malpighi (1628–1694).

maltose $C_{12}H_{22}O_{11}$ Common **disaccharide** sugar, composed of two molecules of **glucose**. It is found in **starch** and **glycogen**, and used in the food and brewing industries.

■ **mammal** Member of the animal class **Mammalia**. [2/3/b] [2/4/b]

■ **Mammalia** Class of tetrapod vertebrates which includes man. General characteristics of mammals include warm-bloodedness, a four-chambered heart, the possession of hair, the possession of middle-ear ossicles, a diaphragm used in respiration, giving birth to live young (nourished in the womb via a placenta), and considerable parental care including feeding of young with secreted milk. [2/3/b] [2/4/b]

■ **mammary gland** Gland for milk-production (lactation) in female mammals. [2/4/b]

mammotrophin Alternative name of **prolactin**.

■ **mandible** *1.* Mouth-part in animals such as crustaceans, insects and cephalopods, used for cutting food. *2.* Lower **jaw** of a vertebrate. [2/5/a]

manganese Mn Metallic element in group VIIA of the Periodic Table, used mainly for making special alloy steels and as a deoxidizing agent. It is also an essential trace element for plants and animals. At. no. 25, r.a.m. 54.94.

manganese dioxide Alternative name for **manganese (IV) oxide**.

manganese (IV) oxide MnO_2 Black amorphous compound, used as an oxidizing agent, catalyst and depolarizing agent in dry batteries. Alternative name: manganese dioxide.

manganic Alternative name for manganese (III) in manganese compounds.

manganin Alloy of copper with manganese and nickel that exhibits high electrical **resistance**, and which is used in resistors.

manganous Alternative name for manganese (II) in manganese compounds.

mannitol *$HO.CH_2(CHOH)_4CH_2OH$ Soluble hexahydric* alcohol that occurs in many plants, used as a sweetener and in medicine as a diuretic.

■ **manometer** Device for measuring fluid pressure. [5/6/a]

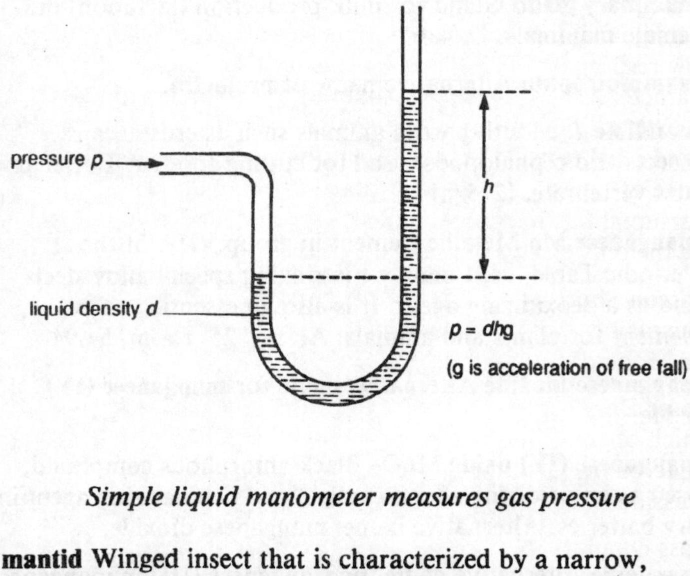

Simple liquid manometer measures gas pressure

■ **mantid** Winged insect that is characterized by a narrow, elongated body and powerful forelegs which are used for seizing insect prey. There are about 1,800 species, found throughout the warmer regions of the world, of which the praying mantis is the best known. [2/3/b] [2/4/b]

mantle *1.* Body part from which the shell of a mollusc or brachiopod develops. *2.* Alternative name for the **carapace** of an adult barnacle.

marble Metamorphic rock consisting of **calcium carbonate** ($CaCO_3$), derived from **limestone**.

mare Adult female horse.

■ **marsupial** *1.* Member of the order **Marsupiala**. *2.* Animal that has a body part that bears a structural or functional resemblance to the pouch possessed by true female marsupials, *e.g.* marsupial frog. [2/3/b] [2/4/b]

■ **Marsupiala** Order of mammals in the subclass Metatheria, now restricted to Australasia and South America, characterized by the development of the **embryo** after birth in an external pouch (marsupium) that surrounds the **mammary glands**. In Australasia, marsupials have evolved into a wide range of types – kangaroo, koala, Tasmanian wolf, marsupial rat, marsupial mouse, etc. In America, opossums are the only surviving members of the order. [2/3/b] [2/4/b]

mascon Short form of mass concentration, one of many unexplained areas of high gravity that have been detected on the Moon's surface.

maser Device that produces narrow **microwave** beams, and which operates on the same principle as a **laser**. The name is an acronym for 'microwave amplification by stimulated emission of radiation'.

■ **mass** Quantity of matter in an object, and a measure of the extent to which it resists acceleration if acted on by a force (*i.e.* its inertia). The SI unit of mass is the **kilogram**. *See also* **weight**.

mass action, law of The driving force of a homogeneous chemical reaction is proportional to the active masses of the reacting substances.

mass decrement *See* **mass defect**.

mass defect *1.* Difference between the mass of an atomic **nucleus** and the masses of the particles that make it up, equivalent to the **binding energy** of the nucleus (expressed in mass units). *2.* Mass of an **isotope** minus its mass number. Alternative names: mass decrement, mass excess.

mass-energy equation Deduction from Einstein's special theory of relativity that all energy has mass; $E = mc^2$, where E is the energy, m is the amount of mass, and c the speed of light.

■ **mass number** (A) Total number of **protons** and **neutrons** in an atomic **nucleus**. Alternative name: nucleon number. *See also* **isotope**. [4/8/d]

mass spectrograph Vacuum system in which positive rays of charged atoms **(ions)** are passed through electric and magnetic fields so as to separate them in order of their charge-to-mass ratios on a photographic plate. It measures **relative atomic masses** of **isotopes** with precision.

mass spectrometer **Mass spectrograph** that uses electrical methods rather than photographic ones to detect charged particles.

mass spectrum Indication of the distribution in mass, or in mass-to-charge ratio, of ionized atoms or molecules produced by a **mass spectrograph**.

maternal twins *See* **twins**.

matrix In biology, extracellular substance that embeds and connects cells, *e.g.* connective tissue.

matter Substance that occupies space and has the property of **inertia**. These two characteristics distinguish matter from energy, the various forms of which make up the rest of the material Universe.

maxilla *1.* Mouthpart of an insect used in feeding. *2.* Upper **jaw** of a vertebrate.

maximum and minimum thermometer *See* **thermometer, maximum and minimum**.

maxwell CGS unit of magnetic flux, the SI unit being the weber. 1 maxwell = 10^{-8} weber. It was named after the British physicist James Clerk Maxwell (1831–79).

Maxwell-Boltzmann distribution Law that describes the distribution of energy among molecules of a gas in thermal equilibrium.

mean free path Average distance that a gas molecule moves between two successive collisions with other molecules.

mean free time Average time between collisions of *1.* gas molecules; *2.* electrons and impurity atoms in a **semiconductor**.

mean life *1.* Average time for which the unstable **nucleus** of a radioisotope exists before decaying. *2.* Average time of survival of an elementary particle, ion, etc., in a given medium or of a charge carrier in a **semiconductor**.

mean solar day Average value over a year of the solar day, the constant time between two transits of the Sun across the meridian. It equals 24 hours.

meatus Duct or channel between body parts, *e.g.* the auditory canal, which leads into the ear.

mechanical advantage (MA) For a simple machine, the ratio of the load L (output force) to the effort E (input force); *i.e.* MA = L/E. Alternative name: force ratio.

■ **mechanics** Study of the interaction between matter and the forces acting on it. Its three divisions are **kinematics** (concerned with acceleration, velocity, etc.), **dynamics** (concerned with forces acting on objects in motion) and **statics** (concerned with forces that do not produce motion).

median lethal dose *See* **LD-50**.

medicine Science that studies diseases and other bodily disorders, how they are caused and how they are treated. In everyday language, a medicine is any **drug** or other remedy used to treat a disorder.

■ **medulla** Central part of an organ or tissue (*e.g.* adrenal medulla). *See also* **cortex**. [2/4/a] [2/5/a] [2/7/a]

medulla oblongata Posterior part of the **brain** of vertebrates, which contains centres that control the frequency of respiration and heartbeat, and is concerned with the co-ordination of nerve impulses from hearing, touch and taste receptors.

medullary ray Strip of **parenchyma** cells that runs through tissues of plants, used for the storage and transport of food. It may be primary (running from the centre to the **cortex**), or secondary (formed by secondary thickening from **cambium**).

medullated nerve fibre *See* **myelin; nerve**.

medusa Gamete-producing form of a **coelenterate**, *e.g.* jellyfish, which thereby reproduces sexually. Other forms in the life-cycle reproduce asexually. *See also* **alternation of generations**.

meerschaum Very fine-grained claylike mineral composed of hydrated magnesium silicate, used for making smoking pipes and ornamental sculptures.

mega- Metric prefix meaning million times; $\times\ 10^6$ (*e.g.* megahertz).

megahertz (MHz) Unit of frequency of one million **hertz**.

megaton Measure of the explosive power of a nuclear explosion or warhead, equivalent to a million tons of TNT.

■ **meiosis** Type of cell division in which the number of **chromosomes** in the daughter cells is halved; thus they are in the **haploid** state. Two successive divisions occur in the process, giving four daughter cells. The first division takes place in four stages: **prophase, metaphase, anaphase** and **telophase**. The second division has three stages: metaphase, anaphase and telophase. In animals meiosis occurs in the formation of **gametes**, *e.g.* eggs and sperm. Alternative name: reduction division. *See also* **mitosis**. [2/8/b]

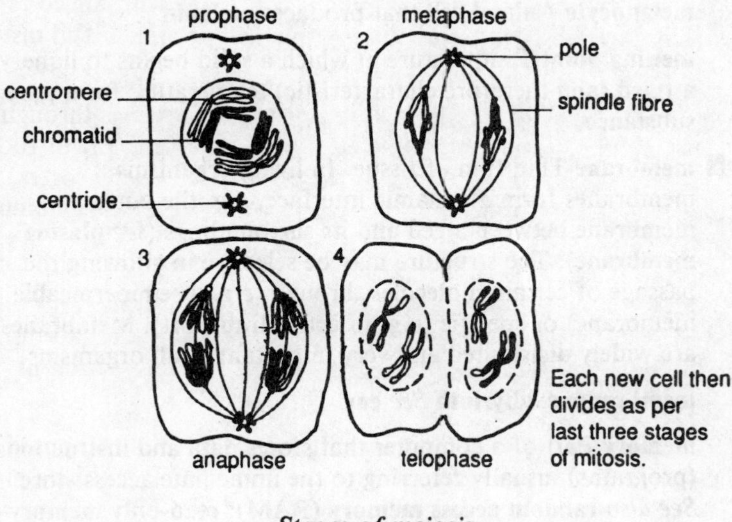

Stages of meiosis

meiotic Describing or referring to **meiosis**.

■ **melamine** $C_3N_3(NH_2)_3$ White solid organic compound which consists of three amino groups ($-NH_2$) attached to a six-membered **heterocyclic** ring of alternate carbon and nitrogen atoms. It condenses with **formaldehyde** (methanal) or other **aldehydes** to form artificial **resins** that have excellent resistance to heat, water and many other chemicals, as well as possessing surface hardness. Alternative name: triaminotriazine. [4/8/b]

■ **melanin** Dark brown or black pigment that gives colour to hair and skin. It is formed in melanocytes by the oxidation of **tyrosine**, induced by the action of sunlight. [2/6/a] [2/7/a]

melanism Occurrence in animal populations of dark-coloured individuals having an excess of **melanin** in their tissues.

melanocyte Animal cell that produces **melanin**.

melting point Temperature at which a solid begins to liquefy, a fixed (and therefore characteristic) temperature for a pure substance.

■ **membrane** Thin film of tissue. In living organisms, membranes form a dynamic interface, *e.g.* the outer membrane between a cell and its surroundings (*see* **plasma membrane**). The structure may be selective in allowing the passage of certain molecules through (*e.g.* a semipermeable membrane) or specific (*e.g.* in active transport). Membranes are widely distributed and very important in all organisms.

membranous labyrinth *See* **ear**.

memory Part of a computer that stores data and instructions (programs), usually referring to the immediate access store. *See also* **random access memory** (RAM); **read-only memory** (ROM).

Mendeleev's law Properties of elements listed in order of increasing relative atomic masses are generally similar every eighth element. The law was named after the Russian chemist Dimitri Mendeleev (1834–1907). Alternative name: periodic law.

mendelevium Md Radioactive element in Group IIIB of the Periodic Table (one of the **actinides**); it has several **isotopes**, with half-lives of up to 54 days. At. no. 101; r.a.m. 258 (most stable isotope).

■ **Mendelism** Study of inheritance, and therefore genetics, as a result of work carried out by the Austrian monk Gregor Mendel (1822–84). *See also* **Mendel's laws**. [2/9/c]

■ **Mendel's laws** Conclusions drawn from work on inheritance carried out by Gregor Mendel in breeding experiments. The first is the law of segregation: an inherited characteristic is controlled by a pair of factors (**alleles**), which separate and become incorporated into different **gametes**. The second is the law of independent assortment: the separated factors are independent of each other when gametes form. [2/7/b]

meninges System of **membranes** that envelop the brain. The innermost is the pia mater, the outermost the dura mater.

meniscus Curved surface of a liquid where it is in contact with a solid. The effect is due to **surface tension**.

menopause Natural cessation of menstruation (*see* **menstrual cycle**).

■ **menstrual cycle Hormone**-regulated cycle of female reproductive behaviour during which **ovulation** occurs; it occurs in some primates, including human beings. The end of the cycle is marked by a monthly shedding of the endometrium (lining of the womb) accompanied by a discharge of blood from the vagina; this is known as

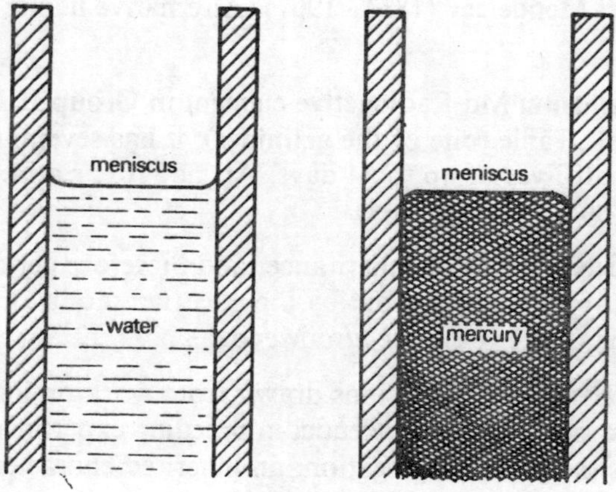

Shape of a meniscus depends on the liquid

menstruation. The cycle begins with the menarche at the beginning of puberty and ends at the menopause. Alternative name: sexual cycle. *See also* **oestrous cycle.**

menstruation *See* **menstrual cycle.**

mensuration Science of measurement.

mercaptan Alternative name for a **thiol.**

mercuric Alternative name for mercury (II) in compounds of mercury.

mercurous Alternative name for mercury (I) in compounds of mercury.

■ **mercury** Hg Dense liquid metallic element in Group IIB of the Periodic Table (a **transition element**), used in lamps, batteries; switches and scientific instruments. It alloys with most metals to form **amalgams**. Its compounds are used in drugs, explosives and pigments. At. no. 80; r.a.m. 200.59.

mercury arc Bright blue-green light obtained from an electric discharge through mercury vapour.

mercury cell *1.* **Electrolytic cell** that has a **cathode** made of mercury. *2.* Dry cell that has a mercury electrode. Alternative name: Mallory cell.

mercury (I) chloride HgCl White crystalline compound, used as an insecticide and in a **mercury cell**. Alternative names: mercurous chloride, calomel.

mercury (II) chloride $HgCl_2$ Extremely poisonous white compound. Alternative names: mercuric chloride, corrosive sublimate.

mercury (II) oxide HgO Red or yellow compound, slightly soluble in water, which reduces to metallic mercury on heating. Alternative name: mercuric oxide.

mercury (II) sulphide HgS Red compound, which occurs naturally as cinnabar, used as a pigment (vermilion) and source of mercury.

mercury-vapour lamp Lamp that uses a **mercury arc** in a quartz tube; it produces **ultraviolet radiation**.

meridian Line of **longitude**, a great circle that passes through the poles.

meristem Region of active cell division and differentiation in plants. The principal meristems in flowering plants occur at the tips of stems and roots.

mescaline Powerful drug derived from the Mexican mescal cactus (peyote), which has a similar effect to **lysergic acid**

diethylamide (LSD), and causes hallucinations and mental disturbance.

mesencephalon Alternative name for **mid-brain.**

mesentery Vertical fold of tissue on the inner surface of the body wall of animals, which supports internal organs or associated structures.

meso- Prefix meaning middle.

mesoderm Tissue in an animal embryo that develops into tissues between the gut and **ectoderm.**

mesoglea Layer of unstructured material that occurs between the **ectoderm** and **endoderm** of **coelenterates** (*e.g.* jellyfish).

mesomerism Phenomenon in which a chemical compound can adopt two or more different structures (called canonical forms) by the alteration of (covalent) bonds while the atoms in the molecules remain in the same relationship to each other (*e.g.* the Kekulé forms of **benzene**). Alternative name: resonance. *See also* **tautomerism.**

meson Member of a group of unstable **elementary particles** with masses intermediate between those of **electrons** and **nucleons**, and with positive, negative or zero charge. Mesons are emitted by **nuclei** that have been bombarded by high-energy electrons.

mesophyll Tissue that forms the middle part of a leaf.

■ **mesophyte** Plant that grows in an environment with an average water supply. *See also* **hydrophyte; xerophyte.** [2/6/a]

■ **messenger RNA** (mRNA) Ribonucleic acid that conveys instructions from **DNA** by copying the code of DNA in the cell **nucleus** and passing it out to the cytoplasm. It is translated into a **polypeptide** chain formed from **amino acids** which join in a sequence according to the instructions in the messenger RNA. *See also* **transcription**.

meta- *1*. Relating to the 1–3 carbon atoms in a **benzene ring**, abbreviated to *m*-; *e.g.* *m*-xylene is 1,3-dimethylbenzene). *2*. Relating to compounds formed by **dehydration**; *e.g.* metaphosphoric acid, made by strongly heating orthophosphoric acid. *3*. Alternative name for **metaldehyde**. *See also* **ortho-**; **para-**. *4*. Prefix that generally indicates the concept of change, and which may mean after or beyond; *e.g.* metastable.

■ **metabolism** Biochemical reactions that occur in cells and are a characteristic of all living organisms. Metabolic reactions are initiated by **enzymes** and liberate energy in a usable form. Organic compounds may be broken down to simple constituents (catabolism) and used for other processes. Simple compounds may be built up to more complex ones (anabolism). [2/5/a] [2/7/a] [2/9/a]

metabolite Molecule participating in **metabolism**, which may be synthesized in an organism or taken in as food. Autotrophic organisms need only to take in inorganic metabolites; heterotrophs also need organic metabolites.

■ **metal** Any of a group of **elements** and their alloys with general properties of strength, hardness and the ability to conduct heat and electricity (because of the presence of free electrons). Most have high melting points and can be polished to a shiny finish. Metallic elements (about 80 per cent of the total) tend to form **cations**. *See also* **metalloid**.

metaldehyde $(CH_3CHO)_4$. White crystalline solid, polymer of **acetaldehyde** (ethanal), used to kill slugs and as a fuel for camping and emergency stoves. Alternative names: meta, ethanal tetramer.

metal fatigue Weakness in a metal caused by a long period of stress, which can cause cracks and disintegration, *e.g.* in the structures of aircraft.

metallic bond Type of interatomic bonding that exists in metals.

metallic crystal Crystalline structure that is held together by **metallic bonds**.

metallocene Member of a group of chemical compounds formed between a metal and an **aromatic compound** in which the oxidation state of the metal is zero; *e.g.* **ferrocene**.

metalloid Element with physical properties resembling those of metals and chemical properties typical of non-metals (*e.g.* arsenic, germanium, selenium). Many metalloids are used in **semiconductors**.

metameric segmentation Alternative name: metamerism. *See* **segmentation**.

metamorphosis Stage in the life cycle of some animals, including marine invertebrates, arthropods and amphibians, in which by hormonal control the **larva** undergoes transformation to the adult form.

■ **metaphase** Second stage of **mitosis** and **meiosis**, in which **chromosomes** are lined up along the equator of the nuclear **spindle**. Alternative name: aster phase. [2/8/b]

Metaphyta The plant kingdom. Alternative name: Plantae.

metaplasia Transformation of one normal tissue type into another as a response to a disease or abnormal condition.

metastasis Process by which disease-bearing cells are transferred from one part of the body to another via the **lymph** and **blood vessels**; the term is usually applied to the spread of **cancers**. The term also applies to the newly diseased area arising from the process.

metastatic Describing **electrons** that leave an orbital shell, either entering another shell or being absorbed into the nucleus.

■ **metatarsal** One of the rod-shaped bones that forms the lower hind limb or part of the hind foot in four-legged animals and the arch of the foot in human beings. [2/3/a]

Metatheria Subclass of mammals that contains the **marsupials**.

■ **Metazoa** Subkingdom consisting of multicellular animals with a body having two or more tissue layers and a co-ordinating nervous system. [2/3/b] [2/4/b]

metazoan Animal that is a member of the **Metazoa**.

methanal HCHO Alternative name for **formaldehyde**.

■ **methane** CH_4 Simplest **hydrocarbon**, an **alkane** and a major constituent of **natural gas** (up to 97() and **coal gas**. It is an end-product of the **anaerobic** decay of plants (hence its occurrence as marsh gas in swamps); it also occurs in coal mines (where it is known as fire damp). Methane can be manufactured by the catalytic **hydrogenation** of **carbon monoxide**. It is used as a fuel, as an industrial source of hydrogen and in chemical synthesis.

methanoic acid HCOOH Alternative name for **formic acid**.

methanol CH_3OH Simplest primary **alcohol**, a poisonous liquid used as a solvent and added to ethanol to make **methylated spirits**. Alternative names: methyl alcohol, wood spirit.

methionine $CH_3S(CH_2)_2CH(NH_2)COOH$ **Sulphur**-containing

amino acid; a constituent of many **proteins**. Alternative name: 2-amino-4-methylthiobutanoic acid.

methyl alcohol Alternative name for **methanol**.

methylamine CH_3NH_2 Simplest primary **amine**, a gas smelling like ammonia, used in making herbicides.

methylaniline Alternative name for **toluidine**.

methylated spirits General solvent consisting of absolute alcohol (ethanol) that has been denatured with **methanol** and **pyridine**; it often has added purple dye. Alternative name: denatured alcohol.

methylation Chemical reaction in which a **methyl group** is added to a chemical compound.

methylbenzene Alternative name for **toluene**.

methylbutadiene Alternative name for **isoprene**.

methyl chloroform Alternative name for **1,1,1-trichloroethane**.

methyl cyanide Alternative name for **acetonitrile**.

methylene blue Blue dye used as a pH **indicator**.

methyl group $-CH_3$ radical.

methyl methacrylate Methyl **ester** of methacrylic acid, used in the preparation of the plastic polymethyl-methacrylate (Perspex) and other acrylic resins, employed to make lenses, spectacle frames, false teeth, etc.

■ **methyl orange** Orange dye used as a pH **indicator**. [4/5/a]

methylphenol $CH_3C_6H_4OH$ Derivative of **phenol** in which one of the hydrogens of the **benzene ring** has been substituted by a **methyl group**. There are three **isomers** (ortho-, meta-

and para-), depending on the positions of the substituents in the ring. Alternative name: cresol.

methylpyridine Alternative name for **picoline**.

methyl red Red dye used as a pH **indicator**.

methyl salicylate Methyl ester of **salicylic acid**, used in medicine. Alternative name: oil of wintergreen.

metre (m) SI unit of length. 1 m = 39.37 inches.

metre-candle Alternative name for **lux**.

metric prefix Any of various numerical prefixes used in the **metric system**. *See* **units**.

metric system Decimal-based system of units. *See* **SI units**.

metric ton Alternative name for **tonne**.

MHD Abbreviation of **magnetohydrodynamics**.

mica Member of a group of minerals consisting of silicates. Micas have low thermal conductivity and high dielectric strength, and are widely used for electrical insulation.

micro- *1*. Metric prefix meaning a millionth; $\times 10^{-6}$ (*e.g.* microfarad). It is sometimes represented by the Greek letter μ (*e.g.* μF). *2*. General prefix meaning small (*e.g.* next three articles).

microbalance Balance capable of weighing very small masses (*e.g.* down to 10^{-5} mg).

■ **microbe** Imprecise term for any **microorganism**, particularly a disease-causing **bacterium**.[2/4/a] [2/5/a] [2/7/a]

microbiology Biological study of **microorganisms**.

microcomputer Small computer. Alternative name: personal computer (PC).

microelectronics Branch of electronics that is concerned with the design, production and application of electronic components, circuits and devices of extremely small dimensions.

micron (μ) Former name for the micrometre; 10^{-6} m.

micronutrient General term for any of the **trace elements** or **vitamins**.

■ **micro-organism** Organism that may be seen only with the aid of a **microscope**. Micro-organisms include microscopic **fungi** and **algae, bacteria, viruses** and single-celled animals (*e.g.* protozoans). [2/8/d]

microphone Device for converting sound energy into electrical energy. There are various types. In a moving-coil microphone, a coil connected to a diaphragm (vibrated by sound waves) moves in a magnetic field, generating a current in the coil by **electromagnetic induction**. In a moving-iron microphone, a small piece of iron is vibrated by a diaphragm in a magnetic field, varying the field and inducing a current in a surrounding coil. In a carbon microphone, a vibrating metallic diaphragm compresses carbon granules, thereby altering their resistance. In a crystal microphone, vibrations are transmitted to a piezoelectric crystal which generates a varying electric field.

microprocessor *See* **computer**.

micropyle Small opening or pore. *1.* Pore in a flower's **ovule** through which the **pollen tube** enters (to bring about pollination). *2.* Pore in a flower's seed through which water enters (to initiate germination). *3.* Pore in an insect's egg through which sperm enter (to bring about fertilization).

microscope Instrument that produces magnified images of structures invisible to the naked eye. There are two major optical types: the simple microscope, consisting of one short focal-length convex lens giving a virtual image, and the compound microscope, consisting of two short focal-length convex lenses which combine to give high magnification. Highest magnifications are produced by an **electron microscope**.

microtome Instrument for cutting thin slices (of the order of a few micrometres thick) of biological materials for microscopic examination.

microtubule Minute cylindrical unbranched tubule composed of globular **protein** subunits found either singly or in groups in the **cytoplasm** of **eukaryotic** cells, in which it has the skeletal function of maintaining their shape. Microtubules are also associated with **spindle** formation, and hence are responsible for chromosomal movement during nuclear division.

microwave Electromagnetic radiation with a wavelength in the approximate range 1mm to 0.3m, *i.e.* between **infra-red radiation** and **radio waves**.

microwave spectroscopy Study of atomic and/or molecular resonances in the **microwave** region.

■ **mid-brain** Part of **brain** that connects the fore-brain to the hind-brain, concerned with processing visual information passed from the fore-brain. Fish, amphibians and birds have a well-developed mid-brain roof, the tectum, which forms the integration centre of their brain. Mammals have a less well-developed mid-brain. Alternative name: mesencephalon. [2/5/a]

■ **middle ear** Air-filled part of the **ear** that is on the inner side of the ear drum and transmits sound waves from the outer ear to the inner ear. Alternative name: tympanic cavity.[2/4/a] [2/5/a] [2/7/a]

middle lamella Thin intercellular cementing pectic substances (*see* **pectin**) that hold together adjacent plant cell walls.

mil *1.* A millilitre. *2.* One-thousandth of an inch, equivalent to 0.0254 mm.

■ **mildew** Any fungus disease of plants in which the **mycelium** is visible as pale patches on external surfaces. [2/2/b] [2/2/d] [2/3/b]

mile Unit of length equal to 1,760 yards or 5,280 feet. 1 mile = 1.60934 kilometres. A nautical mile is 6,080 feet = 1.85318 km).

milk sugar Alternative name for **lactose**.

milk teeth First of two sets of teeth possessed by most mammals. Human beings have 20 milk teeth. Alternative name: deciduous teeth.

milli- Metric prefix meaning a thousandth; $\times 10^{-3}$ (*e.g.* milligram).

millibar (mbar) Thousandth of a **bar**, a common unit of atmospheric pressure in meteorology.

milligram (mg) Thousandth of a gram.

millilitre (ml) Thousandth of a litre, equivalent to a cubic centimetre (cc or cm^3).

millimetre (mm) Thousandth of a metre, equal to a tenth of a centimetre. 1 mm = 0.03937 inches.

millimetre of mercury (mmHg) Unit of pressure, equal to $^{1}/_{760}$ atmospheres.

■ **mimicry** Close resemblance between one animal and another. The mimicking (in colour, sound, habit or structure) of another of a different species gains advantage by its resemblance to the model; *e.g.* palatable insects that mimic the warning coloration of poisonous species are avoided by their predators. Imitation of, *e.g.*, the background is camouflage, not mimicry. [2/4/d]

■ **mineral** Naturally occurring, usually crystalline, inorganic substance of more or less definite chemical composition and physical properties. Mixtures of minerals form rocks. The term is sometimes extended to include fossil fuels (coal, natural gas, petroleum).

mineral acid Inorganic acid such as sulphuric, hydrochloric or nitric acid.

mineral oil Hydrocarbon oil obtained from mineral sources or petroleum (as opposed to an animal oil or vegetable oil).

mineral salts Dissolved salts that occur in soil, derived from weathered rock and decomposed plants. They contain essential nutrients for plant growth, which are in turn utilized by herbivores (and carnivores that feed on them).

minute *1*. Unit of time equal to $^1/_{60}$ of an hour. *2*. Unit of angular measure equal to $^1/_{60}$ of a degree. Both types of minutes are made up of 60 seconds.

mirage False image sometimes observed in deserts and polar regions. It is caused by light reflecting off the upper surface of a layer of very hot (or very cold) air near the ground.

■ **mirror** Optical device that produces reflection, generally having surfaces that are plane, spherical, paraboloidal, ellipsoidal or aspheric. Concave mirrors are hollow, convex mirrors are domed outwards. [5/5/c]

miscarriage Alternative name for a spontaneous **abortion**.

miscible Describing two or more liquids that will mutually dissolve (mix) to form a single **phase**. They can be separated by **fractional distillation**.

■ **mitochondrion** Cell organelle in the **cytoplasm** of **eukaryotic** cells, concerned with **aerobic respiration** and hence energy production from the reduction of **ATP** to ADP. Its shape varies from spherical to cylindrical. Large concentrations of mitochondria are observed in areas of high energy consumption, such as muscle tissue. Alternative name: chondriosome.

■ **mitosis** The usual type of cell division in which the parent nucleus splits into two identical daughter nuclei, which contain the same number of **chromosomes** and identical **genes** to that of the parent nucleus. Alternative name: karyokinesis. *See also* **meiosis**. [2/8/b]

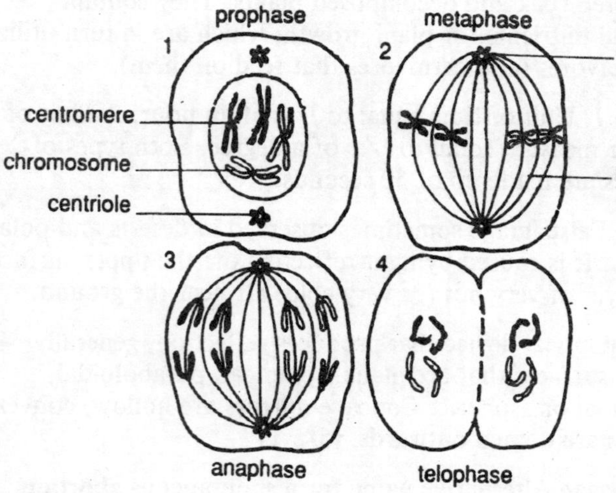

Stages of mitosis

■ **mixture** Combination of two or more substances that do not react chemically and can be separated by physical methods (*e.g.* a **solution**). [4/6/a]

mobility In electronics: *1.* Freedom of particles to move, either randomly, in a field or under the influence of forces. *2.* Average drift velocity of charge carriers (per unit electric field) in a **semiconductor**.

modem Acronym of modulator/demodulator, a device for transmitting computer **data** over long distances (*e.g.* by telephone line).

moderator Material used to slow down **neutrons** in a nuclear reactor (so that they can be captured and initiate **nuclear fission**); *e.g.* water, heavy water or graphite.

modulation In radio transmission, change of amplitude or frequency of a carrier wave by the signal being transmitted. *See* **amplitude modulation**; **frequency modulation**.

modulus of elasticity *See* **elastic modulus**.

moiré pattern **Interference** fringes formed when light passes through two or more fine gratings. A similar effect is created by the tiny dots in a printed (coloured) half-tone photograph when the screen angles are too close.

molality (m) Concentration of a solution given as the number of **moles** of solute in a kilogram of solvent.

molar Describing a quantity of a substance that is proportional to its molecular weight (a **mole**). *See* **molality**; **molarity**.

molar concentration *See* **molality**; **molarity**.

molar conductivity Electrical conductivity of an electrolyte with a concentration of 1 mole of solute per litre of solution. Expressed in siemens cm^2 $mole^{-1}$.

Moiré pattern formed from two sets of regular dots

molar heat capacity Heat required to increase the temperature of 1 mole of a substance by 1 kelvin. Expressed in joules K^{-1} mol^{-1}.

molarity (M) Concentration of a solution given as the number of **moles** of solute in a litre of solution.

molar solution Solution that contains 1 mole of solute in 1 litre of solution.

molar tooth One of the rearmost teeth of a mammal, used for crushing and grinding food. They are absent from the **milk teeth**, and in carnivores are replaced by **carnassial teeth**.

molar volume Volume occupied by 1 mole of a substance under specified conditions.

■ **mole** (mol) SI unit of amount of substance. In chemistry, it is the amount of a substance in grams that corresponds to its molecular weight, or the amount that contains particles equal in number to the **Avogadro constant**. Alternative names: mol, gram-molecule. [4/9/b]

molecular biology Study of biological **macromolecules** (*e.g.* nucleic acids, proteins).

molecular distillation Distillation at extremely low pressures. Alternative name: high-vacuum distillation.

■ **molecular formula** Method of describing the composition of a **molecule** of a chemical compound, using the chemical symbols of the constituent elements with numerical suffixes that indicate the number of atoms of each element in the molecule. *E.g.* H_2O and Na_2SO_4 are the molecular formulae of water and sodium sulphate, respectively. The molecular formula gives no indication how the component atoms are arranged. *See also* **empirical formula; structural formula**.

molecular orbital Region in space occupied by a pair of **electrons** that form a covalent bond in a **molecule**, formed by the overlap of two **atomic orbitals**.

molecular oxygen O_2 Normal **diatomic** molecular form of **oxygen**.

molecular sieve Method of separating substances by trapping (absorbing) the molecules of one within cavities of another, usually a natural or synthetic zeolite. Molecular sieves are used in ion exchange, desalination and as supports for catalysts.

molecular spectrum *See* **spectrum**.

■ **molecule** Group of atoms held together in fixed proportions by chemical bonds; the fundamental unit of a chemical

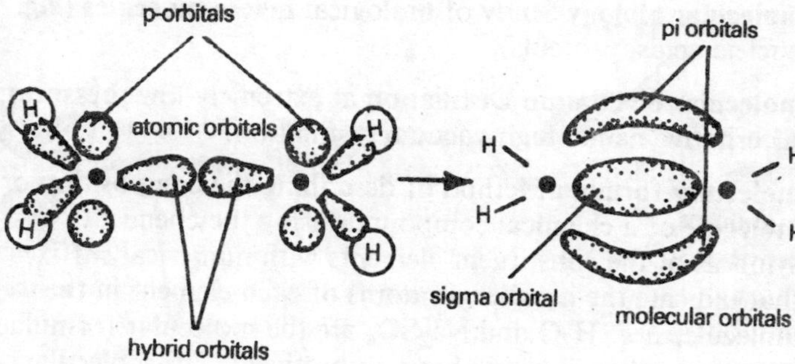

Formation of molecular orbitals

compound. The simplest molecules are diatomic molecules, consisting of two atoms (*e.g.* O_2, HCl); the most complex are biochemicals and macromolecules (such as cellulose, rubber, starch and synthetic plastics).

■ **mollusc** Member of the phylum **Mollusca**. [2/3/b] [2/4/b]

■ **Mollusca** Phylum of invertebrates, all of which have a well-defined body cavity. Molluscs include gastropods, cephalopods, and bivalves (lamellibranchs), such as snails, jellyfish and scallops. Most of the 80,000 species are aquatic, and manufacture a protective shell from dissolved calcium carbonate. [2/3/b] [2/4/b]

molybdenum Mo Metallic element in Group VIB of the

Periodic Table (a **transition element**), used in lamps, vacuum tubes and various alloys. At. no. 42; r.a.m. 95.94.

■ **moment of force** About an axis, the product of the perpendicular distance of the axis from the line of action of the force and the component of the force in the plane perpendicular to the axis. It has a turning effect (torque). *See also* **couple**. [5/7/b]

moment of inertia (I) Sum of the products of the mass of each particle of a body about an axis and the square of its perpendicular distance from the axis (its radius of gyration).

■ **momentum** (p) Product of the mass m and velocity v of a moving object; *i.e.* $p = mv$. It is a vector quantity directed through the centre of mass of the object in the direction of motion. When objects collide, the total momentum before impact is the same as the total momentum after impact. [5/10/a]

monatomic Describing a molecule that contains only one **atom** (*e.g.* the rare gases).

Monera In some classification schemes, the most primitive of the kingdoms of life, consisting of the **prokaryotes** – **cyanophytes** (blue-green algae) and **bacteria**.

■ **mongolism** Alternative name for **Down's syndrome**. [2/9/c]

mono- Prefix meaning one (*e.g.* monobasic, monoxide).

monobasic acid Acid that on solvation produces 1 mole of **hydroxonium ion** (H_3O^+) per mole of acid; an acid with one replaceable hydrogen atom in its molecule (*e.g.* hydrochloric acid, HCl, and nitric acid, HNO_3). It cannot therefore form **acid salts**.

monocarboxylic acid Carboxylic acid with only one **carboxylic group** *e.g.* acetic (ethanoic) acid, CH_3COOH).

monochromatic light Light of a single **wavelength**.

monochromator *See* **spectrometer**.

monoclinic Crystal form in which all three axes are unequal, with one of them perpendicular to the other two, which intersect at an angle inclined at other than a right angle.

■ **monoclonal** Derived from a single parent **clone**. [2/10/a]

■ **monoclonal antibody Antibody** produced by a single-cell clone, and hence consisting of a single **amino acid** sequence. Such cell clones are produced by the artificial fusion of cancerous and antibody-forming cells from the mouse spleen. The hybrid cells are grown in vitro as clones of cells, with each producing only a single type of antibody molecule.

■ **monocotyledon** Member of one of the two subdivisions of the flowering plants, in which the embryo has a single **cotyledon** (seed-leaf); class Liliopsida. *See also* **dicotyledon**. [2/3/b] [2/4/b]

monoculture Describing a form of **agriculture** in which only a single crop is grown continuously.

monocyte Largest phagocytic **leucocyte**, of order of about 10–12 micrometres. Monocytes have monogranulated cytoplasm with a large oval nucleus.

monoecious Describing plants in which separate female and male reproductive bodies are borne on the same individual or flower; *e.g.* maize.

■ **monohybrid** Offspring of parents that have different **alleles** for a particular gene; one parent has two **recessive** alleles and the other has two **dominant** ones. All offspring inherit one recessive and one dominant allele for the gene. [2/7/b]

monohybrid cross Cross between **monohybrids**, giving

offspring in the characteristic 3:1 ratio of those with two dominant **alleles** to those with two recessive alleles for a particular gene. [2/7/b]

monohydrate Chemical compound (a **hydrate**) that contains 1 mole of **water of crystallization** in each of its molecules.

monohydric Describing a chemical compound that has one **hydroxyl group** in each of its molecules (*e.g.* ethanol, C_2H_5OH, is a monohydric alcohol).

monomer Small molecule that can polymerize to form a larger molecule. *See* **polymer**.

monophyodont Describing an animal that has only one set of (irreplaceable) teeth during its whole life cycle.

monopodium Main axis of a plant stem that undergoes indefinite growth.

monosaccharide $C_nH_{2n}O_n$, where n = 5 or 6. Member of the simplest group of **carbohydrates**, which cannot be hydrolysed to any other smaller units; *e.g.* the sugars glucose, fructose.

■ **monosodium glutamate** (MSG) White crystalline solid, a sodium salt of the **amino acid** glutamic acid, made from soya bean protein and used as a flavour enhancer. Eating it can cause an allergic reaction in certain susceptible people.

Monotremata Order of egg-laying mammals in the subclass Prototheria found in Australia and New Guinea. There are only three species of monotremes: the duck-billed platypus and two spiny ant-eaters, or echidnas.

monotreme Member of the order **Monotremata**.

monovalent Having a **valence** of one. Alternative name: univalent.

montmorillonite Alternative name for **fuller's earth**.

mordant Inorganic compound used to fix colours on cloth where the cloth cannot be dyed directly.

■ **morphine** Sedative narcotic **alkaloid** drug isolated from opium, used for pain relief. Alternative name: morphia. [2/1/c]

morphogenesis Origin and development of form and structure of an organism or part of one.

morpholine C_4H_9O **Heterocyclic** secondary **amine**, used as a solvent.

morphology Study of the origin, development and structures of organisms. [2/4/b]

■ **mosaic** *1*. General term for a plant disease caused by viruses that results in patchy leaf coloration. *2*. Organism derived from a single embryo that displays the characteristics of different **genes** in different parts of its body. *See* **chimaera**. *3*. Ordered arrangement that maximizes a functional requirement (*e.g.* positioning of plant leaves that gives maximum exposure to sunlight with a minimum of mutual shading). [2/6/a] [2/10/b]

■ **Moseley's law** The X-ray spectrum of an element can be divided into several distinct line series: K, L, M and N. The law states that for certain elements the square root of the frequency f of the characteristic X-rays of one of these series is directly proportional to the element's **atomic number** Z. It was named after the British physicist Henry Moseley (1887–1915).

■ **moss** Plant that belongs to the class **Musci**. [2/3/b] [2/4/b]

Mössbauer effect Absorption of momentum of an atom by the whole crystal lattice because it is so firmly bound that it cannot recoil after its **nucleus** has emitted a gamma-ray **photon**. It was named after the German physicist Rudolf Mössbauer (1929–).

■ **motile** Describing an organism or structure that can move. [2/3/a] [2/4/a]

motor effect A conductor carrying a direct current (d.c.) in a magnetic field tends to move, which is the principle by which an **electric motor** works. *See* **left-hand rule**.

motor, electric *See* **electric motor**.

motor generator Motor supplied at one voltage frequency-coupled to a generator that provides a different voltage/frequency.

motor neurone Nerve cell that transmits impulses from the spinal cord or the brain to a muscle. Alternative names: motor neuron, motor nerve.

mould Fungal growth usually consisting of a mass of **hyphae**, especially on rotting food.

■ **moult(ing)** *1.* Periodic shedding of hair or feathers by animals. *2.* Widely used term for ecdysis, the process by which an immature insect sheds its **exoskeleton** in order that it may develop and grow in size.

movement Change of position of part of the body, a characteristic of living organisms. *See also* **locomotion**.

moving-iron microphone *See* **microphone**.

mucilage Group of gum-like compounds that produce a slimy solution. Most are highly branched flexible molecules, which may also form part of the cell-wall matrix in plants.

■ **mucin** Any of a number of **glycoproteins** that occur in **mucus**. [2/4/b]

■ **mucous membrane** Moist, **mucus**-lined **epithelium** which itself lines vertebrate internal cavities, including the alimentary, respiratory and reproductive tracts, which are continuous with the outer environment. [2/4/a]

mucus Slimy substance secreted by the **goblet cells** of **mucous membrane**. It lubricates and protects the epithelial layer on which it is secreted.

mule Sterile **hybrid** animal born of a mare and a male donkey.

multicellular Describing plants and animals that have bodies consisting of many cells.

■ **multifactorial inheritance** Existence of more than two **alleles** for one **gene**; *e.g.* as in A, B, O blood grouping. [2/9/c]

multimeter Instrument that can be used as a galvanometer, ammeter and voltmeter.

multiple bond **Chemical bond** that contains more electrons than a single bond (which contains 2 electrons); *e.g.* a double bond (4 electrons) or a triple bond (6 electrons).

■ **multiple proportions, law of** If two elements A and B can combine to form more than one compound, then the different masses of A that combine with a fixed mass of B are in a simple ratio. *E.g.* in carbon monoxide, CO, 16 atomic mass units of oxygen are combined with 12 units of carbon; in carbon dioxide, CO_2, 32 units of oxygen are combined with 12 units of carbon. The ratio of the oxygen masses in the two compounds is 32:16 or 2:1.

multiplication constant In a nuclear reactor, the ratio of the total number of **neutrons** produced by fission in a given time to the number absorbed or escaping in the same period. Alternative name: multiplication factor.

multiplier One of a set of resistors that can be used in series with a **voltmeter** to vary its range.

muon Subatomic particle, a type of **lepton**, that participates in weak interactions.

Musci Class of **spore**-producing plants in the division **Bryophyta**, comprising the mosses. Alternative name: Bryopsida.

■ **muscle** Animal tissue that contracts (by means of muscle fibres) to produce movement, tension and mechanical energy. *See* **involuntary muscle; voluntary muscle.** [2/3/a]

mustard gas $(CH_2ClCH_2)_2S$ Poisonous blistering gas used as a chemical warfare agent. Alternative name: 2,2′-dichlorodiethyl sulphide.

■ **mutagen** Chemical or physical agent that induces or increases the rate of **mutation**; *e.g.* ethyl methanesulphonate, ultraviolet light, X-rays and gamma-rays. [2/9/c]

■ **mutant** Organism that arises by **mutation**. [2/9/c]

mutarotation Change in the **optical activity** of a solution containing photo-active substances, such as sugars.

■ **mutation** Alteration in the sequence of **bases** encoded by **DNA**, resulting in a permanent inheritable change in the **gene** and consequently the **protein** encoded. Mutations may occur in different ways, and may be induced by a **mutagen** or occur spontaneously. A mutation can be detrimental, *e.g.* those thought to be involved in carcinogenesis (formation of cancer). However, some mutations can be advantageous, *e.g.* in **evolution**, where favourable characteristics may be passed on to offspring. [2/9/c]

■ **mutualism** Relationship between two organisms from which each benefits (*e.g.* cellulose-digesting micro-organisms and the animals, such as ruminants, whose gut they inhabit). *See also* **commensalism; symbiosis.**

mycelium Mass of **hyphae** that form the body of a **fungus.**

■ **mycology** Scientific study of **fungi**. [2/4/b]

Mycota Division (phylum) that includes the **fungi**, usually included in the plant kingdom but sometimes accorded a kingdom of its own.

mycotoxin Toxin produced by a **fungus**.

myelin sheath Thin fatty layer of membranes, produced by **Schwann cells**, that covers the **axon** of most vertebrate **neurones** (nerve cells).

myeloid tissue Tissue usually present in **bone marrow** which produces **red blood cells** and other **blood** constituents.

myocardial Relating to **myocardium**.

myocardium Muscle tissue of the vertebrate heart.

myogenic Describing muscular contraction which is not caused by nervous stimulus. *See* **cardiac muscle**.

myoglobin In vertebrate **muscle** fibre, a **haem** protein capable of binding with one atom of oxygen per molecule. It is abundant in the muscles of diving mammals (*e.g.* seals, whales), in which it acts as an oxygen store.

myology Study of muscles.

myopia Short-sightedness, a visual defect in which the eyeball is too long (front to back) so that rays of light entering the **eye** from distant objects are brought to a focus in front of the retina. It can be corrected with spectacles or contact lenses made from diverging (concave) lenses. *See also* **hypermetropia**.

myosin Fibrous **protein** which, with **actin**, makes up muscle. Movement of myosin fibres between actin fibres causes muscle contraction.

■ **myriapod** Member of the animal class **Myriapoda**. As a general term, myriapod is also taken to include centipedes (class Chilopoda). [2/4/b]

Myriapoda Class of terrestrial arthropods characterized by a distinct head bearing antennae, mandibles and maxillae, and segmented bodies with many pairs of walking legs; the millipedes.

myxoedema Disorder caused by lack of hormones from the **thyroid gland**; if present at birth it causes cretinism.

■ **myxomatosis** Disease of rabbits caused by a virus, which has been deliberately introduced in some regions as a form of pest control. [2/4/d] [2/10/e]

N

nacre Mother of pearl, the inner layer of the shell of a mollusc. It is an iridescent substance, composed mainly of **calcium carbonate**.

NAD (nicotinamide adenine dinucleotide) **Coenzyme** form of the vitamin **nicotinic acid**, necessary in certain enzyme-catalysed oxidation-reduction reactions in cells. Its reduced form is a precursor in the fixation of carbon dioxide in chloroplasts during **photosynthesis**.

nail Layer of **keratin** that grows on the upper surface of the fingers of human beings and other primates (except tree-shrews, which have claws).

nano- Prefix meaning a thousand-millionth; $\times 10^{-9}$. *E.g.* 1 nanosecond is 10^{-9} s.

nanometre Thousand-millionth of a metre; 10^{-9} m. It is the usual unit for wavelengths of light and interatomic bond lengths in chemistry.

naphtha Variable mixture of **hydrocarbons**, boiling range approximately 70–160 °C, obtained from coal and petroleum and used as an industrial solvent and a raw material (feedstock) for the chemical industry. Alternative name: solvent naphtha.

naphthalene $C_{10}H_8$ Solid aromatic **hydrocarbon** that consists of two fused **benzene rings**, insoluble in water but soluble in hot ethanol. It is a starting material in the manufacture of dyes.

narcotic **Analgesic** drug that, in addition to killing pain, causes loss of sensation or loss of consciousness (*e.g.* morphine and other opiates).

nasal To do with the nose or sense of smell.

nasal cavity Cavity located in the head of tetrapods, containing the olfactory sense organs.

nastic movement Response by plants to stimuli that do not come from any one direction, *e.g.* temperature, humidity. *See also* **tropism**.

natron Naturally occurring hydrated **sodium carbonate**, found on the bed of dried-out soda-lakes.

natural abundance Relative proportion of one **isotope** to the total of the various isotopes in a naturally occurring sample of an element.

■ **natural frequency** When a vibrating system is displaced from its neutral position it oscillates about that position with a natural frequency characteristic of the system. *See also* **resonance**. [5/8/c]

natural gas Mixture of hydrocarbon fuel gases, rich in **methane**, obtained from deposits that occur naturally underground. It sometimes contains **helium**.

■ **natural selection** One of the conclusions drawn by the British naturalist Charles Darwin (1809–82) from the theory of evolution: certain organisms with particular characteristics are more likely to survive and hence pass on their characteristics to their offspring, *i.e.* survival of the fittest. Thus the characteristics of a population are controlled by this process. [2/6/b] [2/9/c]

nautical mile Distance used at sea. In Britain 1 nautical mile = 6,080 feet; the international definition is 1,852 metres.

Neanderthal man Extinct hominid from the Pleistocene epoch, now usually regarded as a subspecies of *Homo* *sapiens*. It probably possessed the same size of brain as modern man but with a structure that was different.

■ **near infra-red** or **ultraviolet** Parts of the infra-red or ultraviolet regions of the **electromagnetic spectrum** that are close to the visible region. [5/9/a,c]

nearsightedness Alternative name for **myopia**.

■ **nectar** Sticky sweet liquid produced by flowers which attracts insects, small birds and even certain bats (which thus pollinate the flower). It contains up to 80% **sugar** and is used by bees to make honey.

negative feedback *See* **feedback**.

nekton Aquatic organism that actively swims, as opposed to floating passively; *e.g.* fish, jellyfish and aquatic mammals. *See also* **plankton**.

nematocyst Small retractable tentacle, possessed by many species of coelenterates, that is used for defence or to catch prey. The hollow tentacle is inflated by a sac of fluid, and may be used to inject venom.

■ **Nematoda** Phylum of invertebrate animals that contains round, thread and eel worms. They have unsegmented bodies that taper at each end. Many nematodes are parasites (*e.g.* filaria). [2/4/b]

■ **nematode** Worm of the phylum **Nematoda**. [2/4/b]

■ **neo-Darwinism** Modern version of the Darwinian theory of **evolution**, expanded to take into account modern knowledge of genetics. *See* **Darwinism**. [2/9/c]

neodymium Nd Metallic element in Group IIIB of the Periodic Table (one of the **lanthanides**), used in special glass for lasers. At. no. 60; r.a.m. 144.24.

neo-Lamarckism *See* **Lamarckism**.

■ **neon** Ne Gaseous nonmetallic element in Group 0 of the Periodic Table (the **rare gases**) which occurs in trace quantities in air (from which it is extracted). It is used in discharge tubes (for advertising signs) and indicator lamps. At. no. 10; r.a.m. 20.179. [4/8/a] [4/9/a]

neonatal Concerning the newborn.

neon tube Gas-discharge tube that contains neon at low pressure, the colour of the glow being red.

■ **neoplasm** Tumour or group of cells with uncontrolled growth. It may be benign and localized, or if cells move from their normal position in the body and invade other organs the tumour is malignant. *See also* **cancer; metastasis**. [2/7/a] [2/9/c]

Neoprene Commercial name for the synthetic rubber polychloroprene, produced by the **polymerization** of chlorobutadiene.

neotenin Insect **hormone** that suppresses the onset of adult characteristics until the final moult.

■ **neoteny** Presence of the larval or early stage of development in adulthood. It is important in the evolution of some animal groups.

■ **nephridium** Excretory organ possessed by many invertebrates, consisting of a tube leading from the **coelom** to the external body surface. [2/4/a] [2/5/a] [2/7/a]

■ **nephritis** Inflammation of the **kidney**. [2/7/a]

nephron Functional filtering unit of the vertebrate **kidney**, consisting of **Bowman's capsule** and the **glomerulus**.

neptunium Np Radioactive element in group IIIB of the Periodic Table (one of the **actinides**); it has several **isotopes**. At. no. 93; r.a.m. 237 (most stable isotope).

Nernst effect When heat flows through a strip of metal in a magnetic field, the direction of heat flow being across the lines of force, an **e.m.f.** is developed perpendicular to both the flow and the lines. It was named after the German physical chemist Walther Nernst (1864–1941).

Nernst heat theorem If a chemical change occurs between pure crystalline solids at a temperature of absolute zero, the entropy of the final substance equals that of the initial substances.

■ **nerve** *Structure that carries nervous impulses to and from the* central nervous system, consisting of a bundle of **nerve fibres**, and often associated with blood vessels and connective tissue. *See also* **neurone**.

nerve cell Alternative name for a **neurone**.

nerve cord Cord of nervous tissue in invertebrates that forms part of their **central nervous system**.

■ **nerve fibre** Extension of a nerve cell. Nerve fibres may be surrounded by a **myelin sheath** (except at the **nodes of Ranvier**), as in many vertebrates; or they may be unmyelinated and bound by a **plasma membrane**.

nerve gas Chemical warfare gas that acts on the nervous system, *e.g.* by inhibition or destruction of chemicals (neurotransmitters) that convey nerve impulses across synapses between nerves.

nerve impulse Electrical signal conveyed by a **nerve** to carry information throughout the **nervous system**. External stimuli trigger nerve impulses in **receptor** cells, and travel along

afferents towards the central nervous system. Impulses generated within the **central nervous system** travel along **efferents** towards organs and tissues. Intermediate neurones connect sensory afferent neurones to motor efferent neurones and to the brain.

nervous system System that provides a rapid means of communication within an organism, enabling it to be aware of its surroundings and to react accordingly. In most animals it consists of a **central nervous system** (CNS) that integrates the sensory input from peripheral nerves which transmit stimuli from receptors (afferents) to the CNS, allowing the appropriate response from the effectors.

network System of interconnected points and their connections; *e.g.* a grid of electricity supply lines or a set of interconnected terminals on-line to one or more computers.

neural Relating to **nerves** or the **nervous system**.

neuroglia Connective tissue between nerve cells (**neurones**) of the brain and spinal cord.

■ **neurone** Basic cell of the **nervous system** which transmits **nerve impulses**. Each cell body typically possesses a nucleus and fine processes: short **dendrites**, an **axon** and a **dendron**. The axon carries impulses to distant effector cells or other neurones, depending on whether it is a sensory or motor neurone. Neurones also make functional contacts over the surface of shorter, thread-like projections from the cell body (dendrites). Alternative names: neuron, nerve cell.

neurotransmitter Chemical released by **neurone** endings to either induce or inhibit transmission of nerve impulses across a **synapse**. Neurotransmitters are typically stored in small vesicles near the synapse and released in response to arrival of an impulse. There are more than 100 different types (*e.g.* acetylcholine, noradrenalin). Alternative name: transmitter.

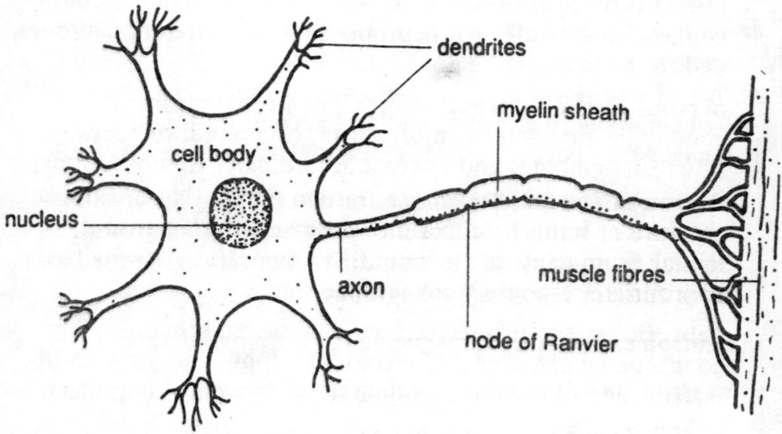

Structure of a neurone

neutral *1*. Describing a solution with **pH** equal to 7 (*i.e.* neither acidic nor alkaline). *2*. Describing a subatomic particle, atom or molecule with no residual electric charge *e.g.* a **neutron** is a neutral subatomic particle.

neutral oxide Oxide that is neither an **acidic oxide** nor a **basic oxide**; *e.g.* dinitrogen oxide (N_2O), water (H_2O).

■ **neutralization** Chemical reaction between an **acid** and a **base** in which both are used up; the products of the reaction are a **salt** and water; *e.g.* the reaction between hydrochloric acid (HCl) and sodium hydroxide (NaOH) to form sodium chloride (NaCl) and water (H_2O). The completion of the reaction (end-point) can be detected by an **indicator**.

neutrino Uncharged subatomic particle with zero rest mass, a type of **lepton**, that interacts very weakly with other particles.

■ **neutron** Uncharged particle that is a constituent of the atomic **nucleus,** having a rest mass of 1.67492×10^{-27} kg (similar to that of a **proton**). Free neutrons are unstable and disintegrate by **beta decay** to a proton and an **electron**; outside the nucleus they have a mean life of about 12 minutes. [4/8/d]

neutron diffraction Technique for determining the crystal structure of solids by diffraction of a beam of **neutrons**. Similar in principle to electron diffraction, it can be used as a substitute for **X-ray crystallography**.

neutron excess Alternative name for **isotopic number**.

neutron flux Product of the number of free neutrons per unit volume and their mean speed. Alternative name: neutron flux density.

■ **neutron number** Number of neutrons in an atomic nucleus, the difference between the **nucleon number** of an element and its **atomic number**. [4/8/d]

neutrophil Type of **leucocyte** (white blood cell), important in the immune system because it ingests bacteria, whose protoplasm can be stained by neutral dyes.

Newlands' law Alternative name for the law of **octaves**.

■ **newton** (N) SI unit of **force,** defined as the force that provides a mass of 1 kg with an **acceleration** of 1 m s^{-2}. It was named after the British mathematician and physicist Isaac Newton (1642–1727).

Newtonian fluid Fluid in which the amount of strain is proportional both to the stress and to the time. The constant of proportionality is known as the coefficient of viscosity.

Newtonian mechanics System of mechanics that relies on **Newton's laws of motion** and is applicable to objects moving at speeds relative to the observer that are small compared to the speed of light. Objects moving near to the speed of light require an approach based on relativistic mechanics (*see* **relativity**), in which the mass of the object changes with its speed.

newton meter Instrument for measuring a **force** in newtons (*e.g.* spring balance).

Newton's formula For a lens, the distances p and q between two conjugate points and their respective foci (f) are related by $pq = f^2$.

Newton's law of cooling Rate of loss of heat from an object is proportional to the excess temperature of the object over the temperature of its surroundings.

■ **Newton's law of gravitation** Force F of attraction between two objects of masses m_1, m_2 separated by a distance x is given by $F = Gm_1m_2/x^2$; G is the gravitational constant and has a value of 6.6732×10^{-11} Nm^2 kg^{-2}. Alternative name: law of universal gravitation.

■ **Newton's laws of motion** Three laws of motion on which **Newtonian mechanics** is based. *1.* An object continues in a state of rest or uniform motion in a straight line unless it is acted upon by external forces. *2.* Rate of change of momentum of a moving object is proportional to and in the same direction as the force acting on it. *3.* If one object exerts a force on another, there is an equal and opposite force, called a reaction, exerted on the first object by the second. Alternative name: Newton's laws of force.

Newton's rings Circular **interference** fringes formed in a thin gap between two reflective media, *e.g.* between a lens and a glass plate with which the lens is in contact. There is a central dark spot around which there are concentric dark rings.

niacin Vitamin B_3, the only one of the B vitamins that is synthesized by animal tissues. Deficiency causes the disease pellagra. Alternative name: nicotinic acid.

niche Status or way of life of an organism (or group of organisms) within an environment, which it cannot share indefinitely with another competing organism. Alternative name: ecological niche.

■ **nickel** Ni Siver-yellow metallic element in Group VIII of the Periodic Table (a **transition element**). It is used in thermionic valves, electroplating and as a **catalyst**, and its alloys (*e.g.* stainless steel, cupronickel, German silver, nickel silver) are used in making cutlery, hollow-ware and coinage. At. no. 28; r.a.m. 58.71.

nickel carbonyl $Ni(CO)_4$ **Co-ordination compound** in which the oxidation state of nickel is zero, used as a starting material in the synthesis of a wide variety of nickel compounds.

nickel-iron accumulator Rechargable **electrolytic cell** (battery) with a positive elecrode of nickel oxide and a negative electrode of iron, in a potassium hydroxide electrolyte. Alternative names: Edison accumulator, NiFe cell.

nickel plating Process in which a thin coating of nickel is electroplated on another metal (*see* **electroplating**).

nickel silver Alloy of nickel and copper, used for coinage and for making cutlery and hollow-ware (which is often silver plated; the abbreviation EPNS stands for electroplated nickel silver). Alternative name: German silver.

nicotine Poisonous **alkaloid**, found in tobacco, which potentially binds to the receptor for the neurotransmitter acetylcholine. It is used as an insecticide.

nicotinic acid Alternative name for **niacin**.

nictitating membrane Protective transparent lid that can be drawn across the eye of many birds and reptiles, and some amphibians and sharks, and which is possessed by a few aquatic mammals, *e.g.* seals.

NiFe cell Alternative name for **nickel-iron accumulator**.

niobium Nb Metallic element in Group VB of the Periodic Table (a **transition element**). Its alloys are used in high-temperature applications and superconductors. At. no. 41; r.a.m. 92.906.

■ **nitrate** Salt of **nitric acid**, containing the NO_3^- anion. Nitrates are commonly used as fertilizers, but their misuse can give rise to pollution of water supplies.

■ **nitration** Chemical reaction in which a nitro group ($-NO_2$) is incorporated into a chemical structure, to make a **nitro compound**; *e.g.* nitration of benzene (C_6H_6) gives nitrobenzene ($C_6H_5NO_2$). Nitration of organic compounds is usually achieved using a mixture of concentrated nitric and sulphuric acids (known as nitrating mixture).

nitre Old term for potassium nitrate, KNO_3, also commonly known as saltpetre.

■ **nitric acid** HNO_3 Strong extremely corrosive **mineral acid**. It is manufactured commercially by the catalytic oxidation of **ammonia** to **nitrogen monoxide** (nitric oxide) and dissolving the latter in water. Its salts are **nitrates**. The main use of the acid is in making explosives and fertilizers.

nitric oxide Alternative name for **nitrogen monoxide**.

nitrification Conversion of **ammonia** and **nitrites** to **nitrates** by the action of nitrifying bacteria. It is one of the important parts of the **nitrogen cycle**, because nitrogen cannot be taken up directly by plants except as nitrates.

nitrile Member of a group of organic compounds that contain the nitrile group (–CN). Alternative name: cyanide.

nitrite Salt of **nitrous acid**, containing the NO_2^- anion. Nitrites are used to preserve meat and meat products.

nitrobenzene $C_6H_5NO_2$ Aromatic liquid organic compound in which one of the hydrogen atoms in **benzene** has been replaced by a nitro group ($-NO_2$). It is used to make aniline and dyes.

nitrocellulose Alternative name for **cellulose trinitrate**.

nitro compound Organic compound in which a nitro group ($-NO_2$) is present in the basic molecular structure. Usually made by **nitration**, some nitro compounds are commercial explosives.

■ **nitrogen** N Gaseous nonmetallic element in Group VA of the Periodic Table. It makes up about 80%of air by volume, and occurs in various minerals (particularly **nitrates**) and all living organisms. It is used as an inert filler in electrical devices and cables, and is an essential plant nutrient (*see* **fertilizer**). At. no. 7; r.a.m. 14.0067.

■ **nitrogen cycle** Circulation of nitrogen and its compounds in the environment. The main reservoirs of nitrogen are **nitrates** in the soil and the gas itself in the atmosphere (formed from nitrates by **denitrification**). Nitrates are also taken up by plants, which are eaten by animals, and after their death the nitrogen-containing **proteins** in plants and animals form **ammonia**, which **nitrification** converts back into nitrates.

Some atmospheric nitrogen undergoes fixation by lightning or bacterial action, again leading to the eventual formation of nitrates (*see* **fixation of nitrogen**). [2/5/c] [2/8/d]

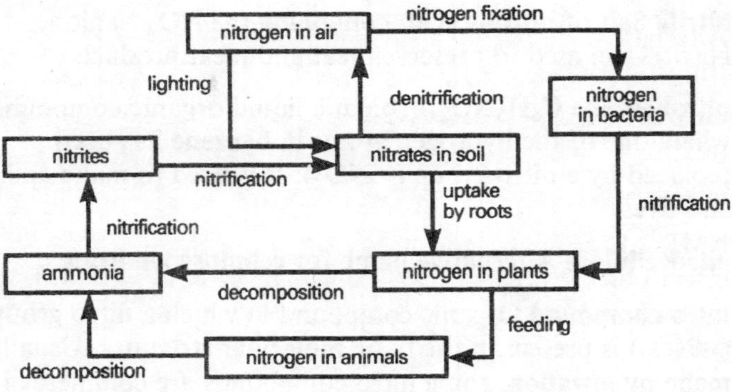

Nitrogen cycle

nitrogen dioxide NO_2 Choking brown gas made by the action of concentrated **nitric acid** on copper or of oxygen on **nitrogen monoxide** (NO). It exists in equilibrium with its dimer, dinitrogen tetroxide (N_2O_4). It is used as an **oxidizing agent** and in the manufacture of nitric acid.

nitrogen fixation *See* **fixation of nitrogen**.

nitrogen monoxide NO Colourless gas made commercially by the catalytic **oxidation** of **ammonia** and used for making **nitric acid**. It reacts with oxygen (*e.g.* in air) to form **nitrogen dioxide**. Alternative name: nitric oxide.

nitrogen oxides Compounds containing nitrogen and oxygen in various ratios, including N_2O, NO, N_2O_3, NO_2, N_2O_4, N_2O_3, N_2O_5, NO_3 and N_2O_6. The most important are dinitrogen oxide, nitrogen dioxide and nitrogen monoxide.

nitroglycerine $C_3H_5(ONO_2)_3$ Explosive oily liquid, which freezes at about 11 °C. When solid, the crystals may explode at the slightest physical shock. It is used to make dynamite. Alternative names: nitroglycerin, glyceryl trinitrate, trinitroglycerine.

nitrous acid HNO_2 Weak unstable mineral acid, made by treating a solution of one of its salts (**nitrites**) with an acid.

nitrous oxide Alternative name for **dinitrogen oxide**.

NMR Abbreviation of Nuclear Magnetic Resonance, an effect observed when radio-frequency radiation is absorbed by matter. NMR spectroscopy is used in chemistry for the study of molecular structure. It has also been introduced as a technique of diagnostic medicine.

nobelium No Radioactive element in Group IIIB of the Periodic Table (one of the **actinides**); it has various **isotopes**. At. no. 102; r.a.m. 255 (most stable isotope).

■ **noble gas** Any of the elements in group O of the Periodic Table: **helium, neon, argon, krypton, xenon** and **radon**. They have a complete set of outer electrons, which gives them great chemical stability (very few noble gas compounds are known); radon is radioactive. Alternative names: inert gas, rare gas. [4/8/a] [4/9/a]

noble metal Highly unreactive metal, *e.g.* gold and platinum.

■ **node** *1.* In plants, point of leaf insertion on the shoot axis. *2.* In animals, thickening or junction of an anatomical structure, *e.g.* lymph node (gland), sinoatrial node, node of Ranvier.

3. In physics, a stationary point (*i.e.* point with zero amplitude) on a **standing wave**.

node of Ranvier One of several regular constrictions along the myelin sheath of a **nerve fibre**. It was named after the French histologist Louis-Antoine Ranvier (1835–1922).

■ **nodule** Lump that occurs on the roots of certain plants (*e.g.* peas, beans and other legumes) which contains bacteria that are able to bring about the **fixation of nitrogen**, an important part of the **nitrogen cycle**. [2/5/c]

noise Sound (or other radiation) consisting of a mixture of random and unrelated frequencies. If all the frequencies in a range are represented, it is called white noise.

non-benzenoid aromatic Aromatic compound in which the number of carbon atoms in the aromatic ring is not equal to six.

non-metal Substance that does not have the properties of a **metal**. Non-metallic **elements** are usually gases (*e.g.* nitrogen, halogens, noble gases) or low-melting point solids (*e.g.* phosphorus, sulphur, iodine). They have poor electrical and thermal **conductivity**.

non-Newtonian fluid Fluid that consists of two or more phases at the same time. The coefficient of viscosity is not a constant but is a function of the rate at which the fluid is sheared as well as of the relative concentration of the phases.

nonstoichiometric compound Chemical whose molecules do not contain small whole numbers of atoms. *See also* **stoichiometric compound**.

noradrenalin Hormone secreted by the medulla of the **adrenal glands** for the regulation of the cardiac muscle, glandular tissue and smooth muscles. It is also a **neurotransmitter** in the sympathetic nervous system, where it acts as a powerful vasoconstrictor on the vascular smooth muscles. In the brain, levels of noradrenalin are related to normal mental function, *e.g.* lowered levels lead to mental depression. Alternative name: norepinephrine.

norepinephrine Alternative name for **noradrenalin**.

normal *1.* In chemistry, describing a solution that contains 1 gram-equivalent of solute in 1 litre of solution. It is denoted by the symbol N and its multiples (thus 3N is a concentration of 3 times normal; N/10 or decinormal is a concentration of one-tenth normal). *2.* In physics, line at right-angles to a surface (*e.g.* a mirror or block of glass), from which angles of incidence, reflection and refraction are measured.

normal body temperature Average temperature of the healthy human body, about 37 °C.

notochord Skeletal rod that lies dorsally beneath the **neural tube** in certain stages of development of an **embryo** of a **chordate**. In some chordates it persists in the adult. In **vertebrates** it is replaced by the **spinal column**.

NTP Abbreviation of normal temperature and pressure. *See also* **standard temperature and pressure** (STP).

***n*-type conductivity** Electrical conductivity caused by the flow of **electrons** in a **semiconductor**.

■ **nucellus** Mass of thin-walled cells in the centre of an **ovule** of a plant, containing the megaspore or egg cell. The simple cell of the nucellus becomes the megasporocyte, while the rest acts as a nutritive tissue for the developing megaspore. In some plants it may be retained to form the **endosperm** for the developing embryo, *e.g.* in maize.

nuclear barrier Region of high potential energy that a charged particle must pass through in order to enter or leave an atomic nucleus.

nuclear division *See* **meiosis; mitosis.**

■ **nuclear energy** Energy released during a **nuclear fission** or **nuclear fusion** process. [3/7/a]

nuclear envelope Double membrane that surrounds the **nucleus** of **eukaryotic** cells.

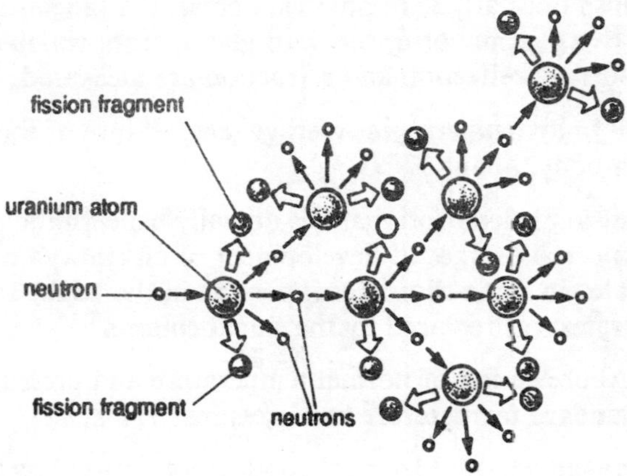

Nuclear fission splits heavy atoms

■ **nuclear fission** Splitting of an atomic nucleus into two or more fragments of comparable size, usually as the result of the capture of a slow, or thermal, **neutron** by the nucleus. It is normally accompanied by the emission of further neutrons or **gamma-rays**, and large amounts of energy. The neutrons can continue the process as a **chain reaction**, so that it becomes the source of energy in a nuclear reactor or an atomic bomb. It may also be the 'trigger' for **nuclear fusion** in a hydrogen bomb. [4/8/d]

nuclear force Strong force that operates during interactions between certain subatomic particles. It holds together the **protons** and **neutrons** in an atomic **nucleus.**

■ **nuclear fusion** Reaction between light atomic nuclei in which a heavier nucleus is formed with the release of large amounts of energy. This process is the basis of the production of energy in stars and in the hydrogen bomb,

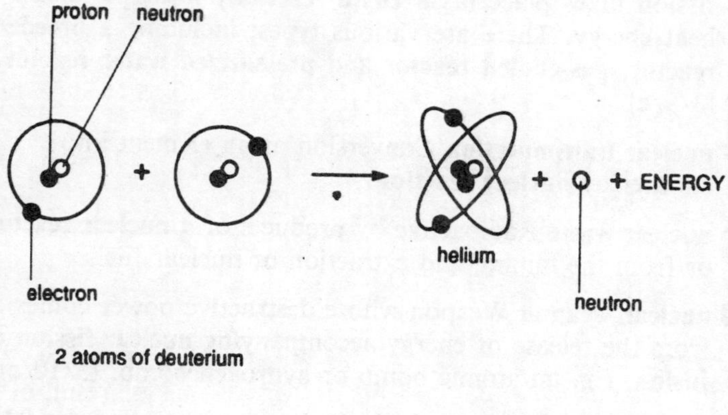

Nuclear fusion combines light atoms

which makes use of the fusion of **isotopes** of hydrogen to form helium. [4/8/d]

nuclear isomerism Property exhibited by nuclei with the same **mass number** and **atomic number** but different radioactive properties.

nuclear magnetic resonance (NMR) *See* **NMR**.

nuclear membrane Membrane that encloses the **nucleus** of a cell.

nuclear power Power obtained by the conversion of heat from a **nuclear reactor**, usually into electrical energy.

■ **nuclear reaction** Reaction that occurs between an atomic **nucleus** and a bombarding particle or photon, leading to the formation of a new nucleus and the possible ejection of one or more particles with the release of energy.

■ **nuclear reactor** Assembly in which controlled **nuclear fission** takes place (as a **chain reaction**) with the release of heat energy. There are various types, including a **breeder reactor**, **gas-cooled reactor** and **pressurized water reactor**. [3/9/c]

nuclear transmutation Conversion of an element into another by **nuclear reaction**.

nuclear waste Radioactive by-products of a **nuclear reactor** or from the mining and extraction of nuclear fuels.

■ **nuclear weapon** Weapon whose destructive power comes from the release of energy accompanying **nuclear fission** or **fusion**; *e.g.* an atomic bomb or hydrogen bomb. [5/10/a]

nuclease Type of **enzyme** that splits the 'chain' of the **DNA** molecule. Nucleases that act at specific sites are called **restriction enzymes**.

■ **nucleic acid** Complex organic **acid** of high molecular weight consisting of chains of **nucleotides**. Nucleic acids commonly occur, conjugated with **proteins**, as **nucleoproteins**, and are found in cell nuclei and **protoplasm**. They are responsible for storing and transferring the **genetic code**. *See* **DNA** (deoxyribonucleic acid); **RNA** (ribonucleic acid).

nucleoid In a **prokaryotic** cell, the DNA-containing region, similar to the nucleus of a **eukaryotic** cell but not bounded by a membrane.

nucleolus Spherical body that occurs within nearly all nuclei of **eukaryotic** cells. It is associated with **ribosome** synthesis and is thus abundant in cells that make large quantities of **protein**. It contains protein DNA and much of the nuclear RNA.

■ **nucleon** Comparatively massive particle in an atomic **nucleus**; a **proton** or **neutron**. [4/8/d]

nucleonics Practical applications of nuclear science and the techniques associated with these applications.

■ **nucleon number** Total number of **neutrons** and **protons** in an atomic nucleus. *See* **mass number**. [4/8/d]

nucleophile Electron-rich chemical reactant that is attracted by **electron-deficient compounds**. Examples include an anion such as chloride (Cl^-) or a compound with a **lone pair of electrons** such as ammonia (NH_3). *See also* **electrophile**.

nucleophilic addition Chemical reaction in which a **nucleophile** adds onto an **electrophile**.

nucleophilic reagent Chemical reactant that contains electron-rich groups of atoms.

nucleophilic substitution **Substitution** reaction that involves a **nucleophile**.

nucleoprotein Compound that is a combination of a **nucleic acid** (DNA, RNA) and a **protein**. *E.g.* in **eukaryotic** cells DNA is associated with histones and protamines; RNA in the cytoplasm is associated with protein in the form of the **ribosomes**.

nucleoside Compound formed by partial hydrolysis of a **nucleotide**. It consists of a base, such as purine or pyrimidine, linked to a sugar, such as ribose or deoxyribose; *e.g.* adenosine, cytidine and uridine.

nucleotide Compound that consists of a sugar (ribose or deoxyribose), base (purine, pyrimidine or pyridine) and phosphoric acid. Nucleotides are the basic units from which **nucleic acids** are formed.

■ **nucleus** *1.* In biology, the largest cell organelle (about 20 micrometres in diameter), found in nearly all **eukaryotic** cells. It is spherical to oval, containing the genetic material **DNA**, and hence controlling all cell activities. It is surrounded by a double membrane that forms the nuclear envelope. A nucleus is absent from mature mammalian **erythrocytes** (red blood cells) and the mature **sieve-tube** elements of plants. *2.* In physics, the most massive, central part of the atom of an element, having a positive charge given by Ze, where Z is the **atomic number** of the element and e the charge on an **electron**. It is composed of chiefly **protons** and (except for hydrogen) **neutrons**, and is surrounded by orbiting **electrons**. *See also* **isotope**. [4/8/d]

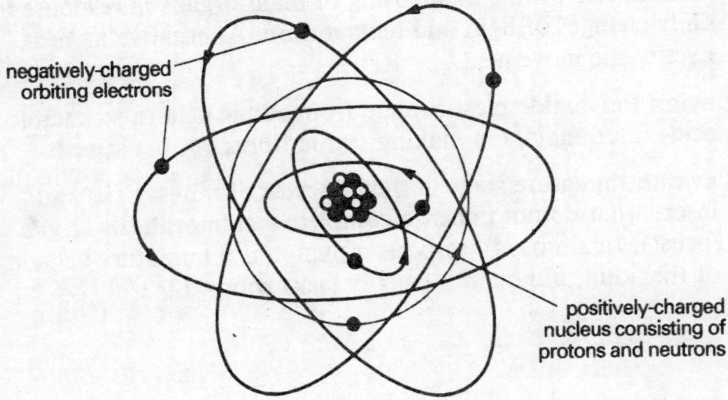

Nucleus is the centre of an atom

nuclide Atomic nucleus, defined by its **atomic number** and **neutron number**.

null method Any measuring system that establishes an unknown value from other known values when a particular instrument registers zero, *e.g.* a **potentiometer**. Alternative name: zero method.

nutation *1*. In botany, the circular swaying movements made by the growing parts of a plant, *e.g.* a shoot. *2*. In physics, a periodic variation in the axis of a spinning object, *e.g.* a **gyroscope**.

■ **nutrient** Useful component of food, such as fats, carbohydrates, proteins, vitamins and mineral salts. *See* **diet**.

■ **nutrition** Feeding. The taking in of food is a characteristic of all living organisms, both plants and animals.

nyctinasty Opening and closing of plant organs in response to daily changes of light and temperature. Alternative name: nyctinastic movement.

nylon Polyamide plastic made from **adipic acid** (hexanedioic acid), used mainly in making textile fibres.

■ **nymph** Immature stage in the life-cycle of mites, ticks and insects that do not undergo complete **metamorphosis** (*e.g.* a locust). In almost all respects a nymph is a miniature version of the adult, although it usually lacks wings. [2/2/c] [2/6/a]

O

object With a mirror, lens or optical instrument, the source of light rays that form an **image**.

objective Lens of an optical system (*e.g.* microscope, telescope) that is nearest the object.

occlusion In biology, closure of an opening (*e.g.* the way an animal's teeth meet when the mouth closes).

ocellus Simple **eye**, possessed by many invertebrates, consisting of a cluster of receptors that are sensitive to changes in the intensity or colour of light.

ochre Mineral of clay and **iron (III) oxide** (Fe_2O_3), used as a light yellow to brown pigment.

OCR Abbreviation of **optical character reader**.

octa-/octo- Prefix meaning eight.

octadecanoic acid Alternative name for **stearic acid**.

octahedral compound Chemical compound whose molecules have a central atom joined to six atoms or groups located at the vertices of an octahedron.

octahydrate Chemical containing eight molecules of **water of crystallization**.

■ **octane** C_8H_{18} Colourless liquid hydrocarbon, the eighth member of the methane series (**alkanes**). It occurs in petroleum, and is used in petrol. It has several **isomers**.

octane number Measure of **knock**-resistance in a motor fuel; the higher the octane number, the less likely it is to cause knock (preignition). Alternative name: octane rating.

octanoic acid $C_7H_{15}COOH$ Colourless oily **carboxylic acid**, used in the manufacture of dyes and perfumes. Alternative name: caprylic acid.

octaves, law of In chemistry, when the elements are arranged in order of their relative atomic masses, any one element has properties similar to those of the element eight places in front of it and eight places behind it in the list. It is a rejected idea of the relationship of elements. Alternative name: Newlands' law. *See also* **Periodic Table**.

octet Stable group of eight **electrons**; the outer electron configuration of most rare gases, and the arrangement achieved by the atoms of other elements as a result of most cases of chemical combination between them. Alternative name: electron octet.

odd-even nucleus Atomic **nucleus** with an odd number of **protons** and an even number of **neutrons**.

odd-odd nucleus Atomic **nucleus** with an odd number of both **protons** and **neutrons**.

Odonata One of the most primitive orders of winged insects, comprising the dragonflies and damselflies.

oersted C.g.s. unit of electromagnetic field strength, replaced in the SI system by the ampere per metre. It was named after the Danish physicist Hans Oersted (1777–1851).

■ **oesophagus** Muscular tube between the **pharynx** and **stomach** (the gullet), through which food passes by **peristalsis**. [2/4/a] [2/5/a] [2/7/a]

■ **oestrogen** Female sex **hormone**, a member of a group of **steroid** hormones that act on the sex organs. The most important is oestradiol, which is responsible for the growth and activity of much of the female reproductive system.

■ **oestrous cycle** Reproductive cycle of most female mammals, during which there is a period of oestrus or 'heat', when ovulation occurs and the female may be successfully impregnated by a male to achieve **fertilization**. The cycle is regulated by hormones. *See also* **menstrual cycle**.

■ **ohm** ω Practical unit of electrical resistance. It is the resistance of a conductor in which the current is 1 ampere when a potential difference of 1 volt is applied across it. It was named after the German physicist Georg Ohm (1787–1854).

Ohm's law Relationship stating that the **voltage** across a conductor is equal to the product of the **current** flowing through it and its **resistance**. It is written $V = IR$, where V is voltage, I current and R resistance.

oil *See* **fats and oils; petroleum**.

oil of vitriol Old name for **sulphuric acid**.

oil of wintergreen Alternative name for **methyl salicylate**.

oil shale Sedimentary rock that has a relatively high content of a bituminous substance and 30–60% organic matter. Heating it in the absence of air yields an oily substance high in sulphur and nitrogen compounds.

oil, synthetic **Hydrocarbon oil** manufactured from **coal**, **lignite** or **natural gas**, often by **hydrogenation**.

-ol Chemical suffix that denotes an **alcohol** or **phenol**.

oleate **Ester** or **salt** of **oleic acid**.

olefin Alternative name for **alkene**.

oleic acid $CH_3(CH_2)_7CH{=}CH(CH_2)_7COOH$ **Unsaturated fatty acid** that occurs in many **fats and oils**. It is a colourless liquid that turns yellow on exposure to air.

oleum $H_2S_2O_7$ Oily solution of **sulphur trioxide** (SO_3) in concentrated **sulphuric acid**. Alternative names: disulphuric acid, fuming sulphuric acid.

■ **olfaction** Process of smelling. In vertebrates, the incoming nerve impulses from the olfactory sense organs are processed in the olfactory lobes of the brain.

olfactory Concerning the sense of smell.

oligo- Prefix that denotes small or few in number.

Oligochaeta Class of annelid worms with few chaetae (bristles); *e.g.* earthworms.

oligochaete Worm of the class **Oligochaeta**.

oligomer **Polymer** formed from the combination of a few **monomer** molecules.

olivine $(Mg,Fe)_2SiO_4$ Mineral which in a certain green form is classed as a gem (peridot).

omega-minus Negatively charged **elementary particle**, the heaviest **hyperon**.

ommatidium One of the individual elements, composed of a lens linked by nerves directly to the brain, that makes up the **compound eyes** of insects and spiders.

■ **omnivore** Animal that eats both plants and animals. *See also* **carnivore**; **herbivore**. [2/3/b] [2/4/b]

oncogenic Cancer-producing. *See also* **carcinogenic**.

oncology Study of **cancer**.

oncovirus Alternative name for **retrovirus**.

onium ion Ion formed by addition of **proton** (H^+) to a molecule; *e.g.* ammonium ion (NH_4^+) from ammonia, hydroxonium ion (H_3O^+) from water.

on-line Describing part of a **computer** (*e.g.* an **input device**) that is linked directly to and under the control of the **central processor**.

ontogeny Developmental history of an organism, from fertilized ovum to sexually mature adult.

■ **oöcyte** Cell at a stage of oögenesis before the complete development of an **ovum**. A primary oöcyte undergoes the first meiotic division to give the secondary oöcyte, which after the second meiotic division produces an ovum. *See* **meiosis**.

oögamy Form of sexual reproduction in which the female egg (ovum) is large and non-motile and is fertilized by a small, motile male **gamete**.

oögenesis Origin and development of an **ovum**. *See* **oöcyte**.

oölite Sedimentary rock consisting of small spherical masses of calcium carbonate (oöliths).

oöspore *1.* Fertilized ovum. *2.* **Zygote** of some algae that has thick walls and food reserves.

oötecha Egg-case formed by some insects, especially cockroaches and mantids.

opal Gemstone composed of non-crystalline **silica** combined with varying amounts of water. The characteristic internal play of colours in reds, blues and greens is caused by reflection of light from different layers within the stone.

opaque Not allowing a wave motion (*e.g.* light, sound, X-rays) to pass; not transmitting light, not transparent.

■ **open-hearth process** Process for the production of **steel** from **pig iron**. Alternative name: Siemens-Martin process. [4/5/c]

■ **operator** In biology, region of **DNA** to which a molecule or repressor may bind to regulate the activity of a group of closely linked structural **genes**.

operculum *1*. Flap that protects the **gill** of bony fishes. *2*. Lid of a moss capsule. *3*. Plate in some **gastropods** that can cover the shell opening.

■ **operon** Groups of closely linked structural **genes** which are under control of an operator gene. The operator may be switched off by a repressor, produced by a regulator gene separate from the operon. Another substance, the effector, may inactivate the repressor.

opium Dried juice from a species of poppy; a bitter nauseous-tasting brown mass with a heavy, characteristic smell. Its narcotic action – for which it is used both medicinally and as an abused drug – depends on the **alkaloids** it contains. *See also* **morphine**.

■ **optic** Concerning the **eye** and vision.

■ **optic nerve** Cranial nerve of vertebrates that transmits stimuli from the eye to the brain. [2/4/a] [2/5/a] [2/7/a]

optical activity Phenomenon exhibited by some chemical compounds which, when placed in the path of a beam of plane-polarized light, are capable of rotating the plane of polarization to the left (**laevorotatory**) or right (**dextrorotatory**). Alternative name: optical rotation.

optical axis Line that passes through the **optical centre** and the **centre of curvature** of a spherical **lens** or **mirror**. Alternative name: principal axis.

optical centre Point at the centre of a **lens** through which a ray continues straight on and undeviated.

optical character reader (OCR) Computer **input device** that 'reads' printed or written **alphanumeric** characters and feeds the information into a computer system.

optical glass Very pure glass free from streaks and bubbles, used for lenses, etc.

optical isomerism Property of chemical compounds with the same molecular structure, but different **configurations**. Because of their molecular asymmetry they are **optically active**.

optical rotation Alternative name for **optical activity**.

optical telescope Instrument used to observe heavenly bodies by the light that they emit. It consists of lenses or mirrors, or both, that make distant objects appear nearer and larger.

optically active Describing a substance that exhibits **optical activity**.

optics Branch of physics concerned with the study of light.

oral Concerning the mouth and, in some contexts, speech.

■ **oral hygiene** Cleanliness and health of the mouth and teeth.

orange oxide Alternative name for **uranium (VI) oxide**.

■ **orbit** *1*. In astronomy and space science, the path of one heavenly body moving round another (*e.g.* the Earth round the Sun, or the Moon round the Earth), or the path of an artificial satellite round a heavenly body. *2*. In atomic physics, the path of motion of an **electron** round the **nucleus** of an **atom**. *3*. In biology, the eye socket.

orbital Region around the **nucleus** of an **atom** in which there is high probability of finding an **electron**. *See* **atomic orbital**; **molecular orbital**.

orbital electron **Electron** that **orbits** the **nucleus** of an **atom**. Alternative name: planetary electron.

■ **order** In biological **classification**, one of the groups into which a class is divided, and which is itself divided into families; *e.g.* Lagomorpha (lagomorphs), Rodentia (rodents). [2/3/b] [2/4/b]

■ **order of reaction** Classification of chemical reactions based on the power to which the concentration of a component of the reaction is raised in the **rate law**. The overall order is the sum of the powers of the concentrations.

ore Rock or mineral from which a desired substance (usually a metal) can be extracted economically.

■ **organ** Specialized structural and functional unit made up of various **tissues**, in turn formed of many cells, found in animals and plants; *e.g.* heart, kidney, leaf. [2/4/a] [2/5/a] [2/7/a]

organ culture Maintenance of an organ in vitro (after removal from an organism) by the artificial creation of the bodily environment.

organelle Discrete membrane-bound structure that performs a specific function within a **eukaryotic** cell; *e.g.* nucleus, mitochondrion, chloroplast, endoplasmic reticulum.

organic acid Organic compound that can give up **protons** to a **base**; *e.g.* a **carboxylic acid**, **phenol**.

organic base Organic compound that can donate a pair of **electrons** to a bond; *e.g.* an **amine**.

organic chemistry Study of **organic compounds**.

■ **organic compound** Compound of **carbon**, with the exception of its oxides and metallic carbonates, hydrogencarbonates and carbides. Other elements are involved in organic compounds, principally **hydrogen** and **oxygen** but also **nitrogen**, the **halogens**, **sulphur** and **nitrogen**.

organ of Corti Organ concerned with hearing, located in the **cochlea** of the **ear**. It was named after the Italian anatomist Alfonso Corti (1822–88).

organometallic compound Chemical compound in which a **metal** is directly bound to **carbon** in an organic group.

organosilicon compound Chemical compound in which **silicon** is directly bound to **carbon** in an organic group.

organ system Functional unit made up of several **organs**; *e.g.* the digestive system.

ornithine $NH_2(CH_2)_3CH(NH_2)COOH$ **Amino acid**, involved in the formation of urea in animals. Alternative name: 1,6-diaminovaleric acid.

■ **ortho-** Prefix that denotes a **benzene** compound with substituents in the 1,2 positions, abbreviated to *o-* (*e.g. o-*xylene is 1,2-dimethylbenzene). *See also* **meta-**; **para-**.

orthoarsenic acid *See* **arsenic acid**.

orthophosphoric acid Alternative name for **phosphoric (V) acid**.

■ **Orthoptera** Large order of insects, characterized by powerful hindlegs, comprising grasshoppers, crickets, locusts and katydids.

orthotropism Growth straight towards or away from a stimulus; *e.g.* primary roots and shoots are orthotropic to gravity and light, respectively.

oscillation Regular variation in position or state about a mean value.

oscillator Device or electronic circuit for producing an **alternating current** of a particular frequency, usually controlled by altering the value of a **capacitor** in the oscillator circuit.

oscilloscope *See* **cathode-ray oscilloscope**.

osmiridium Naturally occurring alloy of **osmium** and **iridium** that is often used to make the tips of pen nibs.

osmium Os Metallic element in Group VIII of the Periodic Table (a **transition element**). The densest element, it is used in hard alloys and as a catalyst. At. no. 76; r.a.m. 190.2.

osmium (IV) oxide OsO_4 Volatile crystalline solid with a characteristic penetrating odour. Its aqueous solutions are used as a catalyst in organic reactions. Alternative name: osmium tetroxide.

■ **osmoregulation** Process that controls the amount of water and electrolyte (salts) concentration in an animal's body. In a saltwater animal, there is a tendency for water to pass out of the body by **osmosis**, which is prevented by osmoregulation by the kidneys. In freshwater animals, osmoregulation by the kidneys (or by **contractile vacuoles** in simple creatures) prevents water from passing into the animal by osmosis. [2/7/a]

osmosis Movement of a solvent from a dilute to a more concentrated solution across a **semipermeable** (or differentially permeable) **membrane**.

osmotic pressure Pressure required to stop **osmosis** between a solution and pure water.

ossicle Alternative name for an **ear ossicle**.

ossification Process by which **bone** is formed, especially the transformation of **cartilage** into bone.

■ **Osteichthyes** Class of animals that comprises the bony fish.

ostracod Member of the subclass **Ostracoda**.

Ostracoda Subclass of **crustaceans** in which the entire body is enclosed in a smooth, rounded **carapace**; *e.g.* mussel shrimps.

outbreeding Mating between members of a species that are not closely related. *See also* **inbreeding**.

■ **outer ear** Part of the **ear** that transmits sound waves from external air to the ear drum. [2/4/a] [2/5/a] [2/7/a]

output device Part of a **computer** that presents **data** in a form that can be used by a human operator; *e.g.* a **printer, visual display unit** (VDU), chart plotter, etc. A machine that writes **data** onto a portable magnetic medium (*e.g.* magnetic disk or tape) may also be considered to be an output device.

■ **oval window** Membranous area at which the 'sole' of the stirrup (stapes) bone of the inner **ear** makes contact with the **cochlea**. Alternative name: fenestra ovalis.[2/4/a] [2/5/a] [2/7/a]

ovarian follicle Alternative name for the **Graafian follicle**.

■ **ovary** Female reproductive organ. In vertebrates, there is a pair of ovaries, which produce the ova (eggs) and sex **hormones**. In plants, the ovary is the hollow base of a **carpel** enclosing one or more ovules borne on a placenta; after fertilization the ovary of a plant becomes the **pericarp** of the fruit.

overtone Harmonic; a note of higher frequency than a fundamental note.

oviduct Tube that conducts released ova (eggs) from the **ovaries** after ovulation. Alternative name (in human beings): Fallopian tube.

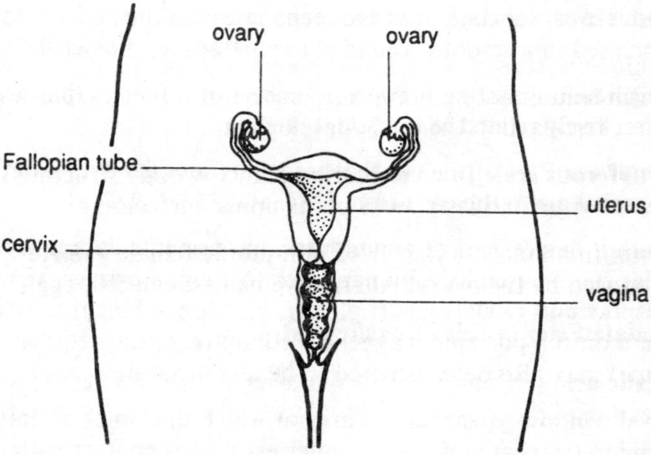

Human ovaries

■ **oviparity** Animal reproduction in which ova (eggs) are laid by the female either before or after fertilization; *e.g.* as in birds and fish.

■ **ovipositor** Egg-laying structure of female insects, formed from a modified pair of appendages at the hind end of the abdomen (*e.g.* in a parasitic wasp it is a modified sting). Some fish too have ovipositors (*e.g.* bitterling).

■ **ovoviviparity** Animal reproduction in which ova (eggs) hatch inside the female body and the embryo is retained for protection. The embryo obtains its nourishment independently from the yolk store. It occurs in some fish and reptiles.

■ **ovulation** In vertebrates, discharge of an ovum (egg) from a mature **Graafian follicle** at the surface of an **ovary**. In mature human females, ovulation occurs from alternate ovaries at about every 28 days until the **menopause** occurs.

ovule Structure in female seed plants. It consists of a nucleus, which contains the embryo sac, surrounded by integuments. After fertilization the ovule develops into the seed.

ovuliferous scale In cone-bearing plants, woody structure that bears **ovules** and later seeds on its upper surface.

■ **ovum** In animals, unfertilized non-motile female **gamete** produced by the **ovary**. Alternative names: egg cell, egg.

oxalate Ester or **salt** of **oxalic acid**.

oxalic acid $(COOH)_2.2H_2O$ White crystalline poisonous **dicarboxylic acid**. It occurs in rhubarb, wood sorrel and other plants of the oxalis group. It is used in dyeing and **volumetric analysis**. Alternative name: ethanedioic acid.

oxatyl Alternative name for **carbonyl group**.

oxidase Collective name for a group of **enzymes** that promote **oxidation** within plant and animal cells.

■ **oxidation** Process that involves the loss of **electrons** by a substance; combination of a substance with **oxygen**. It may occur rapidly (as in combustion and aerobic respiration) or slowly (as in rusting and other forms of corrosion). Oxidation is one of the causes of spoilage in food, sometimes combated by using **additives** called anti-oxidants.

■ **oxidation number** Number of **electrons** that must be added to a **cation** or removed from an **anion** to produce a neutral **atom**. An oxidation number of zero is given to the **elements** themselves. In compounds, a positive oxidation number indicates that an element is in an oxidized state; the higher the oxidation number, the greater is the extent of oxidation. Conversely, a negative oxidation number shows that an element is in a reduced state.

oxidation-reduction reaction Alternative name for a **redox reaction**.

■ **oxide** Compound of **oxygen** and another element, usually made by direct combination or by heating a **carbonate** or **hydroxide**. *See also* **acidic oxide; basic oxide; neutral oxide**.

oxidizing agent Substance that causes **oxidation**. Alternative name: electron acceptor.

■ **oxime** Compound containing the oximino group =NOH, derived by the condensation of an **aldehyde** or **ketone** with hydroxylamine (NH_2OH); *e.g.* acetaldehyde (ethanal), CH_3CHO, forms $CH_3CH{=}NOH$.

■ **oxo process** Reaction that involves the addition of **hydrogen** and the formyl group –CHO, derived from hydrogen and **carbon monoxide**, to an **alkene** in the presence of a **catalyst** (usually cobalt). It is used in the conversion of alkenes into **aldehydes** and **alcohols**. Alternative name: oxo reaction.

2-oxopropanoic acid Alternative name for **pyruvic acid**.

oxy-acetylene burner Device for obtaining a very high-temperature flame (3,480 °C) by the combination of **oxygen** and **acetylene** in the correct proportions. It is used in cutting and welding of metals.

■ **oxyacid Acid** in which the acidic (*i.e.* replaceable) hydrogen atom is part of a **hydroxyl group** (*i.e.* organic **carboxylic acids** and **phenols**, and inorganic acids such as **phosphoric (V) acid** and **sulphuric acid**).

■ **oxygen** O Gaseous nonmetallic element in Group VIA of the Periodic Table. A colourless odourless gas, it makes up about 20%of air by volume, from which it is extracted, and is essential for life in all living organisms (except for a few lower forms such as certain bacteria). In animal tissues its main function is to oxidize **glucose** to release energy. It is the most abundant element in the Earth's crust, occurring in all water and most rocks. It is used in welding, steel-making and as a rocket fuel. It has a triatomic **allotrope, ozone** (O_3). At. no. 8; r.a.m. 15.9994. *See* **aerobic respiration; respiration**.

oxygen debt Physiological condition that induces **anaerobic respiration** in an otherwise **aerobic** organism. It occurs during anoxia, caused *e.g.* by violent exercise.

oxyhaemoglobin Product of **respiration** formed by the combination of **oxygen** and **haemoglobin**. [2/4/c]

■ **ozone** O_3 Allotrope of **oxygen** that contains three atoms in its molecule. It is formed from oxygen in the upper atmosphere by the action of ultraviolet light, where it also acts as a shield that prevents excess ultraviolet light reaching the Earth's surface. It is a powerful **oxidizing agent**, often used in organic chemistry.

■ **ozone layer** Layer in the upper atmosphere at a height of between 15 and 30 km, where **ozone** is found in its greatest concentration. It filters out ultraviolet radiation from the Sun which would otherwise be harmful. If this layer were destroyed, or depleted to a great extent, life on Earth would be endangered. Alternative name: ozonosphere. [2/9/d]

P

pachytene Stage in **prophase** or first division of **meiosis**, in which the paired **chromosomes** shorten and thicken, appearing as two **chromatids**.

packing fraction Difference between the actual mass of an **isotope** and the nearest whole number divided by the **mass number**.

paedogenesis Sexual reproduction in some animals by the immature forms (larvae and pupae).

■ **paint** Suspension of powdered colouring matter (**pigment**) in a liquid, used for decorative and protective coatings of surfaces. In an oil paint, the liquid contains solvents and a drying oil or synthetic resin. In emulsion paint, the liquid is mainly water containing a dispersion of resin.

palaeontology Study and interpretation of **fossils**.

palisade Main photosynthetic tissue of a leaf.

palladium Pd Silver-white metallic element in Group VIII of the Periodic Table (a **transition element**), used as a **catalyst** and in making jewellery. At. no. 46; r.a.m. 106.4.

palmitic acid $C_{15}H_{31}COOH$ Long-chain **carboxylic acid** which occurs in oils and fats (*e.g.* palm oil) as its glyceryl **ester**, used in making **soap**.

palp Jointed sensory structure located on the mouthparts of insects and other invertebrates.

■ **pancreas** Gland situated near the **duodenum** that has digestive and endocrine functions. The enzymes amylase, trypsin and lipase are released from it during digestion. Special groups of cells (the islets of Langerhans) produce the **hormones insulin** and **glucagon** for the control of blood sugar levels. [2/4/a] [2/5/a] [2/7/a]

■ **pancreatic juice** Liquid containing digestive **enzymes** that is secreted by the **pancreas** and passed to the **duodenum**. [2/4/a] [2/5/a] [2/7/a]

■ **pandemic** Describing a disease that affects people or animals throughout the world. *See also* **endemic**; **epidemic**. [2/7/a] [2/9/c]

panicle Branched **racemose inflorescence**, common in grasses.

pantothenic acid Constituent of **coenzyme A**, a carrier of **acyl groups** in biochemical processes. It is required as a B **vitamin** by many organisms, including vertebrates and yeast.

papain Proteolytic **enzyme**, which digests proteins, found in various fruits and used as a meat tenderizer.

■ **paper chromatography** Type of **chromatography** in which the mobile phase is liquid and the stationary phase is porous paper. Compounds are separated on the paper, and can then be identified.

paper tape *See* **punched tape**.

papovavirus Member of a group of double-stranded DNA viruses which infect the cells of higher vertebrates, in which they can cause tumours.

■ **para-** *1.* Prefix that denotes the form of a diatomic molecule in which both nuclei have opposite spin directions. *2.* Referring to the 1,4 positions in the benzene ring (can be abbreviated to *p*-); *e.g.* *p*-xylene is 1,4-dimethylbenzene. *See also* **meta-**, **ortho-**.

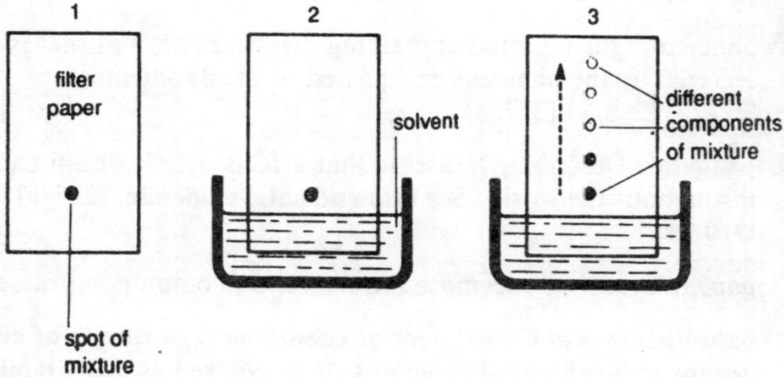

Simple paper chromatography

■ **paraffin Alkane**, a member of series of **hydrocarbons** which include gases as well as liquids and solids, obtained by the distillation of **petroleum**. Paraffins are used as solvents and fuels.

paraffin oil Alternative name for **kerosene**.

paraffin wax Mixture of solid **paraffins** (alkanes) which takes the form of a white translucent solid that melts below 80 °C. It is used to make candles. Alternative name: petroleum wax.

paraformaldehyde Alternative name for **polymethanal**.

paraldehyde $(CH_3CHO)_3$ Cyclic **trimer** formed by the polymerization of **acetaldehyde** (ethanal), used as a sleep-inducing drug. Alternative name: ethanal trimer. *See also* **metaldehyde**.

parallax Apparent change in the position of an object when it is observed from two different viewpoints.

parallel Alternative name for **latitude**.

■ **parallel circuit** Electrical circuit in which the voltage supply is connected to each side of all the components so that only a fraction of the total current flows through each of them. For **capacitors** in parallel, the total capacitance C is equal to the sum of the individual capacitances; *i.e.* $C = C_1 + C_2 + C_3 + \ldots$ For **resistors** in parallel, the reciprocal of the total resistance R is equal to the sum of the reciprocals of the individual resistances; *i.e.* $1/R = 1/R_1 + 1/R_2 + 1/R_3 \ldots$ [5/4/b]

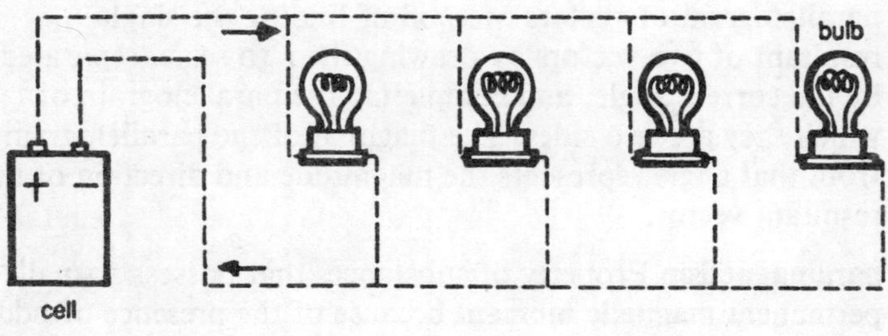

Four lamps connected in parallel

parallelogram of forces Method of finding a resultant of two forces by using the **parallelogram of vectors.**

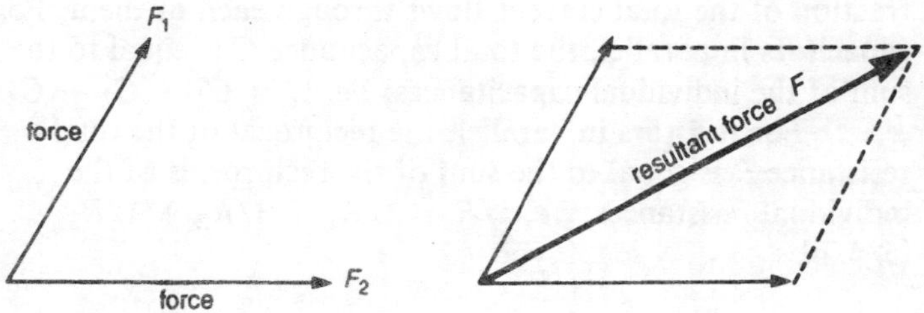

Parallelogram of forces gives the resultant

parallelogram of vectors Method of finding the single resultant of two vectors by drawing them to scale, separated by the correct angle, and completing the parallelogram of which they are two sides. The diagonal of the parallelogram from that angle represents the magnitude and direction of the resultant vector.

paramagnetism Property of substances that possess a small permanent **magnetic moment** because of the presence of odd (unpaired) **electrons**; the substance becomes magnetized in a magnetic field as the magnetic moments align.

■ **parasite** Organism that is the beneficial partner in **parasitism**. *See also* **ectoparasite; endoparasite.**

■ **parasitism** Intimate relationship between two organisms in which one (the parasite) derives benefit from the other (the host), usually to obtain food or physical support. Parasitism can have minor or major effects on the survival of the host. *See also* **commensalism**; **symbiosis**.

parasympathetic nervous system Branch of the **autonomic nervous system** used in involuntary activities, for which **acetylcholine** is the transmitter substance. Effects of the parasympathetic nervous system generally counteract those of the **sympathetic nervous system**.

parathyroid Four **endocrine glands** embedded in the **thyroid** in the neck which release a **hormone** that controls the levels of calcium in the blood.

parenchyma Plant tissue composed of round tightly packed cells, used as packing tissue and for the storage of **starch**.

parsec Unit of astronomical distance equivalent to 3.26 **light-years** (3.086×10^{13} km). At that distance, a length of 1 astronomical unit subtends an angle of 1 second of arc.

■ **parthenogenesis** Development of a new individual from an unfertilized egg, which occurs in certain groups of invertebrates. Eggs may be formed by **mitosis** instead of **meiosis** and are **haploid**. In haploid parthenogenesis, the organism produced is also haploid, *e.g.* drones in the honeybee.

partial pressure *See* **Dalton's law of partial pressures**.

particle Minute portion of matter, often taken to mean an **atom, molecule** or **elementary particle** or **subatomic particle**.

particle accelerator Machine for producing a beam of high-speed **particles**, such as those that are used to disintegrate atomic nuclei.

particle physics Branch of science concerned with the properties of **elementary particles**.

partition coefficient Ratio of the concentrations of a single **solute** in two immiscible **solvents**, at equilibrium. It is independent of the actual concentrations.

■ **parturition** Birth of a full-growth **foetus** at the completion of pregnancy (gestation).

■ **pascal** (Pa) SI unit of pressure, equal to a force of 1 newton per square metre (N m^{-2}). It was named after the French physicist and mathematician Blaise Pascal (1623–62). [5/6/a]

Pascal's law In a fluid, the pressure applied at any point is transmitted equally throughout it. It is the principle by which hydraulic machinery works.

passivity Corrosion-resistance (and unreactivity) of a metal that results from a thin surface layer of oxide. *E.g.* aluminium is passive because of such a layer, which can be thickened by **anodizing**.

■ **pasteurization** Process of heating food or other substances under controlled conditions. It was developed by the French chemist Louis Pasteur (1822–95) to destroy germs. It is widely used in industry, *e.g.* production of milk and wine.

■ **patella** Bone in front of the knee joint. Alternative name: kneecap. [2/3/a]

■ **pathogen** Disease-causing organism, commonly used of micro-organisms such as **bacteria** and **viruses**.

■ **Pauli exclusion principle** No two **electrons** can be assigned the same set of **quantum numbers**; hence there can be only two electrons in any one atomic **orbital**. It was named after the Austrian-born American physicist Wolfgang Pauli (1900–58). Alternative name: exclusion principle.

***p*-block elements** 30 nonmetallic elements that form Groups IIB, IVB, VB, VIB, VIIB and 0 of the Periodic Table (helium is usually excluded), so called because their 1 to 6 outer electrons occupy ***p*-orbitals**.

PD Abbreviation of **potential difference**.

pearl Lustrous, often spherical, accretion formed from the layering of **nacre** (calcium carbonate) on a foreign particle inside the shells of certain molluscs, *e.g.* oysters.

pearl spar Alternative name for **dolomite**.

peat Brown or black fibrous material formed, in surface deposits, by the partial decomposition of plant remains. It is used in making manures and composts, in making charcoal, and as a heat-insulating material. Dried peat can be used as a fuel.

■ **pectin** Complex **polysaccharide** derivative present in plant cell walls, to which it gives rigidity. It can be converted to a gel form in sugary acid solution.

pectoral Concerning the part of the front end of a vertebrate's body which supports the shoulders and forelimbs. [2/3/a]

■ **pedicel** Stem that attaches an individual flower to the main flower stem (peduncle). [2/4/a]

peduncle Main flower stem of a plant (*see also* **pedicel**).

pelagic Describing fish that normally occur in free-swimming shoals in waters less than 200m deep. *See also* **benthic**.

■ **pelvic** Concerning the part of the rear end of a vertebrate's body which supports the hindlimbs. [2/3/a]

■ **pelvis** *1.* Part of the skeleton (pelvic girdle) to which a vertebrate's hindlimbs are joined. *2.* Cavity in the **kidney** which receives urine from the tubules and drains it into the **ureter**. [2/3/a] [2/4/a]

■ **pendulum** A simple pendulum consists of a weight (called a bob) at the end of a string which, over small angles, swings with **simple harmonic motion**. The period of swing t is given by $t = 2\pi(l/g)^{1/2}$, where l is the length of the string and g is the acceleration of free fall. *See also* **compound pendulum**.

■ **penicillin** Member of a class of **antibiotics** produced by moulds of the genus *Penicillium*. It inhibits growth of some **bacteria** by interfering with cell-wall biosynthesis.

■ **penis** Copulatory organ possessed by males or hermaphroditic forms in some animals, which transfers sperm to the female to achieve internal fertilization. In mammals it also houses the **urethra**, through which urine is discharged. [2/4/a] [2/7/a]

pentadactyl Having five digits. It describes the limb typical of most tetrapod vertebrates.

pentahydrate Chemical containing five molecules of **water of crystallization**; *e.g.* $CuSO_4.5H_2O$.

pentane C_5H_{12} Liquid **alkane**, extracted from petroleum and used as a solvent and in organic synthesis. It has three **isomers**.

pentanoic acid $CH_3(CH_2)_3COOH$ Liquid **carboxylic acid** with a pungent odour, used in perfumes. Alternative name: valeric acid.

pentavalent Having a **valence** of five.

■ **pentose Monosaccharide** carbohydrate (sugar) that contains five carbon atoms and has the general formula $C_5H_{10}O_5$; *e.g.* **ribose** and **xylose**. Alternative name: pentaglucose.

pentose sugar Monosaccharide carbohydrate (sugar) that contains five carbon atoms and has the general formula $C_5H_{10}O_5$; *e.g.* **ribose** and **xylose**. Alternative name: pentaglucose.

pentyl group Alternative name for **amyl group**.

■ **pepsin Enzyme** produced in the stomach which, under acid conditions, brings about the partial **hydrolysis** of **polypeptides** (thus helping in the digestion of **proteins**).

■ **peptidase Enzyme**, often secreted in the body (*e.g.* by the intestine), which degrades **peptides** into free **amino acids**, thus completing the digestion of **proteins**.

peptide Organic compound that contains two or more **amino acid** residues joined covalently through peptide bonds (–NH–CO–) by a condensation reaction between the carboxyl group of one amino acid and the amino group of another. Peptides polymerize to form **proteins**.

■ **percentage composition** Make-up of a chemical compound expressed in terms of the percentages (by mass) of each of its component elements. *E.g.* ethane (C_2H_6), ethene (C_2H_4) and ethyne (C_2H_2) all consist of carbon and hydrogen. Their approximate percentage compositions are:

ethane: 80% carbon, 20% hydrogen
ethene: 85% carbon, 15% hydrogen
ethyne: 92% carbon, 8% hydrogen

perdisulphuric (VI) acid H_2SO_5 White crystalline compound, used as a powerful **oxidizing agent**. Alternative names: Caro's acid, persulphuric acid, peroxomonosulphuric acid.

perennating organ Part of a plant that enables it to survive over the winter; *e.g.* a rhizome or tuber.

■ **perennial** Plant that lives for a number of years and may be woody (*e.g.* trees and shrubs) and continuously grow or have herbaceous stems which die at the end of each season and are replaced. *See also* **annual; biennial**.

perfect gas Alternative name for **ideal gas.**

■ **perianth** Outer part of a flower that surrounds the sexual organs, usually consisting of the **sepals** and **petals.**

pericarp Fruit coat formed from the **ovary** wall in plants after fertilization.

perilymph Fluid that surrounds the **cochlea** of the **ear.**

■ **period** *1.* (*t*) Time taken to complete a regular cycle, such as a complete swing of a pendulum or one cycle of any type of simple harmonic motion. *2.* One of the seven horizontal rows of elements in the **Periodic Table.**

periodic function Function that returns to the same value at regular intervals.

periodicity Regular increases and decreases of physical values for elements known to have similar chemical properties.

periodic law Properties of elements are a periodic function of their atomic numbers. Alternative name: Mendeleev's law.

■ **Periodic Table** Arrangement of elements in order of increasing **atomic number**, with elements having similar properties, *i.e.* in the same family, in the same vertical column (group). Horizontal rows of elements are termed periods. [4/7/b]

peripheral nervous system That part of the nervous system that does not include the **central nervous system** (CNS).

peripheral unit Equipment that can be linked to a **computer,** including input, output and storage devices.

periscope Device for viewing objects that are above eye-level, or placed so that direct vision is obstructed. Light rays entering the top of a tube pass into a right-angled prism, which reflects them straight downwards to the bottom of the tube, where a second prism reflects into the eye of the viewer. Angled mirrors may replace the prisms in a simple periscope.

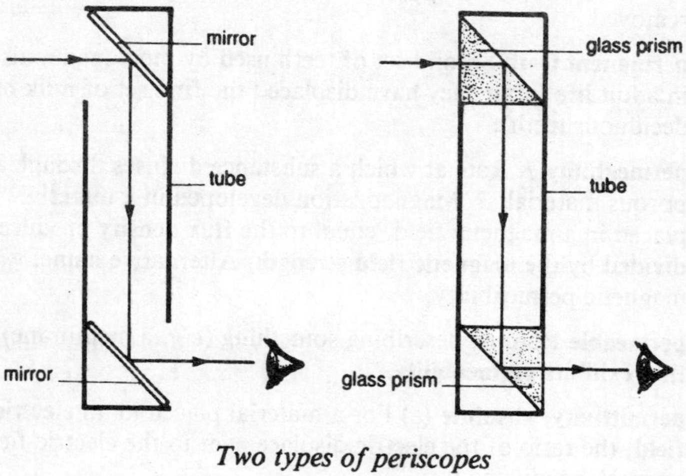

Two types of periscopes

perissodactyl Having an odd number of toes; a member of the **Perissodactyla**.

■ **Perissodactyla** Order of **ungulate** mammals whose members have an odd number of toes (one or three), including horses, rhinoceroses and tapirs. *See also* **Artiodactyla**.

peristalsis Involuntary waves of muscular contraction produced along tubular structures in the body, *e.g.* in the oesophagus or intestine to push food along.

permanent gas Gas that is incapable of being liquefied by **pressure** alone; a gas above its critical temperature.

permanent hardness (of water) Hardness that is not destroyed by boiling the water. It is caused by the presence of calcium or magnesium salts. *See* **hardness of water**.

permanent magnet Ferromagnetic object that retains a permanent **magnetic field** and the **magnetic moment** associated with it after the magnetizing field has been removed.

permanent teeth Second set of teeth used by most mammals in adult life (after they have displaced the first set of milk or deciduous teeth).

■ **permeability** *1.* Rate at which a substance diffuses through a porous material. *2.* Magnetization developed in a material placed in a magnetic field, equal to the **flux density** produced divided by the **magnetic field strength**. Alternative name: magnetic permeability.

permeable Porous; describing something (*e.g.* a membrane) that exhibits **permeability**.

■ **permittivity, absolute** (ε) For a material placed in an **electric field**, the ratio of the electric displacement to the electric field strength producing it.

■ **permittivity, relative** For a material placed in an **electric field,** the ratio of the electric displacement to that which would be produced in free space by the same field. Alternative names: dielectric constant, inductivity, specific capacitance.

Permutit Trade name for a type of **zeolite** used in water softeners.

peroxide *1.* **Oxide** of an element containing more **oxygen** than does its normal oxide. *2.* Oxide, containing the O_2^{2-} ion, that yields **hydrogen peroxide** on treatment with an **acid**. Peroxides are powerful **oxidizing agents**.

peroxomonosulphuric acid Alternative name for **perdisulphuric (VI) acid**.

personal computer (PC) *See* **microcomputer**.

Perspex Trade name for the plastic **polymethyl methacrylate**.

perspiration Watery fluid produced by sweat glands which has a cooling effect as it evaporates from the surface of the skin. Alternative name: sweat.

persulphate Salt of **perdisulphuric (VI) acid**, used as an oxidizing agent.

persulphuric acid Alternative name for **perdisulphuric (VI) acid**.

perturbation theory Method of obtaining approximate solutions to equations representing the behaviour of a system. It is used in **quantum mechanics**.

■ **pesticide** Compound used in agriculture to destroy organisms that can damage crops or stored food, especially insects and rodents. Pesticides include fungicides, herbicides and insecticides. The effects of some of them, *e.g.* organic chlorine compounds such as DDT, can be detrimental to the ecosystem. [2/10/d]

pet Animal that is kept in the home for company or pleasure.

■ **petal** Part of a flower that makes up the **corolla**, often scented and with a bright colour (to attract pollinating insects). Petals are regarded as modified leaves. [2/4/a] [2/5/a] [2/7/a]

■ **Petri dish** Sterilizable circular glass plate with a fitted lid used in microbiology for holding media on which micro-organisms may be cultured. It was named after the German bacteriologist Julius Petri (1852–1921).

petrochemical Any of a range of chemicals derived from **petroleum** or **natural gas**.

■ **petrol** Liquid fuel (for internal combustion engines) consisting of a mixture of **alkanes** in the range pentane (C_5H_{12}) to decane ($C_{10}H_{22}$), made by distillation and **cracking** of **petroleum**. Alternative names: gasoline, motor spirit. *See also* **diesel fuel**. [4/8/b] [4/10/d]

■ **petroleum** Mineral oil, a mixture of **hydrocarbons**, which is usually greenish, brown or black in the crude state. It is the source of various fuels and an important natural raw material in the chemical industry.

petroleum wax Alternative name for **paraffin wax**.

pewter Alloy of **tin** (80–90%) and **lead** (10–20%), used to make tableware and jewellery.

■ **pH** Hydrogen ion concentration (grams of hydrogen ions per litre) expressed as its negative logarithm; a measure of acidity and alkalinity. For example, a hydrogen ion concentration of 10^{-3} grams per litre corresponds to a pH of 3, and is acidic. A pH of 7 is neutral; a pH of more than 7 is alkaline. [4/5/a]

phage *See* **bacteriophage**.

■ **phagocyte** Cell that exhibits **phagocytosis**. It is employed in the defence against invasion by foreign organisms, *e.g.* macrophages in human beings. [2/4/c]

phagocytosis Engulfment of external solid material by a cell, *e.g.* a **phagocyte**. It is also the method by which some unicellular organisms (*e.g.* protozoa) feed.

■ **phalanges** Bones in the digits of the hand or foot. [2/4/c]

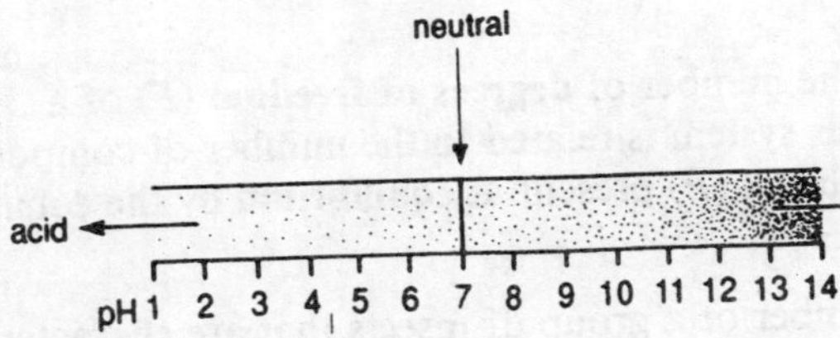

pH scale is neutral at pH 7

pharmacology Study of the properties, manufacture and reactions of drugs.

pharmacy Preparation and dispensing of drugs, and the place where this is done.

■ **pharynx** Area that links the buccal cavity (mouth) to the oesophagus (gullet) and the nares (back of the nostrils) to the trachea (windpipe). Food passes via the pharynx to the oesophagus when the **epiglottis** closes the entrance of the trachea. [2/4/a] [2/5/a] [2/7/a]

■ **phase** *1.* Any homogeneous and physically distinct part of a (chemical) system that is separated from other parts of the system by definite boundaries, *e.g.* ice mixed with water.

Solids, liquids and gases make up different phases. *2.* The part of a periodically varying waveform that has been completed at a particular moment.

phase contrast microscope Microscope that uses the principle that light passing through materials of different **refractive indices** undergoes a change in phase, transmitting these changes as different intensities of light given off by different materials.

phase rule The number of **degrees of freedom** (F) of a **heterogeneous** system is related to the number of components (C) and of phases (P) present at **equilibrium** by the equation $P + F = C + 2$.

phasmid Member of a group of insects that are characterized by extreme elongation of the body and highly effective camouflage, *e.g.* stick insects and leaf insects.

phellem Alternative name for **cork**.

phellogen Layer of **cambium** tissue below the bark of a woody plant. Alternative name: cork cambium.

■ **phenol** C_6H_5OH Colourless crystalline solid which turns pink on exposure to air and light. It has a characteristic, rather sweet odour. It is used as an antiseptic and disinfectant, and in the preparation of dyes, drugs, etc. Alternative names: carbolic acid, hydroxybenzene. Other compounds with one or more **hydroxyl groups** bound directly to a **benzene ring** are also known as phenols. They give reactions typical of **alcohols** (*e.g.* they form **esters** and **ethers**), but they are more acidic and form salts by the action of strong alkalis.

phenolphthalein Organic compound that is used as a laxative and as an **indicator** in **volumetric analysis**. It is red in alkalis and colourless in acids.

■ **phenotype** Outward appearance and characteristics of an organism, or the way **genes** express themselves in an organism. Organisms of the same phenotype may possess different **genotypes**; *e.g.* in a **heterozygous** organism two **alleles** of a gene may be present in the genotype (genetic make-up) with the expression of only one in the phenotype (appearance). [2/6/b] [2/7/b]

phenylalanine $C_6H_5CH_2CH(NH_2)COOH$ Essential amino acid that possesses a **benzene ring**.

phenylamine Alternative name for **aniline**.

phenylethylene Alternative name for **styrene**.

phenyl group C_6H_5- **Monovalent radical** derived from **benzene**.

phenyl methyl ketone Alternative name for **acetophenone**.

phenylpropenoic acid Alternative name for cinnamic acid.

■ **pheromone** Chemical substance produced by an organism which may influence the behaviour of another. *E.g.* in moths, pheromones act as sexual attractants; in social insects such as bees, pheromones have an important role in the development and behaviour of the colony. [2/6/b]

■ **phloem** Type of vascular tissue present in plants, which consists mainly of living cells or **sieve tubes** used for the transport of food material from the leaves to other areas of the plant. *See also* **xylem**. [2/4/a] [2/5/a] [2/7/a]

phosgene $COCl_2$ Colourless gas with penetrating and suffocating smell. It has been employed as a war gas, and is now used in organic synthesis. Alternative name: carbonyl chloride.

■ **phosphate** Salt of **phosphoric (V) acid**, containing the ion PO_4^{3-}. Some phosphates are of enormous commercial and practical importance, *e.g.* ammonium phosphate **fertilizers**, and alkali phosphate **buffers**. Alternative name: orthophosphate. Because phosphoric (V) acid is a tribasic acid, it also forms hydrogenphosphates, HPO_4^{2-}, and dihydrogenphosphates, $H_2PO_4^-$. [2/5/c]

phosphatidyl choline Alternative name for **lecithin**, a **phospholipid** constituent of **plasma membranes.**

phosphide Chemical compound of **phosphorus** and another element.

phosphine PH_3 Colourless poisonous gas, often spontaneously inflammable because of impurities. Alternative name: phosphorus trihydride, phosphorus (III) hydride.

■ **phospholipid** Member of a class of complex **lipids** that are major components of cell membranes. They consist of molecules containing a phosphoric (V) acid **ester** of **glycerol**, the remaining **hydroxyl groups** of the glycerol being esterified by **fatty acids**. Alternative names: phosphoglyceride, phosphatide, glycerol phosphatide.

phosphonium ion Ion PH_4^+.

phosphor Substance capable of **luminescence** or **phosphorescence**, as used to coat the inside of a television screen or fluorescent lamp.

■ **phosphorescence** Emission of light (generally visible light) after **absorption** of light of another wavelength (usually ultraviolet or near ultraviolet) or electrons; unlike **fluorescence**, it continues after the stimulating source is removed. See also: **luminescence**.

phosphoric (V) acid H_3PO_4 Tribasic acid, a colourless crystalline solid, made by dissolving **phosphorus (V) oxide** in water. It is used to form a corrosion-resistant layer on steel. Alternative names: phosphoric acid, orthophosphoric acid.

■ **phosphorus** P Nonmetallic element in Group VA of the Periodic Table. It exists as several **allotropes**, chief of which are red phosphorus and the poisonous and spontaneously inflammable white or yellow phosphorus. It occurs in many minerals (particularly **phosphates**) and all living organisms; it is an essential nutrient for plants (*see* **fertilizer**). Phosphorus is made by heating calcium phosphate (with carbon and sand) in an electric furnace. It is used in matches and for making fertilizers. At. no. 15; r.a.m. 30.9738.

phosphorus (III) bromide PBr_3 Colourless liquid, used in organic synthesis to replace a **hydroxyl group** with a **bromine** atom. Alternative name: phosphorus tribromide.

phosphorus (V) bromide PBr_5 Yellow crystalline solid, used as a brominating agent. Alternative name: phosphorus pentabromide.

phosphorus (III) chloride PCl_3 Colourless fuming liquid, used to make organic compounds of phosphorus. Alternative name: phosphorus trichloride.

phosphorus (V) chloride PCl_5 Yellowish-white crystalline solid, used as a chlorinating agent. Alternative name: phosphorus pentachloride.

phosphorus (III) hydride Alternative name for **phosphine**.

phosphorus (III) oxide P_2O_3 White waxy solid which readily reacts with oxygen to form **phosphorus (V) oxide**. Alternative name: phosphorus trioxide.

phosphorus (V) oxide P_2O_5 **Hygroscopic** white powder which readily reacts with water to form **phosphoric (V) acid**. It is used as a **desiccant**. Alternative name: phosphorus pentoxide.

■ **phosphorylation** Process by which a **phosphate** group is transferred to a molecule of an organic compound. In some substances, *e.g.* **ATP**, a high-energy bond may be formed by phosphorylation, which is essential for energy-transfer in living organisms. It is an important biochemical end-reaction which modifies the conformation of molecules such as enzymes, receptors, etc.

photocell Device that converts light into an electric current. It can be used for the detection and measurement of light intensity. Alternative names: photoelectric cell, photoemissive cell, photovoltaic cell.

■ **photochemical reaction** Chemical reaction that is initiated by the absorption of light. The most important phenomenon of this type is photosynthesis. It is also the basis of photography. [2/3/d]

photochemistry Branch of chemistry concerned with the action of light in initiating chemical reactions.

photochromism Change in colour of a substance through exposure to light; *e.g.* photochromic glass, used in sunglasses.

photoconductivity Change in electrical conductivity of a substance when it is exposed to light; *e.g.* selenium, used in photoelectric light meters. *See also* **light-dependent resistor**.

photoelectric cell Alternative name for **photocell**.

photoelectric effect Phenomenon that occurs with some semi-metallic materials. When **photons** strike them they are absorbed and the energized **electrons** produced flow in the material as an electric current. It is the basis of photocells and instruments that employ them, such as photographers' light meters and burglar alarms.

photoelectron **Electron** produced by the **photoelectric effect** or by **photoionization**.

photoemission Emission of **photoelectrons** by the **photoelectric effect** or by **photoionization**.

■ **photography** Process of taking photographs by the chemical action of light or other radiation on a sensitive plate or film made of glass, celluloid or other transparent material coated with a light-sensitive emulsion. Light causes changes in particles of silver salts in the emulsion which, after development (in a reducing agent), form grains of dark metallic silver to produce a negative image. Unaffected silver salts are removed by fixing (in a solution of ammonium or sodium thiosulphate).

photoionization Ionization of atoms or molecules by light or other **electromagnetic radiation**.

photoluminescence Light emission by a substance after it has itself been exposed to visible light or infra-red or ultraviolet radiation.

photolysis Photochemical reaction that results in the decomposition of a substance.

photometer Instrument for measuring the intensity of light.

photometry Measurement of the intensity of light.

photomicrograph Photograph obtained through a microscope.

■ **photomultiplier Photocell** of high sensitivity used for detecting very small quantities of light radiation. It consists of series of electrodes in an evacuated envelope, which are used to amplify the emission current by electron multiplication. Alternative name: electron multiplier.

photon Quantum of **electromagnetic radiation**, such as light or X-rays; a 'packet' of radiation. The amount of energy per photon is *hn*, where *h* is **Planck's constant** and *n* is the frequency of the radiation.

photoneutron Neutron resulting from the interaction of a **photon** with an atomic **nucleus.**

■ **photoperiodism** Influence of day and night on the activities of an organism. *E.g.* the flowering of **plants** is controlled by the photoperiod and regulated by **phytochrome**. [2/3/d]

■ **photophosphorylation** Process during **photosynthesis** that results in the formation of **ATP** from energy derived from sunlight via **chlorophyll**. The reactions also produce **hydrogen**, which is used in combination with **carbon dioxide** to make **sugars**. Alternative names: photosynthetic phosphorylation. *See also* **phosphorylation; photosynthesis**. [2/3/d]

photoreceptor Receptor consisting of sensory cells that are stimulated by light, *e.g.* light-sensitive cells in the **eye**.

■ **photosynthesis** Type of **autotrophic** nutrition employed by green plants which involves the synthesis of organic compounds (mainly **sugars**) from **carbon dioxide** and water. Sunlight is used as a source of energy, which is trapped by **chlorophyll** present in **chloroplasts**. The process consists of a light stage, in which energy is converted into **ATP** and water is split into **hydrogen** and **oxygen**. Hydrogen is subsequently combined with carbon dioxide in the dark stage to form **carbohydrates**. Photosynthetic **bacteria** use different sources of hydrogen in the process. [2/3/d]

■ **photosynthetic pigment Pigment** involved in the trapping of light energy during **photosynthesis**. The chief photosynthetic pigments are **chlorophylls**. Other accessory pigments also trap energy, *e.g.* **carotenoids**. [2/3/d]

■ **pleura** Double membrane that covers the lungs and lines the chest cavity, with fluid between the membranes. Alternative name: pleural membranes.

pleural To do with the lungs.

plexus Network of interlacing nerves.

plumbic Alternative name for **lead (IV)**.

plumbous Alternative name for **lead (II)**.

plumule *1.* In birds, small soft feather that forms down. *2.* In seed plants, small leafy part of an embryonic shoot.

■ **plutonium** Pu Radioactive element in Group IIIB of the Periodic Table (one of the **actinides**), produced from uranium-238 in a **breeder reactor**. It has several **isotopes**, some of which (*e.g.* Pu-239) undergo **nuclear fission**; all are very poisonous. At. no. 94; r.a.m. 244 (most stable isotope).

pneumatic Operated by air pressure.

Pogonophora Phylum of simple worm-like marine animals; beardworms.

■ **poikilothermic** Describing an animal that is 'cold-blooded' and relies on the heat of the environment to warm its body; *e.g.* all invertebrates and, among vertebrates, fish, amphibians and reptiles. *See also* **homoiothermic**.

■ **poison** *1.* Substance that destroys the activity of a **catalyst**. *2.* Substance that when introduced into a living organism in any way destroys life or causes injury to health; a toxin.

polar body One of a pair of minute cells that divide off from an **ovum** (egg) when the **oöcyte** undergoes **meiosis**. Alternative names: polocyte, polar globule, directive body.

polar bond **Covalent bond** in which the bonding **electrons** are not shared equally between the two **atoms**.

Symbol on containers of poisonous chemicals

polar crystal Crystal that has **ionic bonds** between its atoms. Alternative name: ionic crystal.

polarimeter Instrument for measuring the **optical activity** of a substance. Alternative name: polariscope.

polarimetry Measurement of **optical activity**, used in chemical analysis.

■ **polarization** *1*. In physics, lining up of the electric and magnetic fields of an **electromagnetic wave,** *e.g.* as in **polarized light**. Only transverse waves can be polarized. *2*. Formation of gas bubbles or a film of deposit on an electrode of an **electrolytic cell,** which tends to impede the flow of current. *3*. In chemistry, separation of the positive and negative charges of a molecule. [5/10/c]

■ **polarized light** Light waves (which normally oscillate in all possible planes) with fixed orientation of the electric and magnetic fields. It may be created by passing the light through a polarizer consisting of a plate of tourmaline crystal cut in a special way, by using a **Polaroid** sheet. [5/10/c]

polar molecule Molecule that is polarized (*see* **polarization**) even in the absence of an electric field.

polarography Method of chemical analysis for substances in dilute solution in which current is measured as a function of potential between mercury electrodes in an **electrolytic cell** containing the solution.

■ **Polaroid** Trade name for a thin transparent film that produces plane-**polarized light** when light is passed through it. [5/10/c]

pole *1*. North or south end of the Earth's axis. *2*. **Magnetic pole**. *3*. **Electrode** (particularly of a battery).

■ **pollen** Dust-like microspore of a seed plant produced by microsporangium cones in **gymnosperms** and by anthers in **angiosperms**. Each grain contains male **gametes**. If these gametes are carried by an external agent such as wind, insects or water to the **ovules** of gymnosperms or to the **stigma** of angiosperms, they produce fertilization. In susceptible (*i.e.* sensitive) people, pollen can be a powerful **antigen** that results in a vigorous allergic response (*see* **allergy**). Alternative name: farina. [2/4/a] [2/5/a] [2/7/a]

■ **pollen sac** In **angiosperms**, four sacs (locules) of an **anther** in which **pollen** is produced. In conifer species the microspore of the male **strobilus** bears a variable number of pollen sacs. [2/4/a] [2/5/a] [2/7/a]

pollen tube Fine filamentous process developed from **pollen** grains after they become attached to a **stigma**. It grows towards the **ovule**, directing the male nuclei to the ovule in the embryo sac, where fertilization occurs.

■ **pollination** Transference of **pollen** prior to fertilization by agents such as wind, insects, birds, water etc. from an **anther** to a **stigma** in **angiosperms** and from the male to the female cone in **gymnosperms**. [2/4/a] [2/5/a] [2/7/a]

■ **pollution** Harmful changes or presence of undesirable substances in the environment which result from mankind's industrial or social activities. Pollution of the atmosphere includes the presence of sulphur dioxide, which causes acid rain, and of chlorofluorocarbons (CFCs), which have been linked to the depletion of the **ozone layer** in the stratosphere. River pollution is caused mainly by agricultural run-off (*e.g.* of fertilizers or slurry) or discharge of chemicals. Pollution of the oceans is caused by oil spillage or dumping of untreated sewage, industrial wastes or chemical wastes. [2/3/c]

polonium Po Radioactive metallic element in Group VIA of the Periodic Table, used as a source of alpha-particles. At. no. 84; r.a.m. 209 (most stable isotope).

poly- Prefix meaning many (*e.g.* polyamide, polymer).

polyamide Condensation **polymer** in which the units are linked by **amide** groups (–CONH–); *e.g.* hair, wool fibres, nylon.

polybasic Describing an **acid** with two or more acidic (replaceable) hydrogen atoms in its molecule; *e.g.* phosphorus (V) (orthophosphoric) acid, H_3PO_4, with three replaceable hydrogens, is tribasic.

■ **polycarbonate** Linear low-crystalline **thermoplastic** in which the linking elements are **carbonate** groups. Polycarbonates are used in making electrical connectors and soft-drink bottles.

■ **Polychaeta** Class of **annelid** worms that have many chaetae (bristles). Common polychaetes are the bristleworms and ragworms.

polychaete Worm of the class **Polychaeta**.

polychloroethene Alternative name for **polyvinyl chloride** (PVC).

polycyclic Describing a substance that has more than one ring of atoms in its molecule.

polyester Condensation **polymer** formed from a **polyhydric alcohol** and a **polybasic acid**. Polyesters are used in the manufacture of fibres.

■ **polyethylene** **Thermoplastic** produced by the **polymerization** of **ethene** (ethylene). It is used for making film and sheeting for bags and wrappers and for making moulded articles. Alternative names: polyethene, polythene.

polyhydric Containing a number of **hydroxyl groups**.

polyhydric alcohol **Alcohol** that contains three or more **hydroxyl groups**.

■ **polymer** Long-chain molecule built up of a number of smaller molecules, called **monomers**, joined together by **polymerization**. Natural polymers include starch, cellulose and rubber. Synthetic polymers include all kinds of **plastics**. [4/8/b]

■ **polymerization** Process of joining together of small molecules, called **monomers**, to form larger molecules, **polymers** (often in the presence of a **catalyst**). In condensation polymerization, two types of monomer molecules condense to form long chains, with the elimination of a small molecule (such as water). In addition polymerization, long chains are formed by molecules of a single monomer joining together. [4/8/b]

polymethanal **Polymer** formed from methanal (**formaldehyde**). Alternative name: paraformaldehyde.

■ **polymethyl methacrylate** Transparent colourless **thermoplastic**. Its optical properties of high transmission of light and high internal reflection, coupled with great strength, are responsible for its use in place of **glass**. Alternative name: Perspex.

■ **polymorphism** *1.* In chemistry, occurrence of a substance in more than one crystalline form. *2.* In biology, occurrence of an organism in more than one structural form during its life cycle. [2/6/b]

polynucleotide Polymer of many **nucleotides**.

■ **polyp** *1.* Body type of **Anthozoa** and **Hydrozoa** (*e.g.* corals and hydra), which may reproduce asexually by budding or splitting and sexually to give rise to new polyps or **medusae**. *2.* In some other **Coelenterata**, a sedentary stage in their life cycle in which the body is cylindrical, with a mouth surrounded by tentacles at one end and attached to a fixed surface at the other end. *3.* **Polypus**.

■ **polypeptide** Chain of **amino acids** which is a basic constituent of **proteins**. It may be broken down by **enzyme** action (digestion) to form **peptides**. The linking and folding of polypeptides makes up the three-dimensional structure of a protein.

polyphenylethene Alternative name for **polystyrene**.

■ **polyploidy** Condition in which a cell or organism has three to four times the normal **haploid** or gametic number. It is often made use of in plant breeding because it results in the production of larger and more vigorous crops. Because it disturbs the sex-determining mechanism, polyploidy is rare in animals, and would result in sterility. [2/6/b]

polypropene Thermoplastic produced by the **polymerization** of **propene** (propylene). It is used to produce moulded articles and can be made into a fibre. Alternative name: polypropylene.

polypropylene Alternative name for **polypropene**.

polypus Pendulous but usually benign **tumour** that grows from mucous membrane (*e.g.* in the nose or womb). Alternative name: polyp.

■ **polysaccharide** High molecular weight **carbohydrate,** linked by **glycoside** bonds, that yields a large number of **monosaccharide** molecules (*e.g.* simple sugars) on **hydrolysis** or **enzyme** action. The most common polysaccharides have the general formula $(C_6H_{10}O_5)_n$; *e.g.* starch, cellulose, etc.

■ **polystyrene Thermoplastic** produced by the **polymerization** of **styrene** (phenylethene). It is used to produce moulded articles and, as a foam (expanded polystyrene), for ceiling tiles, insulation and packaging. Alternative name: polyphenylethene. [4/8/b]

■ **polytetrafluoroethene** (PTFE) **Thermoplastic** produced by the **polymerization** of **tetrafluoroethene** (tetrafluoroethylene). It is inert, very stable and has anti-stick properties. It is used in non-stick coatings on cooking utensils and as an insulator. Trade names: Fluon, Teflon. Alternative name: polytetrafluoroethylene.

polythene Alternative name for **polyethylene**.

■ **polyurethane** Member of a family of **polymers** (plastics) in which the formation of the **urethane** group is an important step in **polymerization**. They are used for the manufacture of foams and coatings. Alternative name: urethane resin.

polyvalent *1*. Having a **valence** of more than one. *2*. Having more than one valence.

■ **polyvinyl acetate** (PVA) **Thermoplastic** produced by the **polymerization** of vinyl acetate (ethanoate), $CH_2=CHOOCCH_3$. It is used in adhesives and for coating paper and fabrics.

■ **polyvinyl chloride** (PVC) **Thermoplastic** produced by the **polymerization** of vinyl chloride (chloroethene), ($CH_2=CHCl$). It is used as electrical insulation for wires and cables, for making pipes and gramophone records, and for making waterproof clothing. Alternative name: polychloroethene. [4/8/b]

pome False fruit with a fleshy part which develops from the flower's receptacle and not (as in a true fruit) its ovary; *e.g.* apple, pear.

■ **pons** Thick bundle of nerve fibres that relays impulses between different parts of the brain. It joins the medulla oblongata to the midbrain.

population *1.* Human inhabitants of a country. *2.* Animals or plants in a given area.

■ **population genetics** Study of the theoretical and experimental consequence of Mendelian inheritance on population levels, taking into account the **genotypes, phenotypes, gene** frequencies and the mating systems. [2/10/b]

population inversion Condition in which a higher energy state in an atomic system is more populated with electrons than a lower energy state.

porcelain Ceramic made from **quartz**, white **kaolin, marble** and **feldspar**. The plastic paste is first moulded and then fired.

■ **pore** Any small opening; *e.g.* pores in the skin through which perspiration passes from sweat glands.

porosity Property of substance that allows gases or liquids to pass through it.

porphyrin Member of an important class of naturally occurring organic **pigments** derived from four **pyrrole** rings. Many form complexes with metal ions, as in *e.g.* chlorophyll, haem, cytochrome, etc.

■ **portal vein** Any vein connecting two capillary networks, thus allowing for blood regulation from one network by the other; *e.g.* the hepatic portal vein connects the intestine with the liver. [2/4/a] [2/5/a] [2/7/a]

positive *1*. Describing an **electric charge** or **ion** that is attracted by a negative one. *2*. Describing a north-seeking **magnetic pole**.

positive feedback **Feedback** in which the output adds to the input.

positron **Elementary particle** which has a mass equal to that of an **electron**, and an electrical charge equal in magnitude, but opposite in sign, to that of the electron.

■ **post-actinide** Any of the elements with an atomic number greater than that of lawrencium (103), the last of the **actinides**. The post-actinides 104 (hahnium) and 105 (rutherfordium or kurchatovium) have been prepared in minute quantities by bombarding an actinide with atoms of elements such as carbon, oxygen and neon. Element 106 has also been claimed.

■ **posterior** In bilaterally symmetrical animals, the end of the body directed backwards during locomotion; the rear or hind end. In bipedal animals (*e.g.* human beings), it corresponds to the **dorsal** side of quadrupeds.

potash Substance that contains **potassium**, particularly **potassium carbonate**.

■ **potassium** K Highly reactive silver-white metallic element in Group IA of the Periodic Table (the **alkali metals**). Its compounds occur widely (particularly the **chloride**) and have many uses; potassium is an essential nutrient for plants (*see* **fertilizer**). The metal is used as a coolant in **nuclear reactors**. At. no. 19; r.a.m. 39.102.

potassium bicarbonate Alternative name for **potassium hydrogencarbonate**.

potassium bromide KBr White crystalline salt, used in medicine and photography.

potassium carbonate K_2CO_3 White granular solid, used in the manufacture of **glass** and **soap**. Alternative name: potash.

potassium chloride KCl Colourless or white crystalline salt, used as a **fertilizer** and as a dietary salt (sodium chloride) substitute when **sodium** intake must be limited.

potassium cyanide KCN White poisonous solid, used in metallurgy and electroplating.

potassium ferricyanide $K_3Fe(CN)_6$ Red crystals, used as a chemical reagent and in the manufacture of **pigments** (*e.g.* Prussian blue). Alternative name: potassium hexacyanoferrate (III).

potassium ferrocyanide $K_4Fe(CN)_6.3H_2O$ Yellow crystals. Alternative name: potassium hexacyanoferrate (II).

potassium hydrogencarbonate $KHCO_3$ White granular solid, used in pharmaceuticals. Alternative name: potassium bicarbonate.

potassium hydrogentartrate $HOOC(CHOH)_2\ COOK$ White crystalline powder, used in baking powder. Alternative name: cream of tartar.

■ **potassium hydroxide** KOH Strongly **hygroscopic** white solid. A strong **alkali,** it is used in the manufacture of soft **soaps.** Alternative name: caustic potash.

potassium iodide KI Colourless crystalline salt, used in chemical analysis and organic synthesis. Its solution dissolves iodine.

potassium manganate (VII) Alternative name for **potassium permanganate**.

potassium nitrate KNO_3 Colourless crystalline salt, a powerful **oxidizing agent**. It is used in the manufacture of **glass** and explosives, and as a food preservative. Alternative names: saltpetre, nitre.

potassium permanganate $KMnO_4$ Purple crystals, a powerful **oxidizing agent**. It is used in the manufacture of chemicals, as a disinfectant and fungicide, and in **volumetric analysis**. Alternative names: permanganate of potash, potassium manganate (VII).

potassium thiocyanate KSCN Colourless **hygroscopic** solid, used in solution to test for iron (III) (ferric) compounds, which give a blood-red colour.

■ **potential difference** (PD) Difference in electric potential between two points in a current-carrying circuit, usually expressed in volts (*V*). Alternative name: voltage.

■ **potential energy** Energy possessed by an object because of its position, or because it is stretched or compressed (*e.g.* a spring). For instance, because of its position a mass *m* at a height *h* has potential energy *mgh* (where g is the acceleration of free fall). *See also* **kinetic energy**. [5/9/a]

potentiometer *1*. Instrument for measuring **potential difference** or **electromotive force**. *2*. **Voltage divider**.

■ **potassium hydroxide** KOH Strongly **hygroscopic** white solid. A strong **alkali**, it is used in the manufacture of soft **soaps**. Alternative name: caustic potash.

potassium iodide KI Colourless crystalline salt, used in chemical analysis and organic synthesis. Its solution dissolves iodine.

potassium manganate (VII) Alternative name for **potassium permanganate**.

potassium nitrate KNO_3 Colourless crystalline salt, a powerful **oxidizing agent**. It is used in the manufacture of **glass** and explosives, and as a food preservative. Alternative names: saltpetre, nitre.

potassium permanganate $KMnO_4$ Purple crystals, a powerful **oxidizing agent**. It is used in the manufacture of chemicals, as a disinfectant and fungicide, and in **volumetric analysis**. Alternative names: permanganate of potash, potassium manganate (VII).

potassium thiocyanate KSCN Colourless **hygroscopic** solid, used in solution to test for iron (III) (ferric) compounds, which give a blood-red colour.

■ **potential difference** (PD) Difference in electric potential between two points in a current-carrying circuit, usually expressed in volts (*V*). Alternative name: voltage.

■ **potential energy** Energy possessed by an object because of its position, or because it is stretched or compressed (*e.g.* a spring). For instance, because of its position a mass *m* at a height *h* has potential energy *mgh* (where g is the acceleration of free fall). *See also* **kinetic energy**. [5/9/a]

potentiometer *1*. Instrument for measuring **potential difference** or **electromotive force**. *2*. **Voltage divider**.

pound Unit of weight, abbreviation lb, equal to 16 ounces; 14 lb = 1 stone, 112 lb = 1 hundredweight (cwt). 1 lb = 0.4536 kg.

powder metallurgy Science of producing metal powders and of using them for the production of shaped objects (*see* **sintering**).

■ **power** *1.* In physics, rate of doing **work**. The SI unit of power is the **watt** (equal to 10^7 erg s^{-1} or 1/745.7 horsepower). *2.* In optics, the extent to which a curved mirror, lens or optical instrument can magnify an object. For a simple lens, power is expressed in **dioptres**. [5/6/b] [5/9/b]

poxvirus Member of a group of large DNA-containing viruses that are responsible for smallpox, cowpox and some animal tumours.

praseodymium Pr Metallic element in Group IIIB of the Periodic Table (one of the **lanthanides**). At. no. 59; r.a.m. 140.907.

■ **precipitate** Solid that forms in and settles out from a solution.

■ **precipitation** Process of **precipitate** formation. *See* **double decomposition**. [4/9/a,b]

precursor Intermediate substance from which another is formed in a chemical reaction.

■ **predator** Animal that hunts and eats other animals (the prey). [2/4/d] [2/9/d] [2/10/e]

■ **pregnancy** Time that elapses between fertilization or implantation of a fertilized ovum and an animal's birth; the time that an animal spends as an embryo or foetus. Alternative name: gestation. The time that an embryo reptile

■ **primary colour** *1*. Red, green and violet, which give all other colours when light producing them is combined in various proportions. All three mix to give white. *2*. Pigment colours red, yellow and blue, which can also be combined to give pigments of all other colours. All three mix to give black.

■ **primary growth** Growth of roots and shoots derived from the apical **meristem**, which gives rise to the primary plant body. Alternative name: apical growth.

■ **primary sexual characteristics** Sexual features that are present from birth – *i.e.* excluding **secondary sexual characteristics**.

■ **Primates** Order of mammals that includes tarsiers, pottos, lemurs, monkeys, great apes and man. They are characterized in evolutionary terms by maintaining a generalized limb structure, increasing digital mobility, binocular vision and progressive development of the cerebral cortex of the brain. Primate young undergo a long period of growth and development, during which they learn from their parents. [2/3/b] [2/4/b]

primordium Collection of cells that gives rise to a tissue, organ or a group of associated organs; *e.g.* apical shoot and apical root primordia in plants (*i.e.* at the apex of the growing shoot and root).

print-out Output (**hard copy**) from a computer **printer**.

printer Computer **output device** that produces **hard copy** as a print-out. There are various kinds, including (in order of speed) daisy-wheel, dot-matrix, line, barrel and laser.

prism, optical Transparent solid with triangular ends and rectangular sides, with refracting surfaces at acute angles with each other.

probability distribution of electrons Probability that an

electron within an atom will be at a certain point in space at a given time. It predicts the shape of an atomic **orbital**.

■ **proboscis** Tube-like organ of varying form and functions. In insects, a proboscis is a filamentous structure that projects outwards from the mouthparts, functioning as a piercing and sucking device for obtaining liquid food. In elephants (order Proboscidea), the proboscis is the trunk, and in some marine animals it is a tube-like pharynx that can be protruded.

producer Plant in a **food web** that uses **photosynthesis** to convert light energy to chemical energy (which can be used as food by animals).

producer gas Mixture of the gases **hydrogen, nitrogen** and **carbon monoxide** made by passing steam and air through red-hot coke. Alternative name: air gas.

product Substance formed as a result of a chemical change.

proenzyme Alternative name for **zymogen**.

profile In an ecological survey, method of recording details of the land in terms of the contours and height or depth of vegetation.

■ **progesterone Steroid sex hormone** secreted by the **corpus luteum** of the mammalian ovary, placenta, testes and adrenal cortex. In females it prepares the **uterus** for the implantation of a fertilized ovum (egg) and during **pregnancy** maintains nourishment for the embryo by developing the placenta, inhibiting ovulation and menstruation, and stimulating the growth of the mammary glands.

program Sequence of instructions for a computer. Alternative name: programme.

projectile Object that is projected or thrown by force, often referring to a bullet or shell fired from a gun.

■ **prokaryote** DNA-containing, single-celled organism with no proper **nucleus** or **endoplasmic reticulum**; *e.g.* bacteria, blue-green algae (cyanophytes). *See also* **eukaryote**.

prokaryotic Describing or resembling a **prokaryote**.

■ **prolactin** Protein **hormone** secreted by the anterior **pituitary**. In mammals it stimulates lactation and promotes functional activity of the **corpus luteum**. Alternative names: luteotrophin, mammary stimulating hormone, mammogen hormone, mammotrophin.

proline White crystalline **amino acid** that occurs in most **proteins**.

promethium Pm Radioactive metallic element in Group IIIB of the Periodic Table (one of the **lanthanides**). It has several **isotopes** (none of which occurs naturally), with half-lives of up to 20 years. At. no. 61; r.a.m. 145 (most stable isotope).

promoter Substance used to enhance the efficiency of a **catalyst**. Alternative name: activator.

■ **proof** Measure of the **ethanol** (ethyl alcohol) content of a solution (gunpowder moistened with a 100% proof spirit will just ignite). A 100% proof solution is 57.1% ethanol by volume or 49.3% alcohol by weight.

propagation *1*. In botany, any form of plant reproduction, especially when manipulated by human beings in gardening and agriculture. *1*. In physics, directional transmission of energy in the form of waves, *e.g.* sound, radio waves.

propanal CH_3CH_2CHO Liquid **aldehyde**, used in the manufacture of **plastics**. Alternative name: propionaldehyde, propyl aldehyde.

propane C_3H_8 Gaseous **alkane**, easily liquefied under pressure and used as a portable supply of fuel.

propanoic acid CH_3CH_2COOH Liquid **carboxylic acid**, whose calcium salt is used as a food additive. Alternative name: propionic acid.

■ **propanol Alcohol** that occurs as two isomers. *1.* *n*-propanol C_3H_7OH is a colourless liquid, used as a solvent and in making toilet preparations. Alternative names: *n*-propyl alcohol, propan-1-ol. *2.* Isopropanol $(CH_3)_2CHOH$ is also a colourless liquid, used for preparing **esters, acetone** (propanone), and as a solvent. Alternative names: isopropyl alcohol, propan-2-ol.

2-propanone Alternative name for **acetone**.

propellant *1.* Explosive used to propel bullets and shells, or give thrust to solid-fuel rockets. *2.* Gas used in an **aerosol** to expel the contents through an atomizing jet.

12-propenal Alternative name for **acrolein**.

propene $CH_3CH = CH_2$ Colourless gaseous **alkene** (olefin), used in industry for the preparation of **isopropanol, glycerol, polypropene**, etc. Alternative name: propylene.

propenoic acid Alternative name for **acrylic acid**.

propenonitrile Alternative name for **acrylonitrile**.

■ **prophase** First stage of **cell division** in **meiosis** and **mitosis**. During prophase **chromosomes** can be seen to thicken and shorten and to be composed of **chromatids**. The **spindle** is assembled for division of chromosomes and the **nuclear membrane** disintegrates. In meiosis the first prophase is extended into several stages. [2/6/b] [2/8/b]

propionaldehyde Alternative name for **propanal**.

propionic acid Alternative name for **propanoic acid**.

■ **prostaglandin** Member of a group of **unsaturated** fatty acids that contain 20 carbon atoms. They are found in all human tissue, and particularly high concentrations occur in semen. Their activities affect the nervous system, circulation, female reproductive organs and metabolism. Most prostaglandins are secreted locally and are rapidly metabolized by **enzymes** in the tissue.

■ **prostate gland** Gland located at the base of the urinary bladder that forms part of the reproductive system of male mammals. The size of the gland and the quantity of its secretion are controlled by **androgens**. Its function is secretion of a fluid containing **enzymes** and antiglutinating factor, which contributes to the production of semen. [2/4/a] [2/5/a] [2/7/a]

prosthetic group Non-protein portion of a conjugated **protein**, *e.g.* haem group in haemoglobin.

protactinium Pa Radioactive element in Group IIIB of the Periodic Table (one of the **actinides**).It has several **isotopes**, *with half-lives of up to 2 × 10^4 years. At. no. 91; r.a.m. 231* (most stable isotope).

■ **protease Enzyme** that breaks down **protein** into its constituent **peptides** and **amino acids** by splitting peptide linkages (*e.g.* pepsin, trypsin).

protein Member of a class of high molecular weight **polymers** composed of a variety of **amino acids** joined by **peptide** linkages. In conjugated proteins, the amino acids are joined to other groups. Proteins are extremely important in the physiological structure and functioning of all living organisms.

■ **protein synthesis** Process by which **proteins** are made in cells. A molecule of **messenger RNA** decodes the sequence of copied DNA on **ribosomes** in the cytoplasm. A **polypeptide** chain is generated by the linking of **amino acids** in an order instructed by the base sequence of **messenger RNA**.

prothallus In pteridophytes (ferns), the **gametophyte** generation which consists of a flattened, free-living **haploid** disc of cells bearing sex organs. Homosporous plants produce only one type of prothallus, which bears both the male and female sex organs. Heterosporous plants produce two different types of prothalli, a male prothallus which develops **antheridia** and a female one which develops **archegonia**.

■ **protist** Member of the **Protista**. [2/4/b]

■ **Protista** Kingdom that contains simple organisms such as **algae, bacteria, fungi** and **Protozoa**, although sometimes multicellular organisms are excluded. [2/4/b]

■ **proton** Fundamental **elementary particle** with a positive charge equal in magnitude to the negative charge on an **electron**, and with a mass about 1,850 times that of an electron. Protons are constituents of the **nucleus** in every kind of **atom**. [4/8/d]

proton number Alternative name for **atomic number**. [4/8/d]

protonic acid Compound that releases solvated **hydrogen ions** in a suitable polar solvent (*e.g.* water).

■ **protoplasm** Usually transparent jelly-like substance within a **cell** – *i.e.* the **cytoplasm** (which contains various **organelles**) and the **nucleus**.

■ **Protozoa** Subkingdom or phylum of microscopic unicellular organisms which range from plant-like forms to types that feed and behave like animals. They have no common body shape (and some, such as amoeba, have no fixed shape at all) but all have specialized **organelles**. Their basic mode of reproduction is by **binary fission**, although multiple fission and conjugation occur in some species. Some protozoans are colonial and many are parasitic, inhabiting freshwater, marine and damp terrestrial environments. [2/4/b]

■ **protozoan** Member of the **Protozoa**. [2/4/b]

proventriculus *1*. Anterior glandular part of a bird's stomach, which secretes gastric juice. *2*. In insects and crustaceans, the **gizzard**.

prussic acid Alternative name for **hydrocyanic acid**.

pseudo- Prefix meaning false (*e.g.* **pseudohalogen**).

pseudocarp Fruit that incorporates bracts, an inflorescence or receptacle in addition to the **ovary** of the flowering plant. Alternative name: false fruit.

pseudohalogen Member of a group of volatile chemical compounds that chemically resemble the halogens; *e.g.* cyanogen $(CN)_2$, thiocyanogen $(SCN)_2$.

pseudoparenchyma Fungal or algal tissue in which the filaments or **hyphae** are no longer discrete but have become an interwoven mass, falsely resembling **parenchyma**; *e.g.* stipe of a mushroom, thallus of a red alga.

pseudopodium Part of an **amoeba** or similar **protozoan** that is extruded from its unicellular body. Pseudopodia are used for locomotion and to engulf food particles (for digestion).

pseudopregnancy In some female mammals, physiological state resembling **pregnancy** without the formation of embryos. Alternative name: false pregnancy.

psi particle Meson that has no charge, but a very long lifetime. Alternative name: J-particle.

psychiatry Study and treatment of disorders of the mind (*i.e.* mental disorders).

psychology Scientific study of the mind.

■ **Pteridophyta** Division of the plant kingdom that contains all vascular non-seed-bearing lower plants; *e.g.* clubmosses, ferns, horsetails. Pteridophytes are characterized by having a free-living **haploid gametophyte** generation which produces the male antheridia and the female archegonia. [2/4/b]

■ **pteridophyte** Plant that is a member of the **Pteridophyta**. [2/4/b]

PTFE Abbreviation of **polytetrafluoroethene**.

■ ***p*-type conductivity** Conductivity that results from the movement of positive holes (**lattice** sites of a crystalline **semiconductor** that are occupied by an acceptor impurity atom – *i.e.* an atom with one fewer valence electrons than the semiconductor).

■ ***p*-type semiconductor** Form of **semiconductor** that exhibits *p*-type conductivity.

■ **puberty** Stage in development when a child gradually changes into an adult, during which **sex hormones** are produced (by ovaries or testes) and **secondary sexual characteristics** appear.

pulley Simple machine which changes the direction of an applied force or, if it uses more than one pulley wheel, provides a **mechanical advantage** (force ratio) equal to the ratio of the resultant pulling force to the applied force.

■ **pulmonary** Concerning the lungs and breathing (or the respiratory cavities of molluscs). [2/4/a] [2/5/a] [2/7/a]

■ **pulmonary artery** In mammals, a paired **artery** that carries deoxygenated blood from the right ventricle of the heart to the lungs. It is the only artery that carries deoxygenated blood. [2/4/a] [2/5/a] [2/7/a]

■ **pulmonary vein** In mammals, a paired **vein** that carries oxygenated blood from the lungs to the left atrium of the heart. It is the only vein that carries oxygenated blood. [2/4/a] [2/5/a] [2/7/a]

■ **pulse** *1.* In medicine, regular expansion of the wall of an artery caused by the blood-pressure waves that accompany heartbeats. *2.* In botany, a plant of the pea family; a legume. [2/4/c] [2/5/a] *3.* In physics and telecommunications, brief disturbance propagated in a similar way as a **wave**, but not having the continuous periodic nature of a wave.

pump Mechanical device for transferring liquids or gases, or for compressing gases. A simple lift pump employs atmospheric pressure and cannot pump a liquid vertically more than about 10 m (32 feet); a force pump does not have this restriction.

punched card Computer input or output medium consisting of cards punched with coded holes. The actual input device is a punched card reader; the output device is a card punch.

punched tape Computer input or output medium consisting of paper tape punched with coded holes. The actual input device is a punched tape reader; the output device is a tape punch.

■ **pupa** Inactive stage, characterized by a distinct body form, in the life-cycle of insects that undergo **metamorphosis**, during which a **larva** is transformed into an **imago**. Alternative name: chrysalis (especially in butterflies and moths).

■ **pupil** Hole in the iris of the **eye** (which appears as a black circle in the front of the eye); it allows light to enter and pass through the lens to the retina.

■ **purine** $C_5H_4N_4$ Heterocyclic nitrogen-containing **base** from which the bases characteristic of **nucleotides** and **DNA** are derived; *e.g.* **adenine, guanine**. Other purine derivatives include caffeine and uric acid.

■ **putrefaction** Largely **anaerobic** decomposition of organic matter by microscopic organisms (*e.g.* bacteria, fungi, etc.) which results in the formation of incompletely oxidized products.

putty powder Impure **tin (IV) oxide** (SnO_2), used for polishing glass.

PVC Abbreviation of **polyvinyl chloride** (polychloroethene).

■ **pyramid of numbers** The numbers of animals at each **trophic level** decreases, as can be seen by representing them pictorially as a pyramid. [2/7/d]

pyranose Any of a group of **monosaccharide** sugars (hexoses) whose molecules have a six-membered heterocyclic ring of five carbon atoms and one oxygen atom. *See also* **furanose**.

pyrazine $C_4H_4N_2$ Heterocyclic **aromatic compound** whose ring contains four carbon atoms and two nitrogen atoms. Alternative name: 1,4-diazine.

pyrazole $C_3H_4N_2$ Heterocyclic **aromatic compound** whose ring contains three carbon atoms and two nitrogen atoms. Alternative name: 1,2-diazole.

pyrene $C_{16}H_{10}$ **Aromatic compound** consisting of four **benzene** rings fused together.

pyrenocarp Alternative name for **drupe**.

pyrenoid Spherical **protein** body found in the **chloroplasts** of some **algae** (*e.g.* Chlamydomonas).

Pyrex Trade name for a heat-resistant glass, used for domestic and laboratory glassware.

■ **pyridine** C_5H_5N Heterocyclic liquid organic **base** which occurs in the light oil fraction of coal-tar and in bone oil. It forms salts with acids and is important in organic synthesis.

pyridoxine Crystalline substance from which the active **coenzyme** forms of **vitamin** B_6 are derived. It is also utilized as a potent growth factor for bacteria.

pyrimidine $C_4H_4N_2$ Heterocyclic organic **base** from which bases found in **nucleotides** and **DNA** are derived; *e.g.* **uracil, thymine** and **cytosine**. Its derivatives also include barbituric acid and the barbiturate drugs.

pyro- Prefix that denotes strong heat, fire.

pyroelectricity **Polarization** of certain **crystals** by the application of heat.

pyrolysis Decomposition of a chemical compound by heat.

pyrometallurgy Metallurgy involved in the winning and refining of metals where heat is used, as in roasting and smelting.

pyrometer Instrument for measuring high temperatures, above the range of liquid thermometers.

pyrometry Measurement of high temperatures.

pyrophoric alloy Alloy that catches fire when struck or subjected to friction, used for cigarette lighter flints.

■ **pyrrole** $(CH)_4NH$ Heterocyclic organic compound whose ring contains four carbon atoms and one nitrogen atom. A liquid **aromatic compound,** its derivatives are important biologically; *e.g.* haem, chlorophyll.

pyruvate Ester or **salt** of **pyruvic acid.**

pyruvic acid $CH_3COCOOH$ Simplest keto-acid, important in making energy available from ingested food. Alternative name: 2-oxopropanoic acid.

Q

■ **quadrat** In an ecological survey of an area of ground, a small square (usually 1 square metre) within which all species are recorded or measured. [2/7/d]

quadrate One of a pair of bones in the upper jaw of fish, amphibians, reptiles and birds that has evolved into the incus (an **ear ossicle**) in mammals.

quadrivalent Having a **valence** of four. Alternative name: tetravalent.

qualitative Dealing with the identity, qualities or appearance of something only.

■ **qualitative analysis** Identification of the constituents of a substance or mixture, irrespective of their amount.

quanta Plural of **quantum**.

■ **quantasome** One of the tiny, semi-crystalline particles that occur in disc-shaped arrangements within the **chloroplasts** of plant cells, and which are believed to be the centre of light-processing during **photosynthesis**.

quantitative Dealing with quantities of substances, *e.g.* mass, volume, etc., irrespective of their identity.

■ **quantitative analysis** Determination of the amounts of constituent substances present, often by weighing or manipulating volumes of solutions. *See also* **gravimetric analysis; volumetric analysis.**

■ **quantum** Unit quantity (an indivisible 'packet') of energy postulated in the **quantum theory**. The **photon** is the quantum of **electromagnetic radiation** (such as light) and in certain contexts the **meson** is the quantum of the nuclear field.

quantum electrodynamics Study of **electromagnetic interactions**, in accordance with the **quantum theory**.

quantum electronics Generation or amplification of **microwave** power, governed by **quantum mechanics**.

■ **quantum mechanics** Method of dealing with the behaviour of small particles such as **electrons** and **nuclei**. It uses the idea of the particle-wave duality of matter. Thus an electron has a dual nature, particle and wave, but it behaves as one or the other according to the nature of the experiment.

quantum number Integer or half-integral number that specifies possible values of a quantitized physical quantity, *e.g.* energy level, nuclear spin, angular momentum, etc.

quantum state State of an **atom**, **electron**, **particle**, etc., defined by a unique set of **quantum numbers**.

■ **quantum theory** Theory of radiation. It states that radiant energy is given out by a radiating body in separate units of energy known as **quanta**; the same applies to the absorption of radiation. The total amount of radiant energy given out or absorbed is always a whole number of quanta.

quark Subatomic particle that combines with others to form a **hadron**. Current theory predicts six types of quarks and six antiquarks, but none has yet been observed.

■ **quartz** SiO_2 Natural crystalline **silica** (silicon dioxide), one of the hardest of common minerals. Its crystals (which can generate piezoelectricity) are frequently colourless and transparent. It is used as an abrasive and in mortar and cement.

quaternary ammonium compound Member of a group of white crystalline solids, soluble in water, and completely dissociated in solution. These compounds have the general formula $R_4N^+X^-$, where R is a long-chain **alkyl group**. They have **detergent** properties. Alternative name: quaternary ammonium salt.

■ **quicklime** CaO Whitish powder prepared by roasting **limestone**, used in agriculture and in cements and mortar. Alternative names: calcium oxide, lime.

quinine Colourless crystalline **alkaloid**, obtained from the bark of the cinchona shrub, once much used in the treatment and prevention of malaria.

quinoline C_9H_7N Colourless oily liquid **heterocyclic base**. It is present in coal-tar and bone oil, and was first obtained from **quinine** by alkaline decomposition. It is used in making dyes and drugs.

quinone Member of a group of cyclic **unsaturated** diketones in which the double bonds and keto groups are **conjugated**. Thus they are not **aromatic compounds**.

R

■ **rabies** Virus disease, mainly affecting carnivores but which can afflict any warm-blooded animal, that is usually transmitted by bites from infected animals. Alternative (but erroneous) name: hydrophobia.

race In biological classification, an alternative name for **subspecies**. *See also* **variety**.

racemic acid Racemic mixture of **tartaric acid**.

racemic mixture Optically inactive mixture that contains equal amounts of **dextrorotatory** and **laevorotatory** forms of an **optically active** compound.

racemization Transformation of **optically active** compounds into **racemic mixtures**. It can be effected by the action of heat or light, or by the use of chemical reagents.

racemose inflorescence Type of **inflorescence** in which the youngest flowers are at the growing tip of a flower stalk and the oldest ones are near the bottom.

rad *1*. Unit of absorbed dose of ionizing **radiation**, equivalent to 100 ergs per gram (0.01 J kg^{-1}) of absorbing material. The corresponding SI unit is the gray. *2*. Abbreviation for **radian**.

radar Abbreviation of radio detection and ranging, a method of detecting objects, and their bearing and distance, by transmitting continuous or pulsed radio waves and receiving their echos.

radial symmetry Symmetry about any one of several lines or planes through the centre of an object or organism. *See also* **bilateral symmetry**.

radian (rad) SI unit of plane angle; the angle at the centre of a circle subtended by an arc whose length is equal to the radius of the circle. 1 radian = 57 degrees (approx.).

radiance **Radiant flux** per unit area of a surface.

radiant Describing something that emits **electromagnetic radiation** (*e.g.* light, heat rays).

radiant flux The rate at which **power** is emitted or received by an object in the form of **electromagnetic radiation**.

■ **radiant heat** Heat that is transmitted in the form of **infra-red radiation**. [3/6/c]

radiation Energy that travels in the form of **electromagnetic radiation,** *e.g.* **radio waves, infra-red radiation, light, ultraviolet radiation, X-rays** and **gamma-rays**. The term is also applied to the rays of **alpha** and **beta particles** emitted by **radioactive** substances. Particle rays and short-wavelength electromagnetic radiation may be harmful to tissues because they are **ionizing radiation**.

radiation pressure Minute force exerted on a surface by **electromagnetic radiation** that strikes it.

radiation sickness Illness caused by exposure to excessive ionizing radiation. Alpha particles cannot penetrate skin and are not dangerous internally. Gamma radiation, X-rays and neutrons can penetrate the body and are thus the most harmful.

radiation unit Activity of a **radio-isotope** expressed in units of disintegrations per second, called the **becquerel** in SI units. Formerly it was measured in **curies**.

radiator *1.* Object that emits **radiation**. *2.* Heat exchanger, which is used either to dissipate the heat from a coolant fluid (*e.g.* car radiator) or emit heat from a hot fluid (*e.g.* space heater for buildings).

radical *1.* In chemistry, group of **atoms** within a molecule that maintains its identity through chemical changes that affect the rest of the molecule, but is usually incapable of independent existence; *e.g.* an **alkyl group** such as methyl, CH_3-. *See also* **free radical**. *2.* In botany, relating to a root or stem base (*e.g.* radical leaves).

radicle Young **root** that arises in the **embryo** of plants.

■ **radio** Method of telecommunications that uses **radio waves**. The transmitted signal modulates the **amplitude** (AM) or **frequency** (FM) of a carrier wave, which is picked up by an **aerial** (antenna) at the receiver, which demodulates it and usually drives a **loudspeaker**.

radioactive Possessing or exhibiting **radioactivity**.

■ **radioactive decay** Way in which a **radioactive** element spontaneously changes into another element or **isotope** by the emission of **alpha** or **beta particles** or **gamma-rays**. The rate at which it does so is represented by its **half-life**. [4/7/f] [4/8/d]

radioactive equilibrium Condition attained when a parent **radioactive** element produces a daughter radioactive element that decays at the same rate as it is being formed from the parent.

radioactive series One of three series that describe the **radioactive decay** of 40 or more naturally occurring radioactive **isotopes** of high **atomic number**. They are known (after the element at the beginning of each sequence) as the **thorium** series, **uranium** series and **actinium** series.

radioactive standard **Radio-isotope** of known rate of

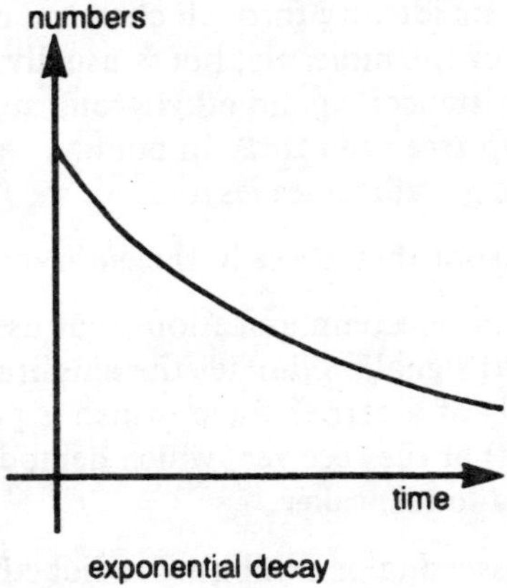

Radioactive substances decay exponentially

radioactive decay used for the calibration of **radiation-measuring** instruments.

radioactive tracing Use of **radio-isotopes** to study the movement and behaviour of an element through a biological or chemical system by observing the intensity of its **radioactivity**.

■ **radioactive waste** Hazardous **radio-isotopes** (fission products) that accumulate as waste products in a **nuclear reactor**. They have to be periodically removed and stored safely or reprocessed. The term is also applied to the waste ('tailings') produced by the processing of uranium ores.

■ **radioactivity** Spontaneous disintegration of atomic **nuclei**, usually with the emission of **alpha particles**, **beta particles** or **gamma-rays**.

radiobiology Study of ionizing radiation in relation to living systems. It includes the effects of radiation on living organisms and the use of radio-isotopes in biological and medical work. *See also* **radiotherapy**.

■ **radiocarbon dating** Method of estimating the ages of carbon-containing (*e.g.* wooden) archaeological and geological specimens that are up to 50,000 years old. A **radio-isotope** of carbon, carbon-14 (C-14), is present in **carbon dioxide** and becomes assimilated into plants during **photosynthesis** (and into animals that eat plants). The C-14 present in 'dead' carbonaceous materials decays and is not replaced (*see* **radioactive decay**). By comparing the **radioactivities** of the 'dead' and 'live' materials, the age of the former can be estimated, because the **half-life** of C-14 is known. A similar technique, potassium-argon dating, is used to determine the age of rocks. Alternative name: radioactive dating.

radiochemistry **Chemistry** of **radioactive** elements and their compounds.

radiodiagnosis Branch of medical **radiology** that is concerned with the use of **X-rays** or **radio-isotopes** in diagnosis.

■ **radio frequency** (RF) Frequency of **electromagnetic radiation** suitable for **radio** transmission, *i.e.* from about 10 kHz to 300,000 MHz.

■ **radiograph** Photographic image that results from uneven absorption by an object being subjected to penetrating **radiation**. An X-ray photograph is a common example.

■ **radiography** Photography using **X-rays** or **gamma-rays**, particularly in medical applications.

■ **radio-isotope** **Isotope** that emits **radioactivity** (ionizing radiation) during its spontaneous decay. Radio-isotopes are useful sources of radiation (*e.g.* in **radiography**) and are used as tracers for **radioactive tracing**.

■ **Radiolaria** Order of single-celled **plankton** animals, spherical in shape, that use **silicon** as a supporting structure rather than the more usual **calcium** compounds.

radiolarian ooze Sediment that contains a high proportion of **silicon** from the skeletons of **Radiolaria**, which covers the sea floor in the deepest equatorial waters of the Indian and Pacific Oceans.

■ **radiology** Study of **X-rays, gamma rays** and **radioactivity** (including **radio-isotopes**), especially as used in medical diagnosis and treatment.

radioluminescence **Fluorescence** caused by **radioactivity**.

radio-opaque Resistant to the penetrating effects of radiation, especially **X-rays**; the term is often used to describe substances injected into the body before a **radiography** examination.

radio telescope **Telescope**, used in astronomy, that can pick up radio signals from extraterrestrial sources. It produces electrical signals which are often recorded graphically.

■ **radiotherapy** Treatment of disorders (*e.g.* **cancer**) by the use of **ionizing radiation** such as **X-rays** or radiation from **radio-isotopes**. [2/9/c]

■ **radio wave** Form of high-frequency **electromagnetic radiation** with a **wavelength** greater than a few millimetres.

■ **radium** Ra Silver-white radioactive metallic element in Group IIA of the Periodic Table (the **alkaline earths**). It has several

isotopes, with half-lives of up to 1,620 years. It is obtained from pitchblende (its principal ore), and used in **radiotherapy** and luminous paints. At. no. 88; r.a.m. 226 (most stable isotope).

radius One of two bones in the forearm or foreleg of a tetrapod vertebrate (the other is the **ulna**).

radius of curvature Of a point on a curve, the radius of a circle that touches the inside of the curve at that point.

radius of gyration Of a rotating object, the distance from the axis of rotation at which the total mass of the object might be concentrated without changing its **moment of inertia**.

radius vector For a point in a plane with the polar co-ordinates (r, θ), the distance from the origin to the point, equal to r.

■ **radon** Rn Radioactive gaseous element in Group 0 of the Periodic Table (the **rare gases**), a **radioactive decay** product of **radium**. It has several **isotopes**, with half-lives of up to 3.82 days. Radon coming out of the ground, particularly in hard-rock areas, is a source of background radiation that has been recognized as a health hazard. At. no. 86; r.a.m. 222 (most stable isotope). [4/8/a] [4/9/a]

radula Organ resembling rows of small teeth, present in plant-eating **molluscs** (*e.g.* snails), used for feeding. Located in the **buccal cavity**, it possesses a serrated edge for rasping plant material.

raffinate Liquid that remains after a substance has been obtained by **solvent extraction**.

raffinose $C_{18}H_{32}O_{16}$ Colourless crystalline **trisaccharide carbohydrate** that occurs in sugar beet, which hydrolyses to the **sugars** galactose, glucose and fructose.

rainbow Arc of spectral colours, seen in the sky opposite the Sun from certain angles only, that results from the reflection and refraction of sunlight by raindrops.

Rajiformes Order of marine fishes comprising the rays, skates, etc., characterized by extremely flattened bodies, and which is most closely related to the sharks.

RAM Abbreviation of **random access memory** of a **computer**.

■ **r.a.m.** Abbreviation of **relative atomic mass** (formerly called atomic weight).

Raman effect Scattering of **monochromatic light**, when it passes through a transparent homogeneous medium, into different **wavelengths** because of interaction with the molecules of the medium.

random access memory (RAM) Part of a computer's **memory** that can be written to and read from. *See also* **ROM**.

Raney nickel Spongy form of **nickel**, made by treating an aluminium-nickel alloy with sodium hydroxide solution, which is used as a **catalyst**, particularly in the **hydrogenation** of **fats and oils**. It was named after the American chemist M. Raney.

■ **Rankine scale** Temperature scale that expresses **absolute temperatures** in degrees Fahrenheit (absolute zero = 0 °R). It was named after the British engineer and physicist William Rankine (1820–72).

Raoult's law The relative lowering of the **vapour pressure** of a **solution** is proportional to the **mole** fraction of the **solute** in the solution at a particular temperature. It was named after the French scientist François Raoult (1830–1901).

rare-earth element Alternative name for a member of the series of elements, in Group IIIB of the **Periodic Table,** known as the **lanthanides.**

■ **rare gas** One of the uncommom, unreactive and highly stable gases in Group 0 of the Periodic Table. They are **helium, neon, argon, krypton, xenon** and **radon**. Alternative names: inert gas, noble gas.[4/8/a] [4/9/a]

raster Display of information in the form of a grid, usually referring to the image produced by the scanning action of a **cathode-ray tube,** *e.g.* in a television receiver.

rate constant Constant of proportionality for the speed of a **chemical reaction** at a particular temperature. It can only be obtained experimentally. Alternative names: velocity constant, specific rate constant.

rate-determining step Slowest step of a **chemical reaction** which determines the overall rate, provided the other steps are relatively rapid. Thus the kinetics and **order of reaction** are basically those of the rate-determining step. Alternative name: limiting step.

rate equation Alternative name for **rate law.**

rate law Equation that relates the rate of a **chemical reaction** to the **concentration** of the individual reactants. It has the form rate $= k[X]^n$, where k is the **rate constant,** X is the **reactant** and n is the **order of reaction.** It can only be obtained experimentally. Alternative name: rate equation.

■ **rate of reaction** Speed of a **chemical reaction,** usually expressed as the change in **concentration** of a reactant or product per unit time. It can be affected by temperature, pressure and the presence of a **catalyst.**

ratite Member of a group of flightless running birds that includes ostriches, rheas, emus and cassowaries.

raw material Substance from which others are made. It may be simple or complex. *E.g.* nitrogen (from air) and hydrogen are raw materials for the synthesis of ammonia; coal and petroleum are raw materials from which a wide range of complex chemicals are made.

■ **ray** Beam of any type of radiation (*e.g.* light, cathode rays). In ray diagrams showing the behaviour of mirrors and lenses, rays of light are drawn as straight lines.

■ **rayon** Man-made fibre manufactured from **cellulose** (obtained from wood pulp), used for making textiles. For acetate rayon, a solution of cellulose acetate in a volatile solvent is extruded through spinnerets into air. For viscose rayon, wood pulp is dissolved in sodium hydroxide and carbon disulphide to produce cellulose xanthate, which is extruded through spinnerets into a bath of sulphuric acid to re-form cellulose fibres. Industrial uses of rayon include conveyor belts and hoses. Former name: artificial silk.

■ **reactance** (X) Property of a **capacitance** or **inductance** in a circuit carrying alternating current (a.c.) that makes it oppose the passage of current; the imaginary part of **impedance**. It is measured in ohms. For a pure capacitance C, $X = ½\pi fC$, where f is the frequency of the a.c.; for a pure inductance L, $X = 2\pi fL$.

reactant Substance that reacts with another in a **chemical reaction** to form new substance(s).

reaction Alternative name for **chemical reaction**.

reactive dye Dye that forms a **covalent bond** with the fibre molecule of the textile being dyed. This provides excellent fastness. Such dyes are used to dye cellulose fibres (*e.g.* rayon).

reactor Alternative name for a **nuclear reactor**.

read head Electromagnetic device on a tape recorder, video recorder or computer that 'reads' signals stored on a magnetic tape, disc or drum.

read-only memory (ROM) Part of a computer's **memory** that can only be read (and not written to). *See also* **random access memory**.

read-write head Electromagnetic device that functions as both a **read head** and a **write head**.

reafforestation Alternative term for **afforestation**.

■ **reagent** *1*. Substance that takes part in a **chemical reaction**; a reactant. *2*. Common laboratory chemical used in chemical analysis and for experiments.

realgar As_2S_2 Natural red arsenic disulphide, used as a pigment and in pyrotechnics.

real gas Gas that never fully achieves 'ideal' behaviour. *See also* **ideal gas**.

recapitulation theory Largely discredited belief based on external observations that the developmental stages undergone by an individual organism – *i.e.* from egg to adult – reflect various stages in the **evolution** of the organism's species.

receptacle Region of a plant that bears flower-parts; the top of a flower stem. It bears the **sepals, petals, stamens, gynaecium** or **carpels**. Its shape varies in different species.

■ **receptor** Sensory **cell**, which may be part of a group that form a **sense organ** capable of detecting stimuli. When a receptor is stimulated (*e.g.* by temperature or light), it produces electrical or biochemical changes that are relayed to the **nervous system** for processing.

■ **recessive** Describing a **gene** that is expressed in the **phenotype** (appearance of an organism) only when it is **homozygous** in a cell; *i.e.* there have to be two recessive genes for their effect to be apparent. The presence of a **dominant allele** masks the effect of a recessive gene; *i.e.* in a combination of a dominant gene and a recessive gene, the dominant gene manifests itself in the phenotype. [2/5/b] [2/6/b] [2/7/b]

recipient Organism that receives material from another, *e.g.* as in the taking up of **DNA** by one **bacterium** from another.

reciprocal cross Hybrid of two plants that results from mutual **pollination**.

reciprocal proportions, law of Alternative name for the law of **equivalent proportions**.

reciprocal wavelength Alternative name for **wave number**.

■ **recombinant DNA** Type of **DNA** that has genes from different sources, genetically engineered using **recombination**.

■ **recombination** Process by which new combinations of characteristics not possessed by the parents are formed in the offspring. It results from **crossing over** during **meiosis** to form **gametes** that unite during fertilization to form a new individual. Genetic engineers have developed techniques for artificially recombining strands of **DNA** (to make recombinant DNA). [2/9/b]

record In computing, a number of elements of data that together form one unit of stored information.

■ **recrystallization** *1.* Change from one crystal structure to another; it occurs on heating or cooling through a critical temperature. *2.* Purification of a substance by repeated **crystallization** from solution.

rectification *1*. In physics, term used to describe conversion of **alternating current** (a.c.) into **direct current** (d.c.) using a **rectifier**. *2*. In chemistry, purification of a liquid using **distillation**.

rectified spirit Solution of **ethanol** (ethyl alcohol) that contains about 5–7% water. It is a constant-boiling mixture and the water cannot be removed by **distillation**.

■ **rectifier** Electrical device, such as a **diode**, for converting an **alternating current** (a.c.) into a **direct current** (d.c.). It may take the form of a plate rectifier, a **diode** valve or a **semiconductor** diode.

■ **rectum** Final part of the **intestine**, through which **faeces** are passed and where they may be stored temporarily after reabsorption of water. [2/4/a] [2/5/a] [2/7/a]

■ **recycle** Use again. In nature, substances are recycled after some form of decomposition; *e.g.* nitrogen is released into the soil after plants and animals decay, to become available for other growing plants. [2/2/d] [2/6/d]

■ **recycling** Method of conserving resources that involves sorting waste materials into their chief components (metals, plastic, paper and glass) and using them to make more of the component materials. *E.g.* up to half the aluminium produced comes from recycled soft drinks cans. [2/6/d]

■ **red blood cell** Alternative name for an **erythrocyte**, also known as a red cell or red corpuscle.

red clay Fine-grained sediment, covering about 50% of the mid-ocean floors, formed from material that originated on land and was carried far out to sea by rivers and the wind.

red lead Pb_3O_4 Bright red powdery oxide of **lead**. An **oxidizing agent**, it is used in anti-rust and priming paints.

Alternative names: minium, dilead (II) lead (IV) oxide, lead tetraoxide, triplumbic tetroxide.

■ **redox reaction Chemical reaction** in which **oxidation** is necessarily accompanied by **reduction**, and vice versa; an oxidation-reduction reaction.

■ **reduced equation of state** Law which states that if any two or more substances have the same reduced pressure π – *i.e.* their pressures are the same fraction or multiple π of their respective **critical pressures** – and are at equal reduced temperatures θ, then their reduced volumes ϕ should be equal.

reduced pressure distillation Alternative name for **vacuum distillation**.

■ **reduced temperature, pressure and volume** Quantities θ, π and ϕ which are the ratios of the temperature, pressure and volume to the **critical temperature**, **critical pressure** and **critical volume** respectively in the **reduced equation of state**.

■ **reducing agent** Substance that causes chemical **reduction**, often by adding hydrogen or removing oxygen; *e.g.* carbon, carbon monoxide, hydrogen. Alternative name: electron donor.

reducing sugar Any **sugar** that can act as a **reducing agent**. *See also* **Benedict's test**; **Fehling's test**.

reductase Enzyme that causes the **reduction** of an **organic compound**.

■ **reduction** Chemical reaction that involves the addition of hydrogen or removal of **oxygen** from a substance, often by the action of a **reducing agent**.

reduction division Alternative name for **meiosis**.

■ **refining** *1.* Purification; the removal of impurities from a substance, particularly a crude metal after it has been extracted from its ore. *2.* Splitting of **petroleum** into its component hydrocarbons, usually by **fractional distillation**. [4/8/b] [4/10/d]

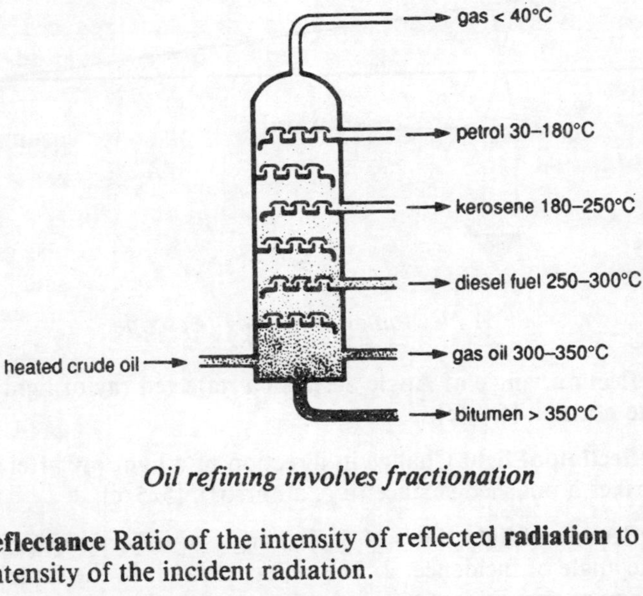

Oil refining involves fractionation

reflectance Ratio of the intensity of reflected **radiation** to the intensity of the incident radiation.

reflecting telescope Astronomical **telescope** that uses mirrors to produce an enlarged image. By multiple reflection, mirrors can 'fold' the light path, so making the telescope shorter than one using only lenses; also mirrors are lighter, and can be made larger, than glass lenses of comparable power. Alternative name: reflector.

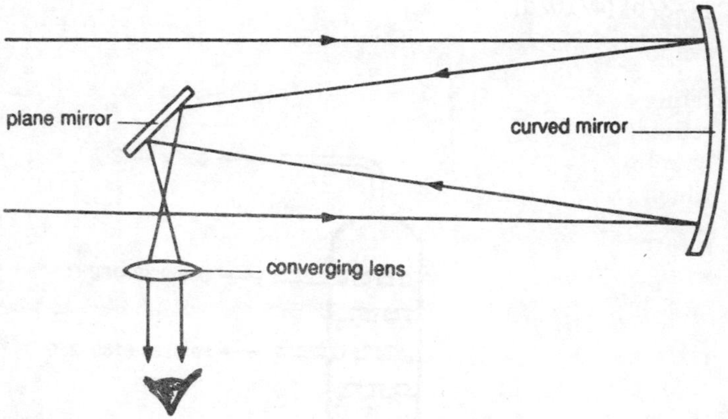

A Newtonian reflecting telescope

■ **reflection, angle of** Angle between a reflected ray of light and the normal. [5/5/c]

■ **reflection of light** Change in direction of a light ray after it strikes a polished surface (*e.g.* a mirror). [5/5/c]

■ **reflection of light, laws of** *1.* The angle of reflection equals the angle of incidence. *2.* The reflected ray is in the same plane as the incident ray and the normal at the point of incidence. [5/5/c]

■ **reflector** *1.* Object or surface that reflects **electromagnetic radiation** (*e.g.* light, radio waves), particularly one around or inside a lamp to concentrate a light beam. *2.* Alternative name for **reflecting telescope**. [5/5/c]

■ **reflex** In biology, sequence of nerve impulses that produces a faster involuntary response to an external stimulus than the corresponding voluntary response (which has to pass via the brain).

reflux Boiling of a liquid for long periods of time. Loss by evaporation is prevented by using a **reflux condenser**.

■ **reflux condenser** Vertical condenser used in the process of refluxing. It is attached to a vessel that contains the liquid to be refluxed and condenses the vapour produced on boiling, which then runs back into the vessel.

reforming Production of branched-chain **alkanes** from straight-chain ones or the production of **aromatic compounds** (*e.g.* benzene) from **alkenes**, using **cracking** or a **catalyst**.

refracting telescope Astronomical **telescope** that has two converging lenses, one an objective of long focal length and the other an eyepiece of short focal length. Alternative name: refractor.

refraction, angle of Angle between a refracted ray of light and the normal.

refraction correction Small correction that has to be made to the observed altitude of a heavenly body because of the **refraction of light** by the Earth's atmosphere. All bodies appear to be higher than they actually are.

■ **refraction of light** Change in direction of a light ray as it passes obliquely from one transparent medium to another of different **refractive index**.

■ **refraction of light, laws of** *1*. For two particular media, the ratio of the sine of the angle of incidence to the sine of the angle of refraction is constant (the **refractive index**). This is a statement of Snell's law. *2*. The refracted ray is in the same

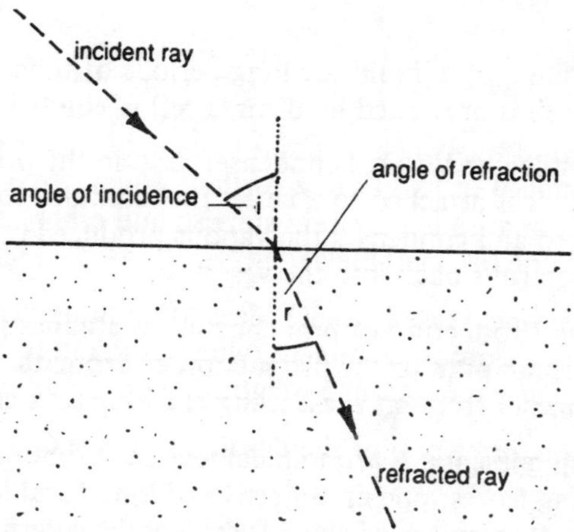

Refraction of light between two media

plane as the incident ray and the normal at the point of incidence. [5/7/e]

refractive constant Alternative name for **refractive index.**

■ **refractive index** μ Ratio of the speed of **electromagnetic radiation** (such as light) in air or vacuum to its speed in another medium. The speed depends on the **wavelength** of the radiation as well as on the **density** of the medium. For a refracted ray of light, it is equal to the ratio of the sine of the angle of incidence *i* to the sine of the angle of refraction *r*; *i.e.* $\mu = \sin i / \sin r$. Alternative name: refractive constant.

refractometer Instrument for measuring the **refractive index** of a substance.

refractor Alternative name for **refracting telescope**.

refrigerant Substance used as the working fluid in a refrigerator (*e.g.* ammonia, fluon, CFCs).

regelation Refreezing of ice that has been melted by pressure when the pressure is removed.

■ **regeneration** Regrowth of tissue to replace that which has been damaged or lost (*e.g.* growth of a plant from a cutting, regeneration of a lost limb in starfish and crabs, and wound healing in higher mammals).

■ **relative atomic mass** (r.a.m.) Mass of an atom relative to the mass of the isotope carbon-12 (which is taken to be exactly 12). Former name: atomic weight.

■ **relative density** Ratio of the **density** of a given substance to the density of some reference substance. For liquids, relative densities are usually expressed with reference to the density of water at 4 °C. Former name: specific gravity.

■ **relative humidity** Ratio of the **pressure** of water vapour present in air to the pressure the water vapour would have if the air were saturated at the same temperature (*i.e.* to the saturated water vapour pressure). It is expressed as a percentage.

relative molecular mass Sum of the **relative atomic masses** of all the atoms in a molecule of a substance. Alternative name: molecular weight.

relative permeability For a material placed in a **magnetic field**, the ratio of its **permeability** to the permeability of free space.

relativity Einstein's theory; scientific principle expressing in mathematical and physical terms the implications of the fact that observations depend as much on the viewpoint as on what is being observed.

■ **relay, electrical** Electric switching device that brings about changes in an independent circuit. [5/5/b]

■ **relict** Species or population that has survived, often with a greatly reduced range, when all related organisms have become extinct. [2/9/c]

reluctance Total **magnetic flux** in a magnetic circuit when a **magnetomotive force** is applied.

reluctivity Reciprocal of magnetic **permeability**. Alternative name: specific reluctance.

rem Abbreviation of röntgen equivalent man, the quantity of **ionizing radiation** such that the energy imparted to a biological system per gram of living material has the same effect as one **röntgen**.

■ **renewable resources** Natural material that can be renewed (*e.g.* timber can eventually be replaced by planting more trees). **Fossil fuels** are not renewable because they cannot be replaced once they have been used up. [2/6/d]

renin Enzyme produced by the kidney that constricts arteries and thus raises blood pressure.

rennin Enzyme found in gastric juice that curdles milk. It is the active ingredient of rennet.

replicase Enzyme that promotes the synthesis of **DNA** and **RNA** within living cells.

■ **reproduction** Procreation of an organism. **Sexual reproduction** involves the fusion of **sex cells** or **gametes** and the exchange of genetic material, thus bringing new vigour to a species. **Asexual reproduction** does not involve gametes, but usually the vegetative proliferation of an organism (*see* **vegetative propagation**).

■ **reptile** Member of the animal class **Reptilia**. [2/3/b] [2/4/b]

■ **Reptilia** Class of **poikilothermic** ('cold-blooded'), principally egg-laying vertebrates characterized by a body covering formed of scales. Reptiles include snakes, lizards, crocodiles, turtles, etc. [2/3/b] [2/4/b]

■ **resin Organic compound** that is generally a viscous liquid or semi-liquid which gradually hardens when exposed to air, becoming an amorphous, brittle solid. Natural resins, found in plants, are yellowish in colour and insoluble in water, but are quite soluble in organic solvents. Synthetic resins (types of plastics) also possess many of these properties. *See also* **rosin**.

■ **resistance, electrical** Property of an electrical **conductor** that makes it oppose the flow of **current** through it. It is measured in **ohms**.

■ **resistivity** Property of a material that makes it oppose the flow of an electric current. For a given specimen of a conductor, it is the product of its resistance R and cross-sectional area A divided by its length l, at a specified temperature (in units of ohm metres). Alternative name: specific resistance.

■ **resistor** Device that provides **resistance** in electrical circuits.

resolution of forces Process of division of **forces** into components that act in specified directions. *See also* **resultant**.

■ **resonance** *1.* Movement of **electrons** from one **atom** of a **molecule** or **ion** to another atom of that molecule or ion. *2.* Phenomenon in which a system is made to vibrate at its natural frequency as a result of vibrations received from another source of the same frequency. [5/8/c]

resorcinol $C_6H_4(OH)_2$ Crystalline dihydric **phenol,** used in the synthesis of drugs and dyes. Alternative names: *m*-dihydroxybenzene; 1,3–benzenediol.

■ **respiration** *1.* Release of energy by living organisms from the breakdown of organic compounds. In **aerobic respiration,** which occurs in most cells, **oxygen** is required and **carbon dioxide** and water are produced. Energy production is coupled to a series of **oxidation-reduction reactions,** catalysed by **enzymes.** In **anaerobic respiration** (*e.g.* **fermentation**), food substances are only partly broken down, and thus less energy is released and oxygen is not required. *2.* Alternative name for breathing.

■ **respiratory movement** Movement by an organism to allow the exchange of respiratory gases, *i.e.* the taking up of **oxygen** and release of **carbon dioxide**. In mammals such as human beings this entails breathing, involving movements of the chest and **diaphragm**. In fish, water is passed over the **gills** for gaseous exchange.

■ **respiratory organ** Organ in which **respiration** (breathing) takes place. In mammals (*e.g.* human beings), the process is carried out in the **lung**; in fish, the **gills**. There gaseous exchange takes place (usually of **oxygen** and **carbon dioxide**).

respiratory pigment Substance that can take up and carry **oxygen** in areas of high oxygen concentration, releasing it in parts of the organism with low oxygen concentrations where it is consumed, *i.e.* by **respiration** in cells. In vertebrates the respiratory pigment is haemoglobin; in some invertebrates it is haemocyanin.

■ **respiratory quotient** (RQ) Ratio of **carbon dioxide** produced by an organism to the **oxygen** consumed in a given time. It gives information about the type of food being oxidized. *E.g.* **carbohydrate** has an RQ of approximately 1, but if the RQ becomes high (*i.e.* little oxygen is available), **anaerobic respiration** may occur.

■ **response** Physical, chemical or behavioural change in an organism initiated by a **stimulus**.

resting potential Potential difference between the inner and outer surfaces of a resting **nerve**, which is about −60 to −80 mV. It happens when the nerve is not conducting any impulse, and is in contrast to the **action potential**, which occurs during the application of a stimulus and brings about a rise in the potential difference to a positive value.

restitution, coefficient of Ratio of the difference in velocity of two colliding objects after impact to the difference before impact. An elastic object that bounces well has a high coefficient of restitution.

■ **restriction enzyme Enzyme** (a **nuclease**) produced by some **bacteria** that is capable of breaking down foreign **DNA**. It cleaves double-stranded DNA at a specific sequence of **bases**, and the DNA of the bacteria is modified for protection against degradation. Restriction enzymes are used widely as tools in **genetic engineering** for cutting DNA. Alternative name: restriction endonuclease.

resultant Of two or more **vectors**, the single vector that has the same effect. It can be found by constructing the **parallelogram of vectors**.

reticulum Second chamber of the stomach of a **ruminant**. *See also* **endoplasmic reticulum**.

■ **retina** Light-sensitive tissue at the back of the vertebrate **eye**, made up of a network of interconnected nerves. The first cells in the network are photo-receptors consisting of **cones** (which are sensitive to colour) or **rods** (which are sensitive to light). They act by means of visual pigments (*e.g.* **rhodopsin**) which cause impulses to be transmitted to the visual centre of the brain via the **optic nerve**. [2/4/a] [2/5/a] [2/7/a]

retinol Fat-soluble **vitamin** found in plants, in which it is formed from **carotene**. Alternative name: vitamin A.

retort Heated vessel used for the **distillation** of substances, as in the separation of some metals.

retrix One of a bird's tail feathers; most species have 12 retrices.

■ **retrovirus** Member of a group of **viruses** that contain **RNA** as their genetic material. They use an RNA-dependent **DNA polymerase** or reverse transcriptase **enzyme** to carry out **transcription**. Many RNA viruses are **carcinogenic** in their hosts, which include mammals. Alternative name: oncovirus.

reverberatory furnace Furnace for **smelting** metal **ores** in which the flame and hot gases are reflected downwards by a curved roof to the material to be heated. The fuel is in one part of the furnace and the ore in another.

■ **reversible process** Process that can theoretically be reversed by an appropriate small change in any of the thermodynamic variables (*e.g.* pressure, temperature). Real natural processes are irreversible.

■ **reversible reaction Chemical reaction** that can go either forwards or backwards depending on the conditions.

Reynolds number (Re) Dimensionless quantity of the form $LV\rho/\mu$ which is proportional to the ratio of inertial force to viscous force in a liquid flowing through a cylindrical tube, where L is diameter of the tube, V is linear velocity, ρ is fluid density and μ is fluid viscosity. The critical Reynolds number corresponds to the change from turbulent flow to laminar flow as the velocity is reduced.

rhabdovirus Member of a group of viruses that can infect multi-cellular animals and plants. One type causes **rabies**.

rhenium Re Rare metallic element of Group VIIB of the Periodic Table (a **transition element**), used in making thermocouples. At no 75; r.a.m. 186.20.

rheostat Variable resistor, used to control voltage (*e.g.* as the volume control in audio equipment).

rhinovirus Member of a group of viruses that infect the respiratory tract of vertebrates and which are one of the main causative agents of the common cold.

■ **rhizoid** Small **root**-like structure, composed of one or few cells, used for anchorage by some lower plants (*e.g.* liverworts, mosses). [2/4/a] [2/5/a] [2/7/a]

■ **rhizome** Underground main **stem** used for food storage by some plants (*e.g.* many grasses, iris). [2/4/a] [2/5/a] [2/7/a]

rhodium Rh Silver-white metallic element in Group VIII of the Periodic Table (a **transition element**), used as a **catalyst** and in making thermocouples. At. no. 45; r.a.m. 102.905.

rhodopsin Protein (derived from vitamin A) in the **rods** of the **retina** of the eye which acts as a light-sensitive pigment; the action of light brings about a chemical change that results in the production of a nerve impulse. Alternative name: visual purple.

■ **riboflavin** Orange water-soluble crystalline solid, member of the vitamin B complex. It plays an important role in growth. Alternative names: riboflavine, lactoflavin, vitamin B_2. [3/9/b]

■ **ribonucleic acid** *See* **RNA**. [2/10/b]

ribose $C_5H_{10}O_5$ Optically active **monosaccharide** sugar, a component of the nucleotides of **RNA**.

■ **ribosome** Particle present in the **cytoplasm** of cells, often attached to the **endoplasmic reticulum**, that is essential in the biosynthesis of **proteins**. Ribosomes are composed of protein and **RNA**, and are the site of attachment for **messenger RNA** during protein synthesis. They may be associated in chains called polyribosomes. [2/10/b]

■ **rickets** Disorder that results from a deficiency of **vitamin** D. It mainly affects children and can cause deformed limbs. [2/7/a]

■ **rickettsiae** Group of micro-organisms, often classified as being part-way between **bacteria** and **viruses**, that are parasitic on the cells of arthropods (lice, mites and ticks) and vertebrates. Some can cause serious disorders (*e.g.* typhus in human beings).

■ **right-hand rule** *1.* Rule that relates the directions of induced current, magnetic field and movement in electromagnetic induction. If the right hand is held with the thumb, first and second fingers at right-angles, the thumb represents the direction of movement, the first finger the direction of the magnetic field and the second finger the direction of the induced current. Alternative name: Fleming's right-hand rule (*see* **left-hand rule**). *2.* Rule that gives the direction of the concentric magnetic field round a current-carrying conductor. If the right hand is held with the thumb directed upwards to represent the direction of the current, the fingers curl round in the direction of the magnetic field. Alternative name: right-hand grip rule. [5/7/c]

rigidity modulus Measure of the resistance of an object or material to a shearing **strain**, equal to the shear force per unit area divided by the angular deformation. Alternative name: shear modulus.

Ringer's fluid/solution Physiological **saline** solution used for keeping **tissues** and **organs** alive outside the body (in vitro). It is similar in composition to the fluid that naturally bathes cells and tissues, maintaining a constant internal environment. It contains chlorides of sodium, potassium and calcium. It was named after the British physiologist Sydney Ringer (1835–1910).

ring main Method of wiring electric power sockets so that they are connected in **parallel** to a supply that forms a ring or chain (typically round one floor of a building), instead of wiring each socket back to a single supply point.

ringworm *See* **tinea**.

ripple tank Apparatus for demonstrating the behaviour of water waves, which in many cases is analogous to the behaviour of other types of wave motion.

■ **RNA** Abbreviation of ribonucleic acid, one of the nucleic acids present in cells, the other being **DNA**. It is composed of nucleotides that contain **ribose** as the sugar. Messenger RNA takes part in transcription or copying of the genetic code from a DNA template. Transfer RNA and ribosomal RNA take part in translation or protein synthesis.

■ **RNA virus** A **virus** that has **RNA** as its genetic material (instead of **DNA**).

■ **rock** Mixture of minerals; any distinguishable portion of the Earth's crust, soft or hard, loose or consolidated, *e.g.* granite, limestone, clay, coal and sand.

rock salt Naturally occurring crystalline **sodium chloride**, an important raw material in the chemical industry.

■ **rod** Type of sensory cell present in the **retina** of the vertebrate **eye.** It is stimulated by light and is concerned with vision in low illumination. The absorption of light energy (photons) by the visual pigment **rhodopsin** present in the rod causes a nervous impulse, which travels along the **optic nerve** to the brain. *See also* **cone.**

roentgen Alternative name for **röntgen.**

ROM Abbreviation of **read-only memory** of a **computer.**

■ **röntgen** (R) Unit of radiation; the amount of **X-rays** or **gamma- rays** that produce a charge of 2.58×10^{-4} coulomb of electricity in 1 cm^3 of dry air. Alternative name: roentgen.

■ **röntgen rays** Alternative name for **X-rays.**

roasting Heating a compound (*e.g.* a metal ore) in air to convert it to an **oxide.**

root Structure in **vascular plants** whose function is anchorage and the uptake of water and **mineral salts** by **osmosis**. Roots are usually partly or completely underground. The vascular tissues form a central core, unlike those in a **stem.**

root hair Single cell in the **outer layer** of a young root which has a very large surface area in relation to its volume, thus increasing its efficiency for absorbing water.

■ **root nodule** Lump that occurs on the roots of leguminous plants (*e.g.* peas, beans, clover) which contains bacteria that can bring about the **fixation of nitrogen.**

■ **root pressure** One of the ways in which water rises up a plant (the others being due to upward 'pull' of **capillarity** and **transpiration**).

rosin Amber brittle translucent **resin** obtained as a residue after distilling **turpentine**. It is used in making paints and

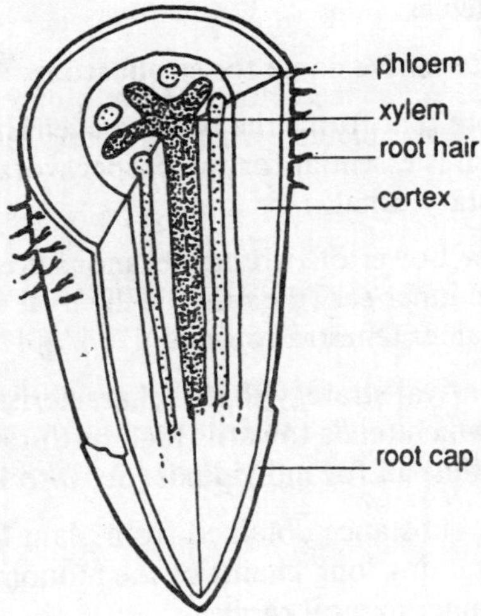

Structure of a root tip

varnishes, as a flux for soldering, and (as a powder) to give grip to boxers' shoes and violin bows.

rostrum *1*. Forward projection of a fish's head beyond its mouth. *2*. Specialized piercing and sucking mouthparts of a **bug**.

rotary converter Electric motor combined with a **dynamo**, used to convert an **alternating current** (a.c.) to a **direct current** (d.c.).

rotatory Optically active; capable of rotating the plane of **polarized light**.

■ **Rotifera** Phylum of small (mostly less than 2 mm in length) aquatic animals, which swim by beating a 'wheel' of cilia.

Most rotifers are found in fresh water. Alternative name: wheel-animalcules.

rotoscope Alternative name for **stroboscope**.

■ **roughage Fibre** that forms the bulking agent in the human diet and which is essential for the proper working and health of the alimentary canal.

■ **round window** Lower of two membranous areas on the **cochlea** of the inner **ear** (the other is the **oval window**). Alternative name: fenestra rotunda. [2/4/a] [2/5/a] [2/7/a]

r-selection Survival strategy that is characteristic of colonizing species, and which tends towards high birth rates combined with a short lifespan for individuals. *See also* **k-selection**.

■ **rubber** Elastic substance obtained from plant **latex**. It is a **polymer**, containing long chains of the monomer **isoprene**. Alternative name: natural rubber.

rubber, synthetic Synthetic compound with a structure resembling that of natural **rubber** in consisting of long-chain **polymer** molecules; these are built of **monomers** such as **acetylene** (ethyne), **isoprene, styrene** and **vinyl** compounds.

rubidium Rb Reactive silver-white metal in Group IA of the Periodic Table (the **alkali metals**), with a naturally occurring radioactive **isotope** (Rb-87). At. no. 37; r.a.m. 85.47.

ruby Natural red impure form of **corundum** (aluminium oxide, Al_2O_3). Imperfect rubies are used as bearings in watches etc. Synthetic rubies are used as gems or in industry.

■ **ruminant** Animal of the suborder Ruminantia, which includes cattle, deer and other mammals that chew the cud. Ruminants are characterized by complex digestive systems, with multi-chambered stomachs, that can break down **cellulose** with the aid of bacterial **fermentation**. [2/4/d] [2/10/e]

■ **rust** Red-brown corrosion product consisting of hydrated iron oxides that forms on the surface of **iron** and **steel** exposed to moisture and air. [4/7/b]

ruthenium Ru Silver-white metallic element in Group VIII of the Periodic Table (a **transition element**), used to add hardness to platinum alloys. At. no. 44; r.a.m. 101.07.

rutherfordium Rf Element no. 104 (a **post-actinide**). It is a radioactive metal with at least three very short-lived isotopes (half-lives up to 70 seconds), made by bombarding an **actinide** with carbon, oxygen or neon atoms. Alternative name: kurchatovium.

rutile Natural form of **titanium (IV) oxide** (TiO_4) and a minor **ore** of **titanium**.

Rydberg constant (R) Constant relating to the **wave number** of atomic spectrum lines. Its value for hydrogen is 1.09677×10^7 m^{-1}. It was named after the Swedish physicist Johannes Rydberg (1854–1919).

Rydberg formula Formula for expressing the **wave number** of a spectral line. The general Rydberg formula is given by the equation $1/\lambda = R(1/n^2 - 1/m^2)$, where n and m are positive integers, and R is the **Rydberg constant**.

S

saccharide Simplest type of **carbohydrate,** with the general formula $(C_6H_{12}O_6)_n$, common to many **sugars.** Alternative name: saccharose. *See also* **disaccharide; monosaccharide; polysaccharide.**

saccharimetry Measurement of the concentration of **sugar** in a solution from its **optical activity,** by using a **polarimeter.**

■ **saccharin** $C_6H_4SO_2CONH$ White crystalline **organic compound** that is about 550 times sweeter than **sugar;** an artificial sweetener. It is almost insoluble in water and hence it is used in the form of its soluble **sodium salt.** Alternative names: 2-sulphobenzimide, saccharine.

saccharose Alternative name for **saccharide.**

■ **sacrificial anode** Method of protecting a steel structure (*e.g.* a bridge, underground or underwater pipeline) by wiring to it plates of an electropositive metal such as magnesium. In a damp or wet environment the magnesium acts as an anode and corrodes in preference to the steel (the cathode); the anodes are periodically replaced with new ones. The technique is called sacrificial protection.

safety lamp Oil lamp designed to be taken into coal mines. It has the flame surrounded by fine wire gauze, so that it can be alight in an atmosphere containing **firedamp** (methane) without causing an explosion. Alternative name: Davy lamp.

sal ammoniac Old name for **ammonium chloride.**

sal volatile Old name for **ammonium carbonate.**

salicylate **Ester** or **salt** of **salicylic acid.**

■ **salicylic acid** $C_6H_4(OH)COOH$ White crystalline organic compound, a **carboxylic acid**. It is used as an antiseptic, in medicine, and in the preparation of **azo dyes**. Its acetyl **ester** is aspirin. Alternative name: 2-hydroxybenzoic acid.

saline Salty; describing a solution of **sodium chloride** (common salt).

■ **saliva** Neutral or slightly alkaline fluid secreted by the salivary glands in the mouth. It lubricates food during chewing and aids digestion. It consists of a mixture of **mucus** and the **enzyme amylase** (ptyalin), which breaks down **starch** to **maltose**.

■ **Salmonella** Genus of anaerobic bacteria that can cause disorders in humans, including enteric fever, paratyphoid, typhoid fever and a common type of food poisoning.

salpingectomy Method of sterilizing a female by surgically cutting (and tying the cut ends) of the oviducts, or Fallopian tubes, thus making it impossible for an ovum (egg) to pass from an ovary to the uterus (womb).

■ **salt** *1.* Product obtained when a **hydrogen** atom in an **acid** is replaced by a **metal** or its equivalent (*e.g.* the ammonium ion NH_4^+). This occurs when a metal dissolves in an acid; *e.g.* zinc (Zn) dissolves in sulphuric acid (H_2SO_4) to form zinc sulphate ($ZnSO_4$) and water (H_2O). A salt is also formed when a **base** or **alkali** neutralizes an acid (*see* **neutralization**), or in a **double decomposition** reaction. *See also* **acid salt**; **basic salt**. 2. Alternative name for common salt, **sodium chloride** (NaCl).

salt bridge Tube that contains a saturated solution of potassium chloride, or an agar gel made with concentrated potassium chloride solution. It is employed to connect two **half-cells**.

saltcake Alternative name for crude **sodium sulphate.**

salting out Precipitation of a **colloid** (*e.g.* gelatine) by the addition of large amounts of a salt.

saltpetre Alternative name for **potassium nitrate.**

samara Winged seed (*e.g.* those produced by sycamore and maple trees).

samarium Sm Metallic element in Group IIIB of the Periodic Table (one of the **lanthanides**). It is slightly radioactive and arises from fission fragments in a **nuclear reactor,** where it acts as a 'poison'. At. no. 62; r.a.m. 150.35.

■ **sand** Separate grains of **quartz**; the indestructible residue from the erosion of rocks. It is used in mortar and concrete and in making glass.

sandstone Consolidated **sand,** cemented together with clay, **calcium carbonate** or iron oxide. It is used as a building stone. [3/2/b]

sandwich compound Organometallic compound whose molecules consist of two parallel planar rings with a **metal** atom centred between them (*e.g.* **ferrocene**).

■ **saponification Hydrolysis** of an **ester**, using an **alkali**, to produce a free **alcohol** and a **salt** of the organic **acid**. It is the process by which **soap** is made.

saponification value Number of milligrams of **potassium hydroxide** required for the complete **saponification** of 1 g of the substance being tested. Alternative name: saponification number.

sapphire Blue gem variety of **corundum** (aluminium oxide, Al_2O_3).

saprolite Naturally occurring deposit of rock that has broken down chemically (*e.g.* by the action of acids in rainwater).

■ **saprophyte** Organism that feeds on dead organic matter (*e.g.* many **bacteria** and **fungi**). Saprophytic activity is the first step in the decomposition of dead animals and plants, and consequently is important in the recycling of elements. Alternative name: saprotroph. [2/2/d]

saprotroph Alternative name for **saprophyte**.

satellite DNA Fraction of **DNA** with significantly different **density** and thus **base** composition from most of the DNA in an organism.

■ **saturated compound** Organic compound that contains only **single bonds**; all the atoms in the compound exert their maximum combining power (valence) with other atoms, so that a chemical change can be effected only by substitution and not by addition. *See* **addition reaction; substitution reaction.**

■ **saturated solution Solution** that cannot take up any more **solute** at a given temperature. *See also* **supersaturated solution.**

■ **saturated vapour Vapour** that can exist in **equilibrium** with its parent **solid** or **liquid** at a given temperature.

■ **saturated vapour pressure Pressure** exerted by a **saturated vapour**. It is temperature dependent.

■ **saturation** Point at which no more of a material can be dissolved, absorbed or retained by another.

sawtooth wave Electronically generated waveform (typically a voltage varying with time) that has a uniform increase of the variable which drops rapidly to the initial value at regular intervals. It is used, *e.g.*, as a **time base** for scanning circuits for a **cathode-ray tube**.

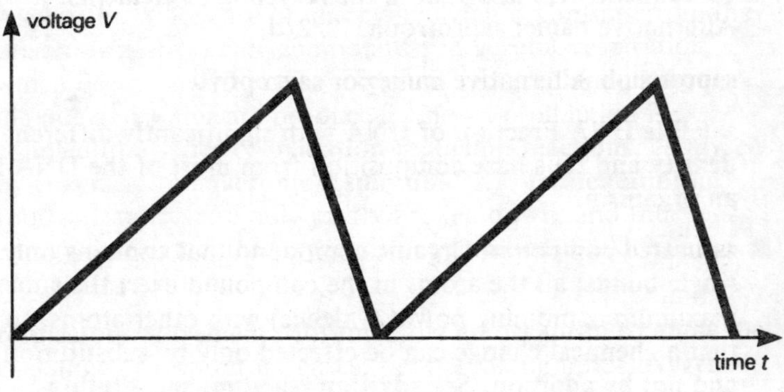

Simple sawtooth wave

***s*-block elements** Metallic elements that form Groups IA and IIA of the Periodic Table, which include the **alkali metals**, **alkaline earths** and the **lanthanides** and **actinides** (together also known as the **rare earths**); hydrogen is usually included as well. They are so called because their 1 or 2 outer electrons occupy ***s*-orbitals**.

scalar Quantity that has magnitude but not direction (unlike a **vector**, which has both).

scaler Electronic device whose circuits measure the amount of charge or energy transfer resulting from radiation. *See also* **counter**.

scandium Sc Silvery-white metallic element in Group IIIB of the Periodic Table (a **rare earth** element); its oxide, Sc_2O_3, is used as a catalyst and to make ceramics. At. no. 21; r.a.m. 44.956.

scanning electron microscope Electron microscope that scans the sample to be examined with a beam of **electrons**.

■ **scapula** Alternative name for the shoulder blade. [2/3/a] [2/5/a]

scattering of light Irregular reflection or diffraction of light rays that occurs when a beam of light passes through a material medium.

■ **Schiff's base** Organic compound formed when an **aldehyde** or **ketone** condenses with a primary aromatic **amine** with the elimination of water. Alternative names: aldimine, azomethine. It was named after the German chemist Hugo Schiff (1834–1915).

■ **Schiff's reagent Reagent** for testing for the presence of aliphatic **aldehydes**, which quickly restore its magenta colour. It is prepared by dissolving rosaniline hydrochloride in water and passing sulphur dioxide through it until the magenta colour is discharged.

schistosomiasis Alternative name for **bilharzia**.

schizogony Form of **asexual reproduction** employed by some single-celled animals, in which a parent cell divides into more than two independent cells. *See also* **binary fission**.

■ **Schrödinger equation** Equation that treats the behaviour of an **electron** in an **atom** as a three-dimensional **stationary wave**. Its solution is related to the probability that the electron is located in a particular place. It was named after the Austrian physicist Erwin Schrödinger (1887–1961). Alternative name: Schrödinger wave equation.

Schwann cell Cell that produces the **myelin sheath** that surrounds a nerve cell (**neurone**). Schwann cells are in close contact with the **axon** of the neurone and are separated by gaps called **nodes of Ranvier**. They were named after the German physiologist Theodor Schwann (1810–82).

sciaphilic Describing a plant that grows only in shady conditions.

scintillation counter Device that counts the incidence of **photons** upon a material by the visible or near-visible light which is emitted.

scintillation spectrometer Scintillation counter capable of measuring the energy and the intensity of **gamma radiation** emitted from a material.

scion Piece of a plant, usually a shoot, that is inserted into the body of another plant when making a **graft**.

sclerenchyma Simple plant tissue composed of cells with thickened walls containing **lignin**, used to give support. The tissue may be composed of fibres or rounded cells.

scleroprotein Member of a group of fibrous **proteins** that provide organisms with structural materials (*e.g.* **collagen**, **keratin**).

sclerotic Outermost of the three layers that form the eyeball (outside the **choroid** and **retina**).

■ **scrotum** Sac present in males of some mammals that contains the **testes**, positioned outside the body cavity so that their temperature is cool enough for **sperm** production.

scurvy Disorder of the skin and gums caused by lack of vitamin C (ascorbic acid) in the diet, once common among sailors because, on long voyages, they had little or no fresh fruit or vegetables in their diet.

seaweed Type of **alga** whose **habitat** is seawater, divided into red, brown and green seaweeds according to which **pigment** is present.

sebaceous gland Small gland found in large numbers in the skin of mammals, usually alongside a hair follicle, that secretes the protective skin oil **sebum**.

sebum Waxy material secreted by **sebaceous glands**, which helps to keep skin waterproof.

■ **second** *1*. **SI unit** of time, defined as the duration of 9,192,631,770 periods of the **radiation** between the two hyperfine levels of the ground state of the caesium-139 atom. Abbreviated to sec or s. *2*. Angle equal to $^{1}/_{60}$ of a minute or $^{1}/_{360}$ of a degree.

secondary cell Electrolytic cell that must be supplied with **electric charge** before use by passing a direct current through it, but it can be recharged over and over again. Alternative names: accumulator, storage cell. *See also* **primary cell**.

secondary colour Colour obtained by mixing **primary colours**.

secondary growth Growth exhibited by some woody plants which takes place in the **stem** and **root**, increasing their girth. It occurs as a result of division of **cambium** located between the **xylem** and **phloem**. Alternative name: secondary thickening.

secondary sexual characteristics Features that develop in some animals after the onset of **puberty**, distinguishing males from females, but not required for sexual function. They result from the actions of **sex hormones**, principally **testosterone** and **oestrogen**.

secondary thickening Alternative name for **secondary growth**.

■ **secretion** Release of a substance by a cell or **gland** with a specialized function, *e.g.* secretion of digestive **enzymes** by cells of the small intestine, or secretion of **hormones** by the pituitary.

■ **sedative** Drug that calms without (in normal doses) causing loss of awareness or consciousness. In larger doses, some sedatives become sleep-inducing drugs. *See also* **tranquillizer**.

■ **sedge** Member of a large group of grasslike plants that have **rhizomes**, and which are found throughout the world, particularly in marshy conditions. Some species are important as animal fodder. [2/1/b] [2/2/c]

sedimentation Removal of solid particles from a **suspension** by gravitational force or in a **centrifuge**.

■ **seed** Structure formed from the **ovule** of a plant after **fertilization**. It contains the developing **embryo** and is a highly resistant structure which can withstand adverse conditions. Dispersal of seeds may be by wind, water or animals and is important in the spreading and colonization of plants. [2/4/a] [2/6/a]

segment One of several adjacent parts of an organism's body that have similar or identical forms (*e.g.* in some worms and insects).

■ **segregation** Separation of a pair of **alleles** in a **diploid** organism during **meiosis** in the formation of **gametes**. A gamete receives one of the two alleles in a diploid organism because it receives only one of a pair of **homologous chromosomes**. [2/5/b]

selection *See* **artificial selection; natural selection**.

selective breeding *See* **artificial selection**.

selectively permeable membrane Alternative name for **semi-permeable membrane.**

selenium Se Nonmetallic element in Group VIA of the Periodic Table, obtained from flue dust in refineries that use **sulphide** ores. One of its **allotropes** conducts electricity in the presence of light, and is used in photocells and rectifiers. At. no. 34; r.a.m. 78.96.

self-absorption Decrease in **radiation** from a large **radioactive** source due to **absorption** by the material itself of some of the radiation produced. Alternative name: self-shielding.

self-fertilization *See* **self-pollination**.

self-induced electromotive force Production of an **electromotive force** (e.m.f.) in an **electric circuit** when the current is varied.

self-induction Resistance to a change in **current** in an **electric circuit** by the creation of a back **electromotive force**.

■ **self-pollination** Type of **fertilization** in plants in which pollen is transferred from the anther to the stigma of the same flower. Many plants have a mechanism which ensures that, if cross-pollination (between two different plants of the same species) does not occur, self-pollination does.

■ **semen** Fluid produced in male reproductive organs of many animals. It contains **sperm**, and in mammals secretions from the accessory sex glands.

semicarbazide $NH_2NHCONH_2$ Organic **base** which forms salts (*e.g.* it forms a hydrochloride with hydrochloric acid) and characteristic condensation products **(semicarbazones)** with **aldehydes** and **ketones**.

semicarbazone Crystalline organic compound containing the group $=C=NNHCONH_2$, formed when an **aldehyde** or a

ketone reacts with **semicarbazide** ($NH_2NHCONH_2$) with the elimination of water. Semicarbazones are used to identify the original aldehyde or ketone.

■ **semicircular canal** Part of the **ear** that is involved in maintaining balance. *See* **labyrinth**. [2/4/a] [2/5/a] [2/7/a]

■ **semiconductor** Substance that is an **insulator** at very low temperatures but which becomes a **conductor** if the temperature is raised or when it is slightly impure. Semiconductors are used in **rectifiers** and photoelectric devices, and for making **diodes** and **transistors**. *See also* **donor**; **doping**; ***n*-type conductivity**; ***p*-type conductivity**.

seminal vesicle Organ in the **testes** that is used for storing **sperm**.

seminiferous tubule One of many tubes within the **testes** in which **sperm** are made.

■ **semipermeable membrane** Porous **membrane** that permits the passage of some substances but not others; *e.g.* **plasma membrane**, which permits entry of small molecules such as water but not large molecules, allowing **osmosis** to occur. Such membranes are extremely important in biological systems and are used in **dialysis**. Alternative name: selectively permeable membrane. [2/7/a]

senescence Processes that mark the final stages of an organism's natural life-span. In plants senescence is usually associated with flowering and fruiting.

■ **sense organ** Group of **receptors** specialized to react to (detect) a certain **stimulus** (*e.g.* the **eye** to light, the **ear** to sound, and **chemoreceptors** in the tongue and nose to tastes and smells).

■ **senses** The five primary senses, common to most vertebrates but sometimes lacking in less highly evolved animals, are sight, hearing, taste, smell and touch, to which may be added the sense of balance. They are effected by various **sense organs**. [2/4/a]

■ **sepal** Part of certain flowers that forms the **calyx**. [2/4/a] [2/5/a] [2/7/a]

septicaemia Disorder that results from the presence of bacteria, or their toxins, in the bloodstream. Alternative name: blood poisoning.

septivalent Having a valency of seven. Alternative name: heptavalent.

septum Dividing wall found in biological systems, *e.g.* between the nostrils or between the two halves of the heart.

sere Any of the characteristic **communities** that occur in sequence during the process of plant **succession**.

■ **series** *1.* In physics, describing the arrangement of components in a **series circuit**. [5/4/b] *2.* In chemistry, systematically arranged succession of chemical compounds (*e.g.* homologous series) or of numbers or algebraic terms (*e.g.* arithmetic series, exponential series, geometric series).

■ **series circuit** Electrical circuit in which the components are arranged one after the other so that the same current flows through each of them. For a series of **resistors**, the total resistance R is equal to the sum of the individual resistances; *i.e.* $R = R_1 + R_2 + R_3 \ldots$ For a series of **capacitors**, the reciprocal of the total capacitance C is equal to the sum of the reciprocals of the individual capacitances; *i.e.* $1/C = 1/C_1 + 1/C_2 + 1/C_3 \ldots$ [5/4/b]

serine $CH_2OHCHNH_2COOH$ White crystalline **amino acid**, present in many **proteins**. Alternative name: 2-amino-3-hydroxypropanoic acid.

serology Branch of **immunology** concerned with reactions between **antibodies** of one organism with **antigens** of the **serum** of another.

serotinal Describing biological activity or events that take place during late summer.

serotonin Substance derived from **tryptophan** and found in blood **serum**, used as a **neurotransmitter** and **vasoconstrictor**. Alternative names: 5H, 5-hydroxytryptamine.

serum Constituent of **plasma** of blood, which contains all the substances in plasma except for **fibrinogen**.

servomechanism Device that uses a small amount of power to control the activity of a much more powerful device (*e.g.* as in power-assisted steering on a motor vehicle).

sessile *1.* Not having a stalk (of leaves). *2.* Non-mobile, describing particularly animals that permanently anchor themselves to a surface such as the sea bed.

■ **sewage** Faeces, urine, washing water and surface water from homes and factories. These waste materials pass along sewers to works where they are treated to make them harmless. Any useful materials are extracted and recycled; treated water is returned to rivers or the sea.

sex cell Alternative name for **gamete**.

■ **sex chromosome** **Chromosome** that carries the **genes** determining sex. In mammals the female possesses two identical sex chromosomes or X- chromosomes, whereas in the male the two sex chromosomes differ, one being an X- and the other a Y-chromosome. *See also* **heterogametic**; **homogametic**; **sex determination**. [2/8/b]

sex determination Inheritance of particular combination of **sex chromosomes**, which is the deciding factor in whether an organism is male or female. Inheritance of a **homologous** pair of sex chromosomes predisposes the organism to one sex (*e.g.* in mammals, the female). Inheritance of a pair of dissimilar sex chromosomes determines the other sex (in mammals, the male). *See also* **heterogametic**; **homogametic**. [2/8/b]

■ **sex hormone** **Hormone** that determines **secondary sexual characteristics** and regulates the reproductive behaviour of an organism. In mammals, the sexual cycle of the female is controlled by such hormones (*see* **oestrous cycle**). In males the **gonads** are regulated. [2/7/a] [2/8/b]

■ **sex linkage** Distribution of **genes** according to the sex of an organism because they are carried on the **sex chromosomes**. In human males a **recessive** gene carried on the **X-chromosome** will be expressed because no corresponding **allele** is present on the **Y-chromosome** to mask it, the Y-chromosome being shorter than the X-chromosome. In the female the corresponding allele will be present on the other X-chromosome, and for this reason human males have a predisposition to recessive sex-linked disorders (*e.g.* haemophilia, colour-blindness), whereas human females are more often **carriers** rather than sufferers of such disorders. [2/8/b]

■ **sex ratio** Ratio of the number of males to the number of females in a population. It may be expressed as the number of males to every 100 females. [2/8/b]

■ **sexual reproduction** **Reproduction** of an organism that involves the fusion of specialized sex cells or **gametes** (which are **haploid**) to form **diploid** progeny. It is important in bringing new vigour to a **species** by the mixing of genetic material from the parents to give a genetically different organism. *See also* **asexual reproduction**; **egg**; **fertilization**; **sperm**. [2/8/b]

shear Deformation in which parallel planes in a material slide over each other (but remain parallel).

shear modulus Alternative name for **rigidity modulus**.

shear stress Shear force per unit cross-sectional area.

■ **shell** In biology, *1*. Protective chalky or leathery outer covering of an egg (*e.g.* a bird's or reptile's egg). *2*. Hard chalky outer covering of some types of **mollusc** (*e.g.* mussel, snail), secreted by the **mantle**. *3*. **Exoskeleton** of an **arthropod** (*e.g.* barnacle, crab, lobster). *4*. **Carapace** of a tortoise or turtle. *5*. Hard outer layer of some fruits (*e.g.* nuts). *6*. In physics, group of **electrons** that share the same principal **quantum number** in an atom. The particles that form an atomic nucleus are also thought to occupy shells. [4/8/d,e]

shock wave Sound wave in air, of exceptionally high intensity, produced by an object travelling faster than the **speed of sound** or by the detonation of a high explosive.

■ **short circuit** Electric circuit through a very low **resistance**, which passes a very high **current**, usually caused accidentally when insulation fails. [5/4/b]

short-sightedness Visual defect in which the eyeball is too long (front to back) so that rays of light entering the **eye** from distant objects are brought to a focus in front of the retina. It can be corrected with spectacles or contact lenses with diverging (concave) lenses. Alternative name: myopia.

shoulder blade Alternative name for **scapula**.

■ **shunt** In medicine, *1*. A surgically implanted tube or vessel that diverts the flow of fluid (*e.g.* to by-pass an obstruction or to drain an area). *2*. Small blood vessel between capillaries that normally carry blood to and from the surface of the skin. When the skin is cold, blood is diverted along the shunt vessels to retain body heat (thus causing very pale, white skin). [2/7/a] *3*. In physics, resistor connected in series with a circuit or device (*e.g.* a meter) that reduces the amount of electric **current** flowing through it. *See also* **multiplier**.

■ **sickle-cell anaemia** Inherited type of **anaemia** that generally affects only black people. An abnormality of the **haemoglobin** in the blood causes red blood cells to take the shape of crescents, which cannot easily pass along narrow blood vessels and tend to be broken down instead (sometimes leading to **thrombosis**). [2/9/c]

side chain Aliphatic radical or group that is attached to one of the atoms in the ring of a **cyclic** compound or to one of the atoms of the longer straight chain of atoms in an **acyclic** compound.

side reaction Chemical reaction that takes place at the same time as the main reaction.

Siemens-Martin process Alternative name for **open-hearth process**.

sieve tube Tubular element that makes up the **phloem** in a vascular plant, through which **translocation** occurs. It is composed of cellulose-walled sieve elements which are joined via pores, and have lost their **organelles**.

silane Hydride of **silicon** with the general formula Si_nH_{2n+2}. The maximum number of bonds possible between silicon atoms is about eight because of the weakness of such bonds. Alternative names: silicon hydride, silicane.

■ **silica** SiO_2 Hard white solid with a high melting point (1,710 °C). It is one of the most abundant natural compounds, occurring as *e.g.* crystobalite, **flint**, **quartz**, **sand**, and combined as **silicates** in rocks. Alternative name: silicon (IV) oxide, silicon dioxide. *See also* **silica gel**. [3/2/b]

silica gel Porous **amorphous** variety of **silica** (SiO_2) which is capable of absorbing large quantities of water and other **solvents**. It is used as a **desiccant** and **adsorbent**.

silicate Derivative of **silica** or a salt of a **silicic acid**. Natural silicates, found in many rocks and minerals, contain varying proportions of silica and of a wide range of metal oxides. *See* **glass; sodium silicate**.

silicic acid Hydrated form of **silica** (SiO_2) made by reacting a soluble **silicate** with an **acid**. It forms a colloidal or gel-like mass.

■ **silicon** Si Nonmetallic element in Group IVA of the Periodic Table, which exists as amorphous and crystalline **allotropes**. It is the second most abundant element, occurring as **silicates** in clays and rocks; sand and quartz consists of **silica** (silicon dioxide, SiO_2). It is used in making refractory materials and temperature-resistant glass. At. no. 14; r.a.m. 28.086.

silicon carbide SiC Very hard black material, used as an abrasive, and in cutting, grinding and polishing instruments. Alternative name: carborundum.

silicon chip Integrated circuit made from a thin wafer of pure crystalline silicon. Alternating insulating and **semiconductor** layers are printed on the wafer. The pattern of an electronic control circuit is also etched onto it.

silicon (IV) chloride Alternative name for **silicon tetrachloride**.

silicon dioxide Alternative name for **silica**.

silicone Member of a group of synthetic **polymers** made from **siloxanes**, used in textile finishing, polishes and lubricants.

silicon (IV) oxide Alternative name for **silica**.

silicon tetrachloride $SiCl_4$ Colourless fuming liquid, a source of pure **silica** for use in the production of silica **glass**. Alternative name: silicon (IV) chloride.

siloxane Member of a group of compounds that contain the linkage Si-O-Si with organic groups bound to the **silicon** atoms. They are used for making **silicones**.

■ **silver** Ag Silver-white metallic element in Group IB of the Periodic Table (a **transition element**). It occurs as the free element (native) and in various **sulphide** ores. It is used in jewellery, electrical contacts, batteries and mirrors. Silver **halides** are used in photographic emulsions. At. no. 47; r.a.m. 107.868.

silver bromide AgBr Pale yellow insoluble crystalline **salt**, used for making light-sensitive photographic emulsions.

silver chloride AgCl White insoluble crystalline **salt**, used in the manufacture of pure **silver** and in photographic emulsions.

silver iodide AgI Pale yellow insoluble crystalline **salt**, used in photographic emulsions.

silver nitrate $AgNO_3$ Colourless crystalline salt, used in **volumetric analysis** and **silver plating**, and as a caustic in medicine (*e.g.* for removing warts).

silver oxide Ag_2O Brown amorphous solid, only slightly soluble in water but soluble in ammonia solution. Alternative name: silver (I) oxide.

silver plating Electrolytic deposition of a layer of **silver** on the surface of another metal, usually from a hot alkaline solution of complex silver (I) cyanides. Alternative name: silvering.

simple harmonic motion Periodic motion of a particle whose **acceleration** (a) is proportional to its distance (x) from a fixed point and is always directed towards that point; *i.e.* $a = -kx$, where k is a constant. The minus sign shows that, as the particle moves to and fro, its acceleration is oppositely directed to its displacement.

simple pendulum *See* **pendulum**.

sine wave Waveform that represents the periodic oscillations of constant **amplitude** as given by the sine of a linear function. Alternative names: sinusoidal wave, sine curve.

single bond Covalent bond formed by the sharing of one pair of **electrons** between two **atoms**. [4/8/d,e]

sintering Method of compacting a powdered solid into a rigid shape by compressing it at a temperature below its melting point.

sinus Irregular cavity or depression that forms part of an animal's anatomy; *e.g.* sinuses in the bones of the face in mammals.

siphon *1.* Device consisting of an inverted U-shaped tube that moves a liquid from one place to another place at a lower level. The tube has to be filled with liquid before being put into position. In biology, *2.* a sucking mouthpart (*e.g.* as in fleas); *3.* a tubular organ through which liquid (often water) passes into or out of an animal's body (*e.g.* as in some molluscs). Alternative name: syphon.

SI units Abbreviation for Système International d'Unités, an international system of scientific units. It has seven basic units: **metre** (m), **kilogram** (kg), **second** (s), **kelvin** (K), **ampere** (A), **mole** (mol) and **candela** (cd), and two supplementary units **radian** (rad) and **steradian** (sr). There are also 18 derived units.

■ **skeleton** Structure that supports the **tissues** and **organs** of an animal and is attached to **muscles** and **ligaments** to allow locomotion. An **endoskeleton** is internal, made of **bone** or **cartilage**, and possessed by **vertebrates**. **Exoskeletons** lie outside the muscles, *e.g.* in **arthropods**. Some **invertebrates** possess a hydrostatic skeleton (water vascular system) which consists of fluid under pressure. [2/3/a]

■ **skin** Organ in mammals that protects the body from invasion by **pathogens** and prevents water from entering. In warm-blooded animals the skin also takes part in temperature regulation, *e.g.* through sweating and the constriction and dilation of its **blood vessels**. It consists of **epithelial tissue** and **connective tissue** arranged in two major layers, the thin outer epidermis and the thicker underlying dermis.

skull Bones that form the head and face, including the **cranium** and **jaws**.

■ **slag** Impurities, mainly silicates, that rise to the surface during the **smelting** of a metal ore. [4/5/c]

slaked lime Alternative name for **calcium hydroxide**.

slow neutron Alternative name for **thermal neutron**.

small intestine Part of the digestive tract in mammals which is composed of the **duodenum** and **ileum**, and is the main site of **digestion** and absorption in the gut. **Bile**, pancreatic juice and intestinal juice are liberated in it, to supply many digestive **enzymes**. *See also* **intestine**.

■ **smell** One of the primary **senses** that enables animals to detect odours, using **chemoreceptors** usually located in mammals in olfactory glands in the nasal cavity. [2/1/a] [2/4/a] [2/5/a]

smelt Material obtained from the **smelting** of a metal ore.

smelting Thermal decomposition of concentrated metal ore to cause the release of the metal. [4/5/c]

smoke Particles of a **solid** dispersed in a **gas**.

■ **smooth muscle** Type of muscle in internal organs and tissues, not under voluntary control. Alternative name: involuntary muscle. [2/5/a] [2/7/a]

■ **soap** Sodium or potassium salt of a high molecular weight **fatty acid** (*e.g.* palmitic acid, stearic acid). Soaps are made by the **hydrolysis** or **saponification** of **fats** with hot **sodium hydroxide** or **potassium hydroxide**, giving **glycerol** as a by-product. They emulsify grease and act as wetting agents. *See also* **detergent**.

soapstone Compact form of **chalk**. Alternative name: steatite.

soda Imprecise term for a compound of **sodium**, usually referring to **sodium carbonate** (washing soda) or **sodium hydrogencarbonate** (baking soda). *See also* **caustic soda; soda ash; soda lime**.

soda ash Common name for anhydrous **sodium carbonate**.

soda lime Solid mixture of **sodium hydroxide** and **calcium oxide**.

soda water Carbon dioxide dissolved in **water** under pressure, used as a fizzy drink.

■ **sodium** Na Soft silvery-white metallic element in Group IA of the Periodic Table (the **alkali metals**). It occurs widely, principally as its **chloride** (common salt, NaCl) in seawater and as underground deposits, from which it is extracted by **electrolysis**. The metal is used as a coolant in some **nuclear reactors**; its many compounds are important in the chemical industry, particularly – in addition to the chloride – **sodium hydroxide** (caustic soda, NaOH) and **sodium carbonate** (soda, Na_2CO_3). At. no. 11; r.a.m. 22.9898.

sodium acetate CH_3COONa White crystalline solid, used in photography and in the manufacture of **ethyl ethanoate** (acetate) and various **pigments**. Alternative name: sodium ethanoate.

sodium aluminate $NaAlO_2$ White solid, used as a coagulant in water purification.

sodium azide NaN_3 White poisonous crystalline solid, used in the manufacture of detonators.

sodium bicarbonate Alternative name for **sodium hydrogencarbonate**.

sodium bisulphate Alternative name for **sodium hydrogensulphate**.

sodium bisulphite Alternative name for **sodium hydrogensulphite**.

sodium borate Alternative name for **borax**.

sodium bromide NaBr White crystalline solid, used in medicine.

■ **sodium carbonate** $Na_2CO_3.10H_2O$ White crystalline solid which exhibits **efflorescence** and forms an alkaline solution in water. It is used in **glass** making, as a water softener, and for the preparation of **sodium** chemicals. Alternative names: washing soda, soda, soda ash.

sodium chlorate $NaClO_3$ White soluble crystalline solid. It is a powerful **oxidizing agent**, used as a weed-killer and in the textile industry. Alternative name: sodium chlorate (V).

■ **sodium chloride** NaCl White soluble crystalline salt, extracted from seawater or underground deposits. It is used for seasoning and preserving food. Industrially, it is used in the manufacture of a wide variety of chemicals, including chlorine, sodium carbonate, sodium hydroxide and hydrochloric acid. Alternative names: common salt, rock salt, salt, sea salt, table salt.

sodium dihydrogenphosphate (V) NaH_2PO_4 White solid, used in detergents and certain baking powders. Alternative name: sodium dihydrogen orthophosphate.

sodium ethanoate Alternative name for **sodium acetate**.

sodium hydrogencarbonate $NaHCO_3$ White soluble powder, used in making baking powder, powder-based fire extinguishers and antacids. Alternative names: sodium bicarbonate, baking soda.

sodium hydrogensulphate $NaHSO_4.H_2O$ White solid, used in the dyeing industry and in the manufacture of **sulphuric acid**. Alternative name: sodium bisulphate.

sodium hydrogensulphite $NaHSO_3$ White powder, used in medicine as an antiseptic, and as a preservative. Alternative name: sodium bisulphite.

■ **sodium hydroxide** NaOH White deliquescent solid; a strong **base**. It is made by the **electrolysis** of brine (**sodium chloride** solution). It is used in the manufacture of **soaps, rayon** and paper and many other **sodium** compounds. Alternative names: caustic soda, soda.

sodium metasilicate Alternative name for **sodium silicate**.

sodium monoxide Na_2O White solid that reacts violently with water to give **sodium hydroxide**; a **strong base**. Alternative name: disodium oxide.

sodium nitrate $NaNO_3$ White crystalline salt, used as a food preservative and in the manufacture of explosives and fireworks. Alternative names: Chilean saltpetre, soda nitre.

sodium peroxide Na_2O_2 Pale yellow powdery solid that reacts readily with water to give **sodium hydroxide** and **oxygen**. It is an **oxidizing agent**, used as a bleach (for cloth and wood pulp).

sodium pump Process by which potassium and sodium **ions** are transported across **membranes** that surround animal cells.

sodium silicate $Na_2SiO_3.5H_2O$ Colourless crystalline solid, used in various types of **detergents** and cleaning compounds, and as a bonding agent in many ceramic cements and in various refractory applications. Alternative name: sodium metasilicate.

sodium silicate solution Concentrated **solution** of **sodium silicate** in water, used to prepare **silica gel** and precipitated silica. Alternative name: water glass.

sodium sulphate $NaSO_4.10H_2O$ White crystalline salt, used in the manufacture of paper, glass, dyes and **detergents**. Alternative name: Glauber's salt, saltcake.

sodium sulphide NaS_2 Reddish-yellow deliquescent amorphous solid, used in the manufacture of dyes.

sodium sulphite Na_2SO_3 White soluble crystalline solid which reacts with an acid to produce **sulphur dioxide** (SO_2), used in photography and as a bleach and food preservative.

sodium thiosulphate $Na_2S_2O_3$ White soluble crystalline solid. It is a strong **reducing agent,** used as photographic fixing agent (when it reacts with unexposed silver halides) and in dyeing and **volumetric analysis**. Alternative name: hypo.

soft iron Iron that has a low content of carbon, unlike steel. It is unable to retain **magnetism**.

soft palate Rear part of the roof of the mouth, consisting of muscle tissue covered by mucous membrane. The uvula hangs from the back of the soft palate.

soft radiation Radiation of relatively long **wavelength** whose penetrating power is very limited.

soft water Water that lathers immediately with **soap**; water from which most of the **calcium** and **magnesium** compounds have been removed. *See* **hardness of water**.

softening of water *See* **hardness of water; ion exchange**.

software Program that can be used on a **computer;** *e.g.* executive programs, operating systems and utility programs. *See also* **hardware**.

solar cell Photocell that converts **solar energy** directly into **electricity**.

■ **solar energy Energy** from the Sun, mainly light and heat radiation. It can be harnessed to make electricity using **photocells**, or the Sun's rays may be used directly to heat water in a radiator or, focused by mirrors, in a solar furnace. Solar energy also provides the energy for photosynthesis, which is an essential process for the existence of life on Earth. [3/7/a]

solar heating Heating from a domestic or industrial heater that uses **solar energy**, generally to heat water.

solar wind Continuous permanent stream of electrically charged particles that are emitted from the Sun.

solder Low-melting-point **alloy**, often containing tin and lead, used for joining various metals (with the aid of a **flux**). It melts readily and remains molten long enough for the joint to be shaped. Alternative name: braze.

solenocyte Alternative name for **flame cell**.

solenoid Cylindrical coil of wire, carrying an electric current, used to produce a **magnetic field**. It may have an iron core that moves, often to work a switch. *See also* **relay**.

■ **solid** *1*. Object having the three dimensions of length, breadth and thickness; a solid figure. *2*. Alternative name for a substance in the **solid state**.

solidifying point Alternative name for **freezing point**.

solid solution Single solid **homogeneous** crystalline **phase** of two or more substances. Many alloys are solid solutions.

solid state Physical **state of matter** that has a definite shape and resists having it changed; the volume of a solid changes only slightly with **temperature** and **pressure**. A true solid state is associated with a definite **crystalline** form, although **amorphous** solids also exist (*e.g.* glass). Alternative name: solid.

solid-state physics Branch of **physics** concerned with matter in the **solid state**, particularly **semiconductors** and **superconductors**.

■ **solubility** Amount of a substance **(solute)** that will dissolve in a liquid **(solvent)** at a given temperature, usually expressed as a mass per unit volume (*e.g.* gm per litre) or a percentage.

solubility product When a **solution** is saturated with an **electrolyte**, the solubility product is the product of the **concentrations** of its constituent **ions**.

soluble Describing a substance **(solute)** that will dissolve in a liquid **(solvent)**.

solute Substance that dissolves in a **solvent** to form a **solution**.

solution Homogeneous mixture of **solute** and **solvent**.

■ **solvation** Attachment between **solvent** and **solute** molecules. The greater the polarity of the solvent, the greater is the attraction between solute and solvent molecules. It is what causes ionic compounds to dissolve, and water is a good polar solvent. Solvation by water is called **hydration**.

Solvay process Process for the commercial preparation of **sodium carbonate** (by reacting sodium chloride solution containing ammonia with carbon dioxide made by heating calcium carbonate). It was named after the Belgian chemist Ernest Solvay (1838–1922). Alternative name: ammonia-soda process.

solvent Substance (usually a liquid) in which a **solute** dissolves; the component of a **solution** which is in excess.

solvent extraction Removal of a substance from a (usually aqueous) solution by dissolving it in a (usually organic) **solvent**. Alternative name: liquid-liquid extraction.

solvent naphtha Alternative name for **naphtha**.

solvolysis Chemical reaction between **solvent** and **solute** molecules. *See also* **hydrolysis**.

somatic Of the body. Somatic cells include all cells of an organism except for the **gametes** or sex-cells.

somatotrophin Alternative name for **growth hormone**.

sonar Abbreviation of sound navigation and ranging, an apparatus for echo sounding that employs **ultrasonic** waves. It can detect echoes from underwater objects and so give their bearing and range. Alternative name: underwater **radar**.

■ **sonometer** Apparatus for investigating stretched wires or strings as sources of sound. [5/8/c]

sorbitol Hexahydric **alcohol**, an **isomer** of **mannitol**, formed by the **reduction** of **glucose**, used as a sweetening agent.

soredium Structure formed during the **asexual reproduction** of some **lichens** that contains both fungal and algal cells.

soret effect *See* **thermal diffusion**.

■ **sound** Periodic vibrations that travel as pressure waves through media and to which the ears of human beings and other animals may be sensitive. There must be a medium; sound will not travel through a vacuum. Human hearing can detect sound in the approximate frequency range 20 Hz to 20 kHz. [5/4/c] [5/5/a]

■ **sound, speed of** *See* **speed of sound**. [5/4/c] [5/5/a]

space Parts of the Universe beyond the Earth's atmosphere, often distinguished as interplanetary, interstellar and intergalactic space.

spartalite Alternative name for **zincite**.

■ **species** Smallest group commonly used in biological **classification** and into which a genus is divided. Species are sometimes further divided into subspecies (*e.g.* races, varieties). Generally, no more than one type of organism is present in one species. Members of a species may breed with one another, but usually cannot breed with members of another species. Rarely, very closely related species interbreed to produce a **hybrid**. *See also* **binomial nomenclature**. [2/3/b] [2/4/b]

specific activity Number of disintegrations of a **radio-isotope** per unit time per unit mass.

specific capacitance *See* **permittivity, relative**.

specific charge Ratio of **electric charge** to unit mass of an **elementary particle**.

specific gravity Former name for **relative density**.

■ **specific heat capacity** (c) Measure of the capacity for heat of a material; the quantity of heat needed to raise the **temperature** of 1 kilogram of material through 1 degree. In general, if *W* joules of heat energy raise the temperature of a mass *m* from t_1 to t_2, $c = W/m(t_2 - t_1)$. Its **SI** units are J kg^{-1} K^{-1}. Alternative names: specific heat, specific thermal capacity.

specific latent heat Amount of **latent heat** per unit mass of a substance.

specific resistance Alternative name for **resistivity**.

specific surface Surface area per unit mass or unit volume of a solid substance.

specific volume Volume of a substance per unit mass; the reciprocal of **density**.

spectra Plural of **spectrum**.

spectral line Particular **wavelength** of light in a **line spectrum**.

spectral series Sequence of lines in a **line spectrum**.

spectrochemistry Branch of **chemistry** concerned with the study of spectra of substances.

spectrometer **Spectroscope** that has some form of photographic or electrical detection device.

spectrometry Measurement of the intensity of **spectral lines** or **spectral series** as a function of **wavelength**.

spectrophotometer Instrument that measures the intensity of **electromagnetic radiation** absorbed or transmitted by a substance as a function of **wavelength**, usually in the visible, infra-red and ultraviolet regions of the **electromagnetic spectrum**.

spectroscope Instrument for splitting various **wavelengths** of **electromagnetic radiation** into a **spectrum**, using a **prism** or **diffraction** grating.

spectroscopy Study of the properties of light, using a **spectroscope**; the production and analysis of **spectra**.

■ **spectrum** *1*. Band, continuous range, or lines of **electromagnetic radiation** emitted or absorbed by a substance under certain circumstances. *2*. Coloured band of light or bands of colours produced by splitting various **wavelengths** of **electromagnetic radiation**, using a **prism** or **diffraction** grating. [5/9/a,c]

■ **spectrum colours** Visible colours that are observed in the **spectrum** of white light (*e.g.* in a **rainbow**). [5/9/a,c]

■ **speed** Rate at which an object moves, expressed as the distance travelled in a given time; it is measured in units such as m s^{-1}, km/h or mph. Unlike **velocity**, it is independent of the direction of travel. Speed is a scalar quantity; velocity is a vector. [5/7/a]

■ **speed of light** Velocity of light waves. Its mean value is 2.997925×10^8 m s^{-1} in a vacuum. Alternative name: velocity of light. [5/9/a,c]

■ **speed of sound** Velocity of **sound** waves. In general, sound travels more rapidly through liquids than through gases, and fastest of all through solids. As temperature increases, so does the speed of sound. In dry air at 0 °C, its value is 330 m s^{-1}. Alternative name: velocity of sound. [5/4/c] [5/5/a]

■ **sperm** Abbreviation of **spermatozoon**, the **gamete** (sex-cell) produced by the male in many species of animals (in mammals, in the **testes**). It consists of a head containing genetic material in the **nucleus** (which is **haploid**), and usually possesses **cilia** or a **flagellum** for movement.

spermaceti Waxy oil that occurs in tissues located behind the forehead of sperm whales. Its function is uncertain, but it is believed to be involved in echo-location. It can be used in the manufacture of medicines and cosmetics.

spermatocyte Cell from which **sperm** (spermatozoa) are derived through **spermatogenesis**. Primary spermatocytes are **diploid**; after **meiosis,** secondary spermatocytes which are **haploid** are formed. These further divide to produce spermatids, which differentiate to form sperm.

spermatogenesis Formation of **sperm** in the **testis**. It commences with the repeated **mitosis** of primordial germ cells to form spermatogonia, which grow to form a primary **spermatocyte**. This divides by meiotic (reduction) division to produce four **haploid** spermatozoa.

Spermatophyta Division of the plant kingdom that contains all seed-bearing plants. Spermatophytes may be divided into **angiosperms** (conifers, etc.) and **gymnosperms** (flowering plants) which can be further subdivided into **dicotyledons** (class Magnoliopsida) and **monocotyledons** (Liliopsida)).

spermatophyte Member of the plant division **Spermatophyta**.

spermatozoon *See* **sperm**.

spherical aberration Type of **aberration** in a lens or mirror.

spherometer Instrument for determining the curvature of spherical objects.

sphincter Circular muscle that controls the flow of a liquid or semi-solid through an orifice (*e.g.* the anal sphincter, round the anus).

sphygmomanometer Instrument for measuring blood pressure.

spin Spinning motion (angular momentum) of an **atom**, **nucleus** or **subatomic particle**, assigned a quantum number of ±½, ±1, ±1½, ±2,... Particles with whole-number spins are **bosons**.

spinal column *See* **spine**.

spinal cord Part of the **central nervous system** (CNS) in vertebrates that is enclosed within the spine. It consists of a hollow nerve tube containing many interconnecting **neurones** and connected to the **spinal nerves**.

spinal nerve Any of several peripheral nerves arising from the **spinal cord** which are connected to **receptors** and **effectors** in other parts of the body.

spindle Structure composed of **protein** fibres that is formed in the **cytoplasm** during **cell division**. It is used for the attachment of **chromosomes** and thought to assist in their movement to the poles of the cell. *See also* **meiosis; mitosis**.

■ **spine** Backbone; dorsally situated bony column composed of **vertebrae**, which enclose the **spinal cord**. Alternative name: spinal column. [2/3/a]

spin, electron Spinning motion of an **electron** in an **atom**.

spiracle *1*. In cartilaginous fish, one of a pair of openings on the head near the gills. *2*. In insects, an opening that leads into a system of **tracheae**, used for **respiration**.

spirit *1*. Volatile liquid obtained by **distillation**; a volatile distillate (*e.g.* aviation spirit). *2*. Solution that consists of a volatile substance dissolved in **ethanol** (ethyl alcohol). *See also* **methylated spirits**.

spirochaete Member of a group of **bacteria** characterized by a spiral shape, some of which are parasitic and cause disorders (*e.g.* syphilis, yaws, infectious jaundice).

spirogyra Commonest filamentous **alga** (genus *Spirogyra*) found in freshwater ponds, which has one or more spiral-shaped **chloroplasts** in each cell.

splanchnology Branch of biology and medicine concerned with the organs within the central body cavity of vertebrates.

■ **spleen** Organ present in the abdomen of some vertebrates that aids in defence against invading organisms. It produces **lymphocytes** and also stores and removes red blood cells (erythrocytes) from the blood system. [2/4/a] [2/5/a] [2/7/a]

sponge Member of the animal phylum **Porifera**.

spontaneous combustion Self-ignition of a substance of low **ignition temperature** (*e.g.* damp straw or oily rags), which results from the accumulation of heat within the substance because of slow **oxidation**.

■ **spontaneous generation** Theory (now disproved) that living matter can arise from non-living matter. *See also* **abiogenesis; biogenesis**. [2/9/c]

sporangium Plant structure that produces **spores**.

■ **spore** *1.* **Haploid** reproductive structure formed in the **sporophyte** of plants. It can be widely dispersed by wind etc. and germinates to form a **gametophyte**. *2.* Resting form of an organism that is highly resistant to adverse conditions, *e.g.* in **bacteria**.

sporophyll Plant part that bears **sporangia**; *e.g.* leaves of ferns, cones of pteridophytes.

sporophyte Stage in the life cycle of a plant that is **diploid** but produces **haploid spores** which generate the **gametophyte**. *See also* **alternation of generations**.

■ **sport** In biology, an organism that possesses abnormal

characteristics as a result of a naturally occurring **mutation**. The term is most often applied to plants. [2/9/c]

square wave Wave motion with very rapid rise and fall times, usually describing a voltage that varies regularly and rapidly between two fixed values.

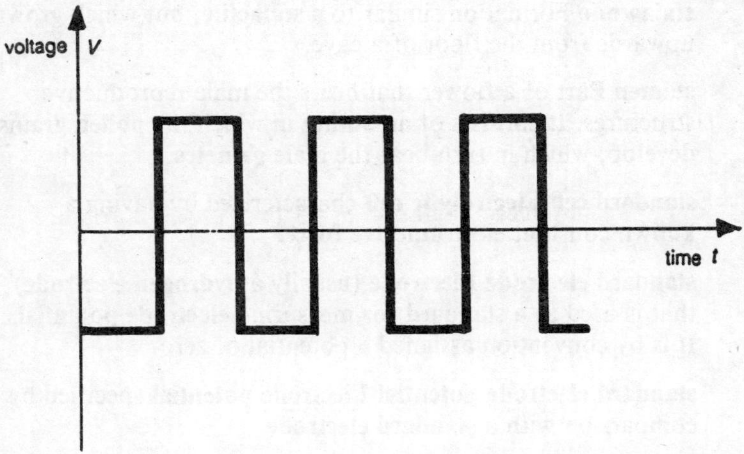

Square wave

stability Property of an object in equilibrium that opposes an attempt to move it. Any movement sets up an opposing force if an object is stable (*i.e.* if movement raises its **centre of mass**). For an object in unstable equilibrium, the slightest movement lowers the centre of mass.

stable Relatively inert; hard to decompose.

stainless steel Member of a group of **chromium**-containing **steels** characterized by a high degree of resistance to corrosion. They are used for cutlery, chemical plant equipment, ball-bearings, etc.

stalactite Long icicle-like formation of **calcium carbonate** (from dripping water containing dissolved calcium carbonate) which grows very slowly downwards from the roof of a limestone cave. *See also* **stalagmite**.

stalagmite Formation similar to a **stalactite**, but which grows upwards from the floor of a cave.

stamen Part of a flower that bears the male reproductive structures. It consists of an **anther** in which the **pollen** grains develop, which in turn bear the male **gametes**.

standard cell **Electrolytic cell** characterized by having a known constant **electromotive force**.

standard electrode **Electrode** (usually a **hydrogen electrode**) that is used as a standard for measuring **electrode potentials**. It is by convention assigned a potential of zero.

standard electrode potential **Electrode potential** specified by comparison with a **standard electrode**.

standard solution **Solution** of definite **concentration,** *i.e.* having a known weight of **solute** in a definite volume of solution, as used in **volumetric analysis**. The concentration is commonly given in terms of normality (*see* **normal**).

standard state Element in its most stable physical form at a specified **temperature** and a **pressure** of 101,325 pascals (760 mm Hg).

■ **standard temperature and pressure** (STP or s.t.p.) Set of standard conditions of **temperature** and **pressure**. By convention, the standard temperature is 273.16 K (0 °C) and the standard pressure is 101,325 pascals (760 mm Hg).

standing wave Stationary wave caused by **interference** when two waves of the same wavelength move in opposite directions (*e.g.* when a single wave is reflected back along itself).

stannane Alternative name for **tin (IV) hydride**.

stannic Tetravalent tin. Alternative names: tin (IV), tin (4^+).

stannic acid H_2SnO_3 White amorphous solid, used for polishing glass and metal. Alternative name: tin hydroxide oxide.

stannic chloride Alternative name for **tin (IV) chloride**.

stannous Bivalent tin. Alternative names: tin (II), tin (2^+).

stannous chloride Alternative name for **tin (II) chloride**.

■ **stapes** One of the **ear** ossicles. Alternative name: stirrup. [2/4/a] [2/5/a] [2/7/a]

■ **Staphylococcus** Type of Gram-positive **bacterium** characterized by its shape, which takes the form of irregular clusters (resembling a miniature bunch of grapes). Staphylococci cause various inflammatory disorders (*e.g.* boils, impetigo, osteomyelitis). *See also* **coccus**.

■ **starch** $(C_6H_{10}O_5)_n$ Complex **polysaccharide carbohydrate**, a **polymer** of **glucose**, that occurs in all green plants, where it serves as a reserve energy material (*e.g.* in roots, tubers and cereal seeds). It forms glucose on complete **hydrolysis**. Alternative name: amylum. [2/3/d]

stasigenesis Process of non-evolution in which the characteristics of a species remain completely unchanged over very long periods of time, and the species becomes a 'living fossil' (*e.g.* coelacanth, tuatara).

state of matter Alternative name for **physical state of matter**.

static electricity Accumulation of **electric charge** on an object; the **electricity** produced by the removal of **electrons** from an **atom** by friction.

statics Branch of **mechanics** concerned with the study of the action of forces on stationary objects.

stationary state Characteristic energy levels of a system, as allowed by **quantum theory**.

stationary wave Alternative name for **standing wave**.

statistical mechanics Branch of statistics concerned with the study of the properties and behaviour of the component microscopic particles of a macroscopic system.

statocyst Sensory organ found in some aquatic invertebrates (*e.g.* certain **crustaceans**) that consists of a cavity containing grains of a hard substance, and which is used to monitor changes in the animals' direction of movement.

stator Non-moving part of an electrical machine such as a motor or dynamo.

■ **steam Water** (H_2O) in the vapour or gaseous state; water above its **boiling point** (100 °C). The white clouds near boiling water often called steam are in fact minute water droplets that have condensed out as steam cools.

steam distillation Distillation of a substance by bubbling **steam** through the heated **liquid**. It is used to purify water-insoluble volatile substances.

steam point Normal **boiling point** of water; it is taken to be a temperature of 100 °C (at normal pressure).

stearate Ester or **salt** of **stearic acid**. *See also* **soap**.

stearic acid $CH_3(CH_2)_{16}COOH$ Long-chain **fatty acid** that occurs in most **fats and oils.** Its sodium and potassium salts are constituents of **soaps**. Alternative name: octadecanoic acid.

steatite Alternative name for **soapstone.**

■ **steel Iron** containing small but controlled quantities of **carbon** (0.1–1.5() and free from silicon, sulphur and phosphorus. Its properties depend on the percentage of carbon; the more carbon, the harder the steel. It is used to make ships, bridges, frameworks of buildings, tools and reinforced concrete, etc. *See also* **stainless steel**. [4/5/c]

stele Vascular parts (**phloem** and **xylem**) of the roots and stems of a vascular plant.

■ **stem** Axis of a plant which is usually above ground, bearing leaves and flowers in vascular plants. Its arrangement of vascular tissues differs from that of a **root**. *See also* **corm; phloem; rhizome; secondary growth; tuber; xylem**. [2/1/a] [2/4/a] [2/5/a] [2/7/a]

steradian (sr) Supplementary SI unit of solid angle which is equal to the angle at the centre of a sphere subtended by part of surface whose area is equal to the square of the radius.

stereochemistry Branch of **chemistry** concerned with the study of the spatial arrangement of the **atoms** within a molecule and the way that these affect the properties of that molecule.

stereoisomer One of two or more **isomers** with the same **molecular formula**, but different **configurations**; any functional groups remain the same. *E.g.* 1-bromopropane, C_3H_7Br, and 2-bromopropane, $CH_3CH(Br)CH_3$, are stereoisomers.

stereoisomerism Isomerism of compounds of the same

molecular formula that results when the spatial arrangement of the atoms within the molecules is different. Any functional groups remain the same but may be in different positions. *See also* **geometric isomerism; optical isomerism**.

stereophony Sound reproduction that uses two or more transmission channels, to give the hearer a similar sound to that of the original.

stereoregular Describing a compound that has a regular spatial arrangement of **atoms** within its **molecule**.

stereoregular rubber Member of a group of stereoregular synthetic rubbers which are produced by a solution **polymerization** process using special **catalysts**.

stereoscope Optical instrument that gives a three-dimensional illusion of depth, normally from a pair of flat photographs.

stereospecific Describing a **chemical reaction** in which one product is formed from each **geometric isomer** of the **reactant**.

steric effect Phenomenon in which the shape of a molecule affects its **chemical reactions**.

steric hindrance Repulsion of an approaching potential **reactant** by a sterically congested compound. *See* **steric effect**.

sterilization *1*. Treatment of an apparatus or substance (*e.g.* food) so that it contains no micro-organisms that could cause disease or spoilage, usually by means of high temperatures, gamma radiation, etc. *2*. Surgical treatment of an animal so that it cannot have offspring (*e.g.* in mammals by cutting or tying the Fallopian tubes of the female or vasa deferentia of the male; removing the ovaries or testes is a more drastic way of achieving the same effect).

sternum Bone in tetrapod vertebrates on the ventral side of the **thorax**, parallel to the **spine**, to which most of the ribs are attached. Alternative name: breastbone.

■ **steroid** Any of a group of naturally occurring tetracyclic organic compounds, widely found in animal tissues. Most have very important physiological activities (*e.g.* adrenal hormones, bile acids, sex hormones, sterols). [2/1/c]

sterol Subgroup of **steroids** or steroid **alcohols**. They include **cholesterol**, abundant in animal tissues, which is the precursor of many other steroids.

stethoscope Instrument for listening to sounds made by internal organs (*e.g.* heart, lungs).

■ **stigma** *1.* In a flower, swelling at the apex of the **style** in the carpel onto which **pollen** is transferred. *2.* In some **flagellates**, pigment spot sensitive to changes in light. [2/4/a] [2/5/a] [2/7/a]

still Apparatus for the **distillation** of liquids.

■ **stimulant** Any **drug** that produces a stimulating effect; *e.g.* alcohol and nicotine in very small quantities (larger quantities are depressants), amphetamines, caffeine. Many stimulants can be addictive (*see* **addiction**).

stimulated emission Process by which a **photon** causes an **electron** in an atom to drop to a lower energy level and emit another photon. It is the principle of the **laser**.

stimulus Environmental factor that is detected by a **receptor** and induces a **response** from an **effector**.

stoichiometric compound Chemical whose molecules have the component **elements** present in exact proportions as demanded by a simple **molecular formula**.

stoichiometric mixture Mixture of **reactants** that in a chemical reaction yield a **stoichiometric compound** with no excess reactant.

stoichiometry Branch of **chemistry** that deals with the relative quantities of atoms or molecules taking place in a reaction.

stoke C.g.s unit of **kinematic viscosity**, equal to 10^{-4} m^2 s^{-1}. It was named after the British physicist George Stokes (1819–1903).

Stokes' law When a small sphere of radius *r* moves through a viscous fluid of viscosity coefficient η, the viscosity *v* is given by $v = 2gr^2(d_1 - d_2)/9\eta$, where d_1 is the density of the sphere, d_2 is the density of the medium, and *g* is the **acceleration of free fall**.

stolon *1.* In botany, stem that grows parallel with the ground; a runner (*see* **vegetative propagation**). *2.* In zoology, stalk-like body part that anchors many types of aquatic invertebrates and which bears polyps.

■ **stoma** Specialized pore found on the stems and leaves (many on the undersurfaces) of vascular plants which allows the exchange of gases and loss of water during **transpiration**. When the stomatal guard cells are turgid, the stomata are open; when the guard cells are flaccid, they are closed. [2/4/a] [2/5/a] [2/7/a]

■ **stomach** Muscular sac present in vertebrates in which food is partly digested and stored after passage through **oesophagus** (gullet). Hydrochloric acid is secreted in the stomach, as is the **enzyme pepsin**, which begins the digestion of **proteins**.

stomata Plural of **stoma**.

■ **stomatal guard cell** One of a pair of sausage-shaped cells within the thickness of cellulose in the cell wall of a leaf.

When the guard cells are turgid they curve, and the pore between them **(stoma)** opens. [2/3/d] [2/7/a]

storage cell Alternative name for **secondary cell**.

storage device Alternative name for a computer **memory** or store. *See also* **random access memory** (RAM); **read-only memory** (ROM).

STP or **s.t.p.** Abbreviation of **standard temperature and pressure**.

straight chain Molecule in which the **atoms** are linked in a long chain, with no **side chains** attached.

strain Deformation of an object due to stress. As a quantity, it is equal to the amount of deformation divided by the original dimension. Strain has no units. *See also* **stress**.

strangeness Property of **hadrons** (subatomic particles), some of which have zero strangeness, whereas others possess non-zero strangeness because they decay slower than expected and are therefore described as strange. *See also:* **spin**.

Streptococcus Type of Gram-positive **bacterium** characterized by its shape, which takes the form of a chain. Streptococci cause various disorders (*e.g.* erysipelas, scarlet fever).

streptomycin Antibiotic that works by inhibiting **protein synthesis** in bacterial cells.

■ **stress** *1*. In physics, any force acting upon an object in such a way as to alter its size or shape. Mechanical stress is measured in units of force per unit area. *See also* **strain**. [5/2/a] *2*. In biology, any environmental factor, or combination of factors, that has adverse effects on the structure or behaviour of an organism.

■ **striated muscle Muscle** that contains well-aligned threads of

protein, which enable it to contract strongly in a particular direction. Alternative name: voluntary muscle. [2/5/a]

stridulation Process by which many insects (*e.g.* grasshoppers, crickets) produce sound by rubbing body parts together. Typically, a toothed 'file' on a wing or leg is moved rapidly against the hardened edge of another wing.

strigil Cluster of hairs on the forelegs of some insects (*e.g.* butterflies, wasps, bees) that is used for grooming.

strobilus Alternative name for a (plant) **cone**.

stroboscope Instrument consisting of a rapidly flashing lamp, employed for measuring speeds of rotation. It can also be used, by controlling the rate of flashing, to view objects that are moving rapidly with periodic motion and to see them as if they were at rest. Alternative name: rotoscope.

stroke Sudden loss of part of the brain's function due to lack of oxygen caused by the blocking of a blood vessel serving the brain by a clot (**thrombosis**) or by a vessel bursting (cerebral haemorrhage). Alternative name: apoplexy.

stroma Matrix in which the **lamellae** of a green plant's **chloroplasts** are embedded, and the site of dark reactions of **photosynthesis** (which build up **carbohydrates**).

strong acid **Acid** that is completely dissociated into its component **ions**.

strong base **Base** that is completely dissociated into its component **ions**.

strontia Alternative name for **strontium oxide**.

strontium Sr Silvery-white metallic element in Group IIA of the Periodic Table (one of the **rare earth** elements). Strontium compounds impart a bright red colour to a flame, and are used in flares and fireworks. At. no. 38; r.a.m. 87.62.

strontium nitrate $Sr(NO_3)_2$ Colourless crystalline solid, used in fireworks and flares to produce a bright red flame.

strontium oxide SrO Grey amorphous solid, which dissolves in water to form strontium hydroxide. Alternative names: strontia, strontium monoxide.

strontium unit (S.U.) Measure of the concentration of the **radio-isotope** strontium-90 in substances such as bone, milk or soil relative to their calcium content.

structural formula Shorthand description of a chemical compound that indicates the arrangement of the atoms in its molecules as well as its composition. *E.g.* the molecular formula of acetone (propanone) is C_3H_6O; its structural formula is CH_3COCH_3. *See also* **empirical formula**; **molecular formula**.

structural isomerism **Isomerism** of chemical compounds that have the same **molecular formula** but different **structural formulae**. Structural isomers have different physical and chemical properties. *E.g.* propanol (propyl alcohol), C_3H_5OH, and acetone (propanone), CH_3COCH_3, have the same molecular formulae, C_3H_6O, but different structures and are therefore structural isomers.

strychnine White crystalline insoluble **alkaloid** with a bitter taste, one of the most powerful **poisons** known. Alternative name: vauqueline.

style Part of the **carpel** of a flowering **plant** that bears the **stigma**.

styrene $C_6H_5CH=CH_2$ Colourless liquid **aromatic compound** that occurs in coal-tar. It polymerizes slowly on standing, rapidly when exposed to sunlight, and is used for making **polymers** (*e.g.* **polystyrene**). Alternative names: phenylethylene, vinylbenzene, ethenylbenzene.

styrene-butadiene rubber Synthetic rubber made by the **polymerization** of **butadiene** with **styrene**. It is used for making vehicle tyres. Alternative names: buna-S, SBR, GR-S.

subatomic particle Particle that is smaller than an **atom** or forms part of an atom (*e.g.* electron, neutron, proton). Sometimes also called an elementary particle.

subcritical Describing a **chain reaction** in a **nuclear reactor** that is not self-sustaining.

subcutaneous tissue Layer of **tissue** below the **dermis** of the **skin** in vertebrates. It often contains deposits of **fat**.

sub-imago Pre-adult winged insect form that is unique to the mayflies (order Ephemeroptera).

sublimate Solid formed by the process of **sublimation**.

sublimation Direct conversion of a solid substance to its vapour state on heating without melting taking place (*e.g.* solid carbon dioxide (dry ice), iodine). The vapour condenses to give a **sublimate**. The process is used to purify various substances.

sub-shell Subdivision of an electron **shell**.

subsonic *1.* Speed less than that of the **speed of sound** in a medium. *2.* Speed that is less than Mach 1. *See* **Mach number**.

■ **subspecies** Group of organisms within a **species** that have certain characteristics not possessed by other members of the species. Breeding may occur between members of different subspecies. *See also* **race; variety**. [2/4/b]

substituent **Atom** or group that replaces another atom or group in a molecule of a compound.

substitution Direct replacement of an **atom** or group in a molecule of a compound by some other atom or group.

substitution product Product formed from **substitution**.

substitution reaction Chemical reaction that involves the direct replacement of an **atom** or group in a **molecule** of a compound by some other atom or group. *E.g.* the reaction in which an atom of chlorine replaces an atom of hydrogen in a molecule of benzene (C_6H_6) to form a molecule of chlorobenzene (C_6H_5Cl). Alternative name: displacement reaction. *See also* **addition reaction**.

substrate *1.* Molecule on which an **enzyme** exerts its catalytic action. *2.* Pure crystal of a **semiconductor** used for making integrated circuits.

■ **succession** Progressive change in the structure of a community when colonizing a **habitat** until a stable **climax community** is established. [2/7/c]

succinic acid $(CH_2COOH)_2$ White crystalline **dicarboxylic acid**, used in the manufacture of dyes and in organic synthesis. Alternative names: butanedioic acid, ethylenedicarboxylic acid.

succulent Plant with swollen leaves or stems, adapted for living in dry habitats or similar conditions in which there is little fresh water (*e.g.* a salt marsh). *See also* **halophyte**; **xerophyte**.

sucrase **Enzyme** that breaks down **sucrose** into simpler sugars. Alternative name: invertase.

■ **sucrose** $C_{12}H_{22}O_{11}$ White **optically active** soluble crystalline disaccharide which is obtained from sugar cane and sugar-beet; ordinary sugar, used to sweeten food. It is hydrolysed to **fructose** and **glucose**. Alternative names: cane-sugar, beet-sugar, sugar. [2/3/d]

sugar *1*. Crystalline soluble **carbohydrate** with a sweet taste; usually a **monosaccharide** or **disaccharide**. *2*. Common name for **sucrose**.

sulphate **Ester** or **salt** of **sulphuric acid**.

sulphation Conversion of a substance into a **sulphate**.

sulphide Binary compound containing **sulphur**; a salt of **hydrogen sulphide**. Organic sulphides are called **thioethers**.

sulphite **Ester** or **salt** of **sulphurous acid**.

2-sulphobenzimide Alternative name for **saccharin**.

sulphonamide Member of a class of synthetic anti-bacterial drugs which act by **enzyme** inhibition. They are **amides** derived from **sulphonic acids**.

sulphonate **Ester** or **salt** of a **sulphonic acid**.

sulphonation **Substitution reaction** that involves the replacement of a **hydrogen** atom by the **sulphonic acid** group.

sulphonic acid **Acid** that contains the group $-SO_3H$. Organic sulphonic acids are used in the manufacture of dyes, detergents and drugs.

sulphonium compound Compound of the general formula R_3SX, where R is an organic **radical** and X is an electronegative element or radical.

sulphoxide Compound of general formula $RSOR'$, where R and R' are organic **radicals**.

■ **sulphur** S Yellow nonmetallic solid element in Group VIB of the Periodic Table, which forms several **allotropes** including alpha- (rhombic) sulphur and beta- (monoclinic) sulphur. It occurs as the free element in volcanic regions and as underground deposits (extracted by the **Frasch process**), and as **sulphates** and **sulphides**, which include important minerals (*e.g.* galena, PbS, and pyrites, FeS_2). Chemically it behaves like oxygen, and can replace it in organic compounds (*e.g.* **thioethers** and **thiols**). Sulphur is used to make **sulphuric acid**, matches, gunpowder, drugs, fungicides and dyes, and in the **vulcanization** of rubber. At. no. 16; r.a.m. 32.06.

sulphur dichloride oxide Alternative name for **thionyl chloride**.

sulphur dioxide SO_2 Colourless poisonous gas with a strong pungent odour, made by burning sulphur, roasting **sulphide** ores or by the action of acids on **sulphites**. It is used to make **sulphuric acid** and, in aqueous solution, as a bleach (*e.g.* for straw and paper). Alternative name: sulphur (IV) oxide.

sulphur dye Dye made by heating certain **organic compounds** with **sulphur** or alkali polysulphides, used for dyeing industrial fabrics.

sulphur (IV) oxide Alternative name for **sulphur dioxide**.

sulphur (VI) oxide Alternative name for **sulphur trioxide**.

sulphur trioxide SO_3 Volatile white solid made by the catalytic oxidation of **sulphur dioxide**, usually stored in sealed tubes. It reacts with water to form **sulphuric acid**. Alternative name: sulphur (VI) oxide.

sulphuric acid H_2SO_4 Corrosive colourless oily liquid **acid**, made mainly from **sulphur dioxide** by the **contact process**. It is a **desiccant**, and when hot a powerful **oxidizing agent**. It is produced in large quantities and used in the manufacture of other acids, fertilizers, explosives, accumulators, petrochemicals, etc. Its salts are **sulphates**. Alternative names: vitriol, oil of vitriol, hydrogen sulphate.

sulphurous acid H_2SO_3 Colourless aqueous solution of **sulphur dioxide**, used as a bleach and a **reducing agent**. Its salts are **sulphites**.

superconductivity Large increase in electrical **conductivity** exhibited by certain metals and alloys at a temperature a few degrees above **absolute zero**. Alternative name: supraconductivity.

supercooling Metastable state of a liquid in which its temperature has been brought below the normal **freezing point** without any solidification or crystallization occurring.

supercritical Describing a **chain reaction** of a **nuclear reactor** which is self-sustaining.

superfluid Fluid that has practically zero **viscosity**; it forms thin films on surfaces and flows without friction (*e.g.* liquid helium).

superheated steam Steam (under pressure) at a temperature above the boiling point of water (100 °C). *See* **superheating**.

superheating Heating of a liquid or gas to above its **boiling point** in the liquid state, by increasing the pressure above that of the atmosphere. Alternative name: overheating.

superheterodyne receiver Radio receiver in which the frequency of the incoming signal is reduced by combining it with another generated in the receiver (the heterodyne principle). The resulting intermediate frequency (IF) is then amplified and demodulated.

supernatant liquid Clear liquid that lies above a sediment or **precipitate**.

superoxide *1.* Compound that yields the **free radical** O_2^- which is highly toxic to living cells. *2.* **Oxide** that yields both **hydrogen peroxide** and **oxygen** on treatment with an **acid**. *See also* **peroxide**.

superphosphate Important **fertilizer** made by reacting calcium phosphate (from slag or bone ash) with **sulphuric acid**.

superplasticity Property of certain materials that are able to stretch several hundred times before failing.

■ **supersaturated solution** Unstable **solution** that contains more **solute** than a **saturated solution** would contain at the same temperature. It easily changes to a saturated solution when the excess solute is made to crystallize.

supersonic Describing anything travelling through a medium faster than the speed of sound in that medium, *e.g.* above 1,160 km/h in the Earth's atmosphere at sea level. Alternative name: ultrasonic. *See also* **Mach number**.

suprarenal gland Alternative name for **adrenal gland**.

surface active agent Alternative name for **surfactant**.

■ **surface tension** Force per unit length acting along the surface of a **liquid** at right angles to any line drawn in the surface. It has the effect of making a liquid behave as if it has a surface skin (which can support *e.g.* small aquatic insects), and is responsible for **capillarity** and other phenomena. It is measured in newtons per metre (Nm^{-1}).

surfactant Substance that reduces the **surface tension** of a liquid, used in detergents, wetting agents and foaming agents. Alternative name: surface active agent.

susceptibility, magnetic *See* **magnetic susceptibility**.

suspension Mixture of insoluble small solid particles and a fluid through which the insoluble substance stays evenly distributed (because of molecular collisions and the fluid's viscosity, which prevents precipitation). Alternative name: suspensoid.

suspensory ligament One of the structures that hold the lens of the **eye** in position.

■ **sweat** Watery fluid containing salts secreted from glands in the **skin. Evaporation** of sweat aids in cooling the body. Alternative name: perspiration. [2/7/a]

swim bladder Structure present in bony fish, used for controlling buoyancy by filling or emptying with air.

■ **switch** Usually mechanical device for opening and closing an electric circuit. A solid-state switch has no moving parts. [5/5/b]

■ **symbiosis** Association between two organisms of different species in which both partners benefit. *See also* **commensalism**; **epiphyte**; **parasitism**; **saprophyte**. [2/2/c] [2/6/a]

symbol, chemical Letter or letters that represent the name of a **chemical element**, or one atom of it in a chemical **formula**; *e.g.* C is carbon, Ca is calcium and Co is cobalt. The symbol may be based on the element's Latin, not English, name; *e.g.* Au for gold (Latin *aurum*), Fe for iron (*ferrum*).

symmetry The property of being symmetrical, *i.e.* having the same shapes on each side of or around a point, axis or plane. Most animals have bilateral symmetry; many echinoderms (*e.g.* starfish) have radial symmetry.

sympathetic nervous system Branch of the **autonomic nervous system** that is structurally different from the **parasympathetic nervous system. Noradrenaline** is produced at the end of sympathetic nerve fibres, unlike the parasympathetic system. Effects produced by each system are generally antagonistic.

symplast All the living **protoplasm** within a plant, consisting of the individual cells, and the network of filaments that pierce the non-living cell walls and link adjoining cells.

synapse Point of connection between **neurones** (nerve cells). It consists of a gap between the **membranes** of two cells, across which **impulses** are passed by a transmitter substance *e.g.* **acetylcholine**). Specialized synapses occur at nerve-**muscle** junctions. *See also* **neurotransmitter**.

synergy Collective action of two or more things (*e.g.* drugs, muscles) that is more effective than it would be if they acted on their own. Alternative name: synergism.

■ **synovial fluid** Liquid secreted by a **synovial membrane.** [2/3/a] [2/5/a]

■ **synovial membrane** Lining of the capsule that encloses a joint between bones. It secretes synovial fluid, which acts as a lubricant to prevent friction in the joint during movement. [2/3/a] [2/5/a]

synthesis Formation of a chemical compound by combining elements or simpler compounds; the building of compounds through a planned series of steps.

synthetic Formed by artificial means; describing a chemical compound that has been produced by synthesis. Alternative names: artificial, man-made.

synthetic gas Mixture of **hydrogen** and **carbon monoxide** made by **reforming** natural gas with steam.

syphon *See* **siphon**.

syrinx Sound-producing organ of a bird.

■ **systole** Phase of contraction of the heartbeat during which blood is forced into the **arteries**. *See also* **diastole**. [2/4/c]

T

tachyon Theoretical subatomic particle that travels faster than light.

tadpole Larval stage of an amphibian (*e.g.* of a frog, toad or newt).

talc $3MgO.4SiO_2.H_2O$ Soft white or grey-green mineral. Its purified form is a white powder, used in talcum powder, in medicine and in ceramic materials. Alternative names: French chalk, magnesium silicate monohydrate.

tall oil Complex mixture of mainly **resin** acids and **fatty acids**, a by-product of wood-pulp manufacture, used in **soaps**, paints and varnishes.

talus The ankle bone.

tannic acid White amorphous solid organic acid, a member of the class of compounds called **tannins**. It is used in **tanning** and for making inks and dyes. Alternative name: tannin.

tannin Yellow substance that is a member of a class of organic compounds, of vegetable origin (*e.g.* in tree bark, oak galls and tea), that are derivatives of **polyhydric benzoic acids**. They are used in tanning hides to make leather. *See also* **tannic acid**.

tanning Process of converting hides and skin into leather by the action of substances such as **tannic acid** or **tannins**.

tantalum Ta Hard blue-grey metallic element in Group VA of the Periodic Table (a **transition element**), used in electronic and chemical equipment, and for making surgical instruments. At. no. 73; r.a.m. 180.948.

tape deck Unit that records audio or video signals onto, or plays back signals from, magnetic tape in cassettes or on reels.

tape punch Machine that produces punched paper tape, either operated by a **keyboard** as part of a computer **input device** or automatically as an **output device** driven by a computer.

tape reader **Input device** that feeds **data** off punched paper tape into a computer.

■ **tar** Dark viscous liquid obtained by the **destructive distillation** of **coal** in the absence of air. It is a complex mixture of organic compounds, mainly **hydrocarbons** and **phenols**. Alternative name: coal-tar. Similar substances obtained from **petroleum** are also called tars. *See also* **bitumen**.

tarsal Bone that occurs in the feet of tetrapods. Human beings have seven tarsals in each foot, one of them modified to form the heel bone (calcaneum).

tarsus *1*. Region of the segmented leg of an **insect**. *2*. The ankle region of the hindlimb of a tetrapod vertebrate.

tartar emetic Potassium antimonyl tartrate, $K.SbO.C_4H_4O_6.\frac{1}{2}H_2O$, a poisonous compound used as an insecticide and as mordant in dyeing.

tartaric acid HOOC.CH(OH)CH(OH).COOH White crystalline hydroxycarboxylicacid that occurs in grapes and other fruits. It is used in dyeing and printing. Its salts, the **tartrates**, are frequently used in medicine. Alternative names: 2,3-dihydroxybutanedioic acid, dihydroxysuccinic acid.

tartrate **Ester** or **salt** of **tartaric acid**.

taste Sense that enables animals to detect flavours, which in mammals involves **taste buds**.

taste bud Small sense organ containing **chemoreceptors** for the sense of **taste**, located in the mouth (particularly on the upper surface of the **tongue**).

tau-particle Subatomic particle, a **lepton** and its associated anti-lepton that have unusually high mass.

tautomerism Equilibrium between two organic **isomers**. It usually involves a shift in the point of attachment of a mobile hydrogen atom and a shift in the position of a **double bond** in a molecule. Alternative name: dynamic isomerism. *See* **keto-enol tautomerism**.

taxis Orientation of an organism, involving movement, with respect to a **stimulus** from a specific direction. *See also* **chemotaxis**; **geotaxis**; **phototaxis**; **tropism**.

■ **taxonomy** Study of the **classification** of living organisms. [2/4/b]

tear gas Volatile substance, usually a **halogen**-containingorganiccompound, that causes irritation of the eyes and is used in crowd control. Alternative name: lachrymator.

tears Slightly bactericidal watery liquid secreted by the **lacrimal gland** near the eye.

technetium Tc Artificial radioactive metallic element in Group VIIA of the Periodic Table (a **transition element**), which occurs among the fission products of uranium. It has several isotopes, with half-lives of up to 2.12×10^5 years. At. no. 43; r.a.m. 99 (most stable isotope).

teeth *See* **tooth**.

Teflon Trade name for **polytetrafluoroethene** (PTFE).

telecommunications Electrical and electronic methods of communicating information (as aural or visual signals, or data) between two distant places; *e.g.* telegraph, telephone, radio, radar, television, facsimile transmission (fax), data transmission.

telegraph Apparatus for transmitting messages over a long distance using electrical impulses sent along wires.

telemeter Device for transmitting information over long distances in modulated form, *e.g.* by means of **radio** or **telephone**.

telemetry Technique of transmitting information by **telemeter**.

telephone Device for converting sounds into electrical impulses (which are then transmitted along wires or by **radio**) and reconverting them to sounds. It uses a **microphone** in the mouthpiece and a small **loudspeaker** in the earpiece.

telephony Transmission of **sound** by electrical means, particularly using a **telephone**.

telescope Optical instrument that forms an enlarged image of a distant object. *See also* **radio telescope**; **reflecting telescope**; **refracting telescope**.

television Method of converting visual images into electrical impulses (which are then transmitted by **radio** or along wires) and reconverting them to images.

telluride **Binary compound** that contains **tellurium**.

tellurium Te Silvery-white semi-metallic element in Group VIB of the Periodic Table, obtained as a by-product of the extraction of gold, silver and copper. It is used as a catalyst and to add hardness to alloys of lead or steel. At. no. 52; r.a.m. 127.60.

■ **telophase** Final phase of cell division that occurs in **mitosis** and **meiosis**. During telophase **chromatids**, on reaching the poles of the cell, become densely packed and the cell divides. In animal cells, the **plasma membrane** constricts. In plants, a wall divides the cell in two. [2/6/b]

telotaxis Animal movement in response to a stimulus detected by any form of specialized sense organ.

telson Lowest segment of the abdomen in some **arthropods** (*e.g.* the tail of a crab), but which is found in insects only at the **embryo** stage.

temperature Degree of hotness or coldness of something, a measure of the average **kinetic energy** of its atoms or molecules. *See also* **temperature scale**.

temperature coefficient Change in a physical quantity with change in **temperature**, usually given as per degree rise of temperature.

temperature gradient Degree or measured rate of the **temperature** change between two points of reference in a substance or in an area.

temperature scale Method of expressing **temperature**. There are various scales, based on different **fixed points**. The **Fahrenheit scale** has largely been replaced by the **Celsius scale** (formerly centigrade), all of which express temperatures in degrees (°F and °C). In science, temperatures are frequently expressed in **kelvin** (abbreviation K, with no ° sign), the thermodynamic unit of temperature in **SI units**.

tempering Method of changing the physical properties of a metal or alloy by heating it for a time and then cooling it gradually or quickly.

temporary hardness *See* **hardness of water**.

■ **tendon** Strong **connective tissue** that attaches **muscles** to **bones**. It consists of **collagen** fibres. *See also* **ligament**. [2/3/a] [2/5/a]

■ **tendril** Long thin extension of a stem or leaf, which usually grows in a spiral, developed by many plants as an aid to climbing. [2/4/a] [2/5/a] [2/7/a]

tension Stretching force. *See also* **stress**; **surface tension**.

teratogen Any substance, natural or artificial, that causes abnormal development of an **embryo**.

terbium Tb Silvery-grey metallic element in Group IIIA of the Periodic Table (one of the **lanthanides**), used in making **semiconductors** and **phosphors**. At. no. 65; r.a.m. 158.92.

terephthalic acid $C_6H_4(COOH)_2$ White crystalline **aromatic compound**, used in the manufacture of **plasticizers** and **polyester** fibres and films (*e.g.* Dacron, Terylene). Alternative name: benzene-1,4-dicarboxylic acid. *See also* **phthalic acid**.

terminal In computing, an **input device** or **output device** that can handle **data**.

terminal velocity For an object falling through a fluid (gas or liquid) under the influence of gravity, the constant velocity it reaches when there is no resultant force acting on it (*i.e.* frictional resistance equals the gravitational pull). Alternative name: terminal speed.

terpene Member of a class of **unsaturated hydrocarbons** found in **essential oils**, with a structure based on the **isoprene** unit.

terrestrial magnetism Earth's **magnetism**. Alternative name: geomagnetism.

tervalent Alternative name for **trivalent**.

Terylene Trade name for a **polyester** produced by the **condensation** of **terephthalic acid** with ethane-1,2-diol (ethylene glycol). It is used for making textiles.

tesla SI unit of magnetic flux density, named after the Croation-born American physicist Nikola Tesla (1857–1943).

testa Protective covering of a **seed**. Dry and fibrous, it is formed from the **nucellus** and **integuments**.

testes Plural of **testis**.

testis Usually paired organ in males of many animals, responsible for the production of **sperm**. In vertebrates, the testes also produce **hormones** that induce development of **secondary sexual characteristics**.

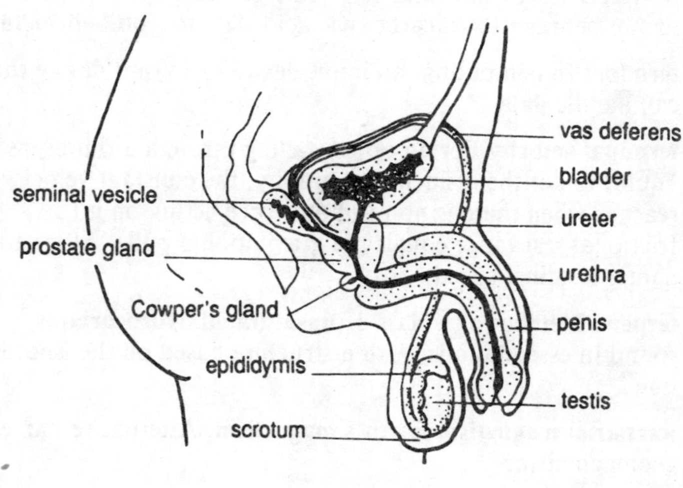

Human testis

testosterone **Sex hormone** produced in the **testes** that induces the development of **secondary sexual characteristics.**

tetrachloroethene $CCl_2=CCl_2$ Liquid halogenoalkene, used as a solvent (especially as a de-greasing agent). Alternative names: ethylene tetrachloride, tetrachloroethylene.

tetrachloroethylene Alternative name for **tetrachloroethene.**

tetrachloromethane CCl_4 Colourless liquid **halogenoalkane,** used as a solvent, cleaning agent and in fire extinguishers. Alternative name: carbon tetrachloride.

tetracycline Member of a class of **antibiotics** (which act by inhibiting **protein** synthesis in bacteria) that have a wide range of applications.

tetrad *1.* Four cells formed after the second division in meiosis is complete. *2.* Four **spores** formed by **meiosis** in a spore mother cell, often seen in **fungi**.

tetraethyllead $Pb(C_2H_5)_4$ Colourless liquid **organometallic compound**, used as an anti-**knock** agent in petrol. Alternative name: tetraethyl plumbane.

tetrafluoroethene $CF_2=CF_2$ Gaseous **fluorocarbon** from which **polytetrafluoroethene** (PTFE, Fluon) is made.

tetrahedral compound Compound that has a central **atom** joined to other atoms or groups at the four corners of a tetrahedron. It is the configuration of carbon in many **saturated compounds** (*e.g.* in methane, CH_4).

tetrahydrate Chemical containing four molecules of **water of crystallization**.

■ **tetraploid** Describing an organism that has four sets of homologous **chromosomes**. *See also* **diploid**; **haploid**. [2/8/b]

tetrapod Animal with four (usually **pentadactyl**) limbs, including all vertebrates with a locomotory system adapted to living on land. Alternative name: quadruped.

tetravalent Having a **valence** of four. Alternative name: quadrivalent.

thalamus Part of the **forebrain** of vertebrates that is concerned with the routing of nervous impulses to and from the **spinal cord**.

thalassaemia Inherited form of **anaemia**, common in Mediterranean countries and Africa, caused by a defect in the blood pigment **haemoglobin**.

thalidomide **Sedative** drug that is now banned in many countries because it can cause deformities in human embryos.

thallium Tl Silver-grey metallic element in Group IIIB of the Periodic Table, used in electronic equipment and to make pesticides. At. no. 81; r.a.m. 204.39.

■ **Thallophyta** Division of the plant kingdom that contains **algae, bacteria** and **fungi**. The simplest form of plant life, very varied in size and structure, thallophytes all possess a **thallus**. [2/4/b]

■ **thallus** Main body of simple plants (*e.g.* algae, lichens and mosses) that do not have true **roots, stems** or leaves. [2/4/a] [2/5/a] [2/7/a]

thanatosis Defensive behaviour exhibited by some animals (*e.g.* opossums and certain reptiles) that feign death as a means of evading the attention of predators.

thelytoky Sexual adaptation in which **parthenogenesis** is the only form of reproduction in populations that are composed entirely of females, as in *e.g.* the great green cricket.

theodolite Instrument used in surveying to measure angles accurately. It is fitted with a special **telescope**.

theorem General conclusion in science or mathematics which makes certain assumptions in order to explain observations.

thermal *1*. Concerning or using heat. *2*. Heat-induced up-current of air, used by soaring birds and glider pilots.

thermal analysis *1*. Method of chemical analysis that depends on a change of weight of a compound as a function of temperature. Alternative name: thermographic analysis. *2*. Method of analysing metals and alloys by studying their heating and cooling curves.

thermal cross-section Probability of a **nucleus** interacting with a **neutron** of thermal energy (*i.e.* a slow neutron).

thermal diffusion Process of forming a concentration gradient in a **fluid** mixture by the application of a **temperature gradient**. Alternative name: Soret effect.

thermal dissociation Temporary reversible decomposition of a chemical compound (into component molecules) resulting from the application of heat.

thermal equilibrium State in which there is no net heat flow in a system.

thermalization Slowing-down of fast **neutrons** to thermal energies (*i.e.* converting them to slow neutrons, capable of initiating **nuclear fission**), achieved by a **moderator** in a **nuclear reactor**.

thermal neutron **Neutron** that has an energy of about 0.025 electron volts (eV), and which can be captured by an atomic **nucleus** (particularly to initiate **nuclear fission**). Alternative name: slow neutron. *See also* **fast neutron**.

thermal reactor **Nuclear reactor** that has a **moderator**, and in which a **chain reaction** is sustained by **thermal neutrons**.

thermal spike Transient rise in **temperature** above the normal.

thermionic emission Emission of **electrons** from a heated material (*e.g.* the **cathode** in a **thermionic valve**).

thermionics Branch of **electronics** concerned with the study of the emission of **electrons** or **ions** from heated substances.

thermionic valve **Valve** (vacuum tube) with a heated **cathode** that emits **electrons**, an **anode** that collects them and often control **grids** (*e.g.* one in a triode, two in a tetrode, etc.). Before the introduction of **semiconductor** devices, such valves were much used in electronics (*e.g.* in amplifiers).

thermistor Temperature-sensitive **semiconductor** device whose resistance decreases with an increase in temperature, used in electronic thermometers and switches.

thermite Mixture of powdered aluminium and iron (III) oxide (ferric oxide) which produces a great amount of heat when ignited, because the reaction between the two substances is highly **exothermic**. It is used in incendiary bombs and for local welding of steel (*e.g.* to mend broken tram-lines). Alternative name: Thermit.

thermobarograph Instrument for measuring and recording the **temperature** and **pressure** of the atmosphere.

thermochemistry Branch of **chemistry** that deals with the study of heat changes in relation to **chemical reactions**. The measurement of **heats of reaction**, heat capacities and **bond energies** falls within its scope.

thermocouple Device for measuring **temperature** which relies on the principle that a heated junction between two dissimilar metals produces a measurable **electromotive force** (e.m.f.) which depends on the temperature of the junction.

thermodynamic temperature Alternative name for **absolute temperature**.

thermodynamics Branch of **physics** concerned with the study of the effects of **energy** changes in physical systems and the relationship between various forms of energy, principally heat and mechanical energy.

thermodynamics, laws of *1*. First law: **energy** can be neither created nor destroyed. Alternative name: Law of conservation of energy. *2*. Second law can be expressed in two ways: **entropy** of a closed system increases during a spontaneous process, or heat will not transfer spontaneously from an object to a hotter object. *3*. Third law: entropy of a perfect crystal at **absolute zero** is equal to zero. *4*. Zeroth law: if two objects are in thermal equilibrium with a third object, then all three are in thermal equilibrium.

thermoelectricity **Electricity** produced from heat energy, as in a **thermocouple**.

thermograph **Thermometer** that records variations in **temperature** over a period in time on a graph; a self-registering thermometer.

thermographic analysis Alternative name for **thermal analysis**.

thermometer Device for measuring **temperature**. The common liquid-in-glass thermometer relies on the expansion of the liquid (*e.g.* mercury or dyed alcohol) in a calibrated sealed glass capillary tube.

thermometer, clinical Mercury thermometer used in medicine for the accurate measurement of body temperature. It measures only a small range of temperatures. A constriction in the capillary near the bulb breaks the mercury thread to retain the temperature reading; before use, the thermometer has to be shaken to return all the mercury to the bulb.

thermometer, maximum and minimum **Thermometer** that records the maximum and minimum **temperatures** attained during a given period of time.

thermonuclear reaction Process that releases **energy** by the **fusion** of atomic nuclei (*e.g.* of hydrogen nuclei in the **hydrogen bomb**). Alternative name: fusion reaction.

thermopile Temperature-measuring device that consists of several **thermocouples** connected in series, with one set of junctions blackened so as to absorb thermal radiation.

■ **thermoplastic** Describing a **polymer** (plastic) that softens on heating and hardens on cooling without changing its properties (*i.e.* it softens again on reheating). The polymer itself may be described as a thermoplastic. *See also* **thermosetting**. [4/8/b]

■ **thermosetting** Describing a **polymer** (plastic) that becomes rigid on heating because of the formation of extra chemical bonds, and therefore has new properties. It does not soften again on reheating – heat will decompose it. The polymer may be described as a thermosetter. *See also* **thermoplastic**. [4/8/b]

■ **thermostat** Device for automatically controlling the **temperature** of an appliance or apparatus. Thermostats for electrically heated appliances work by switching off the current when the required temperature is exceeded (and switching it on again when the temperature falls). [3/7/c]

thiamine White water-soluble crystalline B **vitamin,** found in cereals and yeast. Deficiency of thiamine causes the disorder

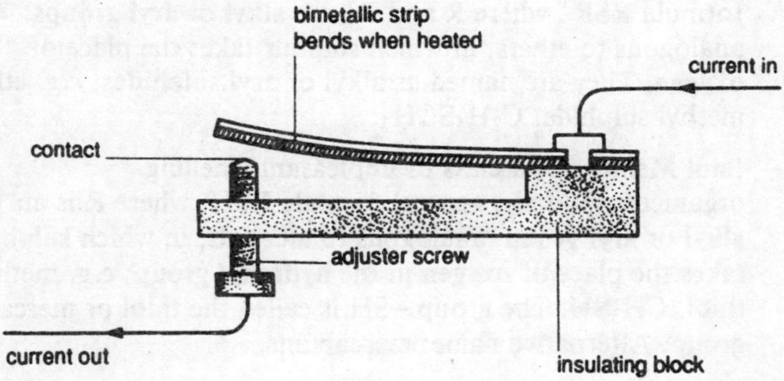

Bimetallic strip thermostat

beri-beri in human beings. Alternative names: thiamin, aneurin, vitamin B_1.

thiazine Member of a group of **heterocyclic compounds** that contain **sulphur** and **nitrogen** (in addition to four carbons) in a six-membered ring.

thigmotropism Tropism in plants (*e.g.* the leaves of *Mimosa pudica* or the tendrils of climbing plants) in response to the **stimulus** of touch.

thin layer chromatography (TLC) Method for separating substances by allowing a **solution** containing them to rise up a thin plate of solid **adsorbent** material, with the different substances moving at different rates. Once separated, they can be identified.

thio- Prefix denoting the presence of **sulphur** in a compound.

thiocarbamide Alternative name for **thiourea**.

thioether Member of a class of organiccompounds, general formula RSR′ where R and R′ are **alkyl** or **aryl groups**, analogous to **ethers**, in which **sulphur** takes the place of **oxygen**. They are named as alkyl or aryl sulphides; *e.g.* ethyl methyl sulphide, $C_2H_5SCH_3$.

thiol Member of a class of unpleasant-smelling organiccompounds, general formula RSH, where R is an **alkyl** or **aryl group**, analogous to **alcohols**, in which **sulphur** takes the place of **oxygen** in the **hydroxyl group**; *e.g.* methane thiol, CH_3SH. The group –SH is called the thiol or mercapto group. Alternative name: mercaptans.

thionyl chloride $SOCl_2$ Colourless fuming pungent liquid, used in organic synthesis (to convert an –OH group to –Cl). Alternative name: sulphur dichloride oxide.

thiophene C_4H_4S Colourless liquid that occurs as an impurity in commercial **benzene**. A **heterocyclic** compound, with four carbon atoms and one sulphur atom in a five-membered ring, it is used in organic synthesis and as a solvent.

thiosulphate Ester or **salt** of **thiosulphuric acid**. The alkali metal salts, *e.g.* sodium thiosulphate, $Na_2S_2O_3$, and ammonium thiosulphate, $(NH_4)_2S_2O_3$, are used in photography as fixing agents (where they dissolve unexposed silver halides from the emulsion on developed film or paper).

thiosulphuric acid $H_2S_2O_3$ Unstable **acid** which readily decomposes to **sulphurous acid** and **sulphur**. Its salts (**thiosulphates**) are used in photography.

thiourea NH_2CSNH_2 Colourless crystalline organic compound, the **sulphur** analogue of **urea**. Its conversion to its

isomer ammonium thiocyanate, NH_4CNS, on heating was the first demonstration of an organic compound being changed directly into an inorganic one. It is used in medicine and as a photographic sensitizer. Alternative name: thiocarbamide.

thixotropy Property of certain **fluids** that decrease in **viscosity** when placed under increased stress. 'Jelly' paints are thixotropic.

Thomson effect Phenomenon in which heat is evolved when current flows along a conductor whose ends are at different temperatures; the rate of heat production is proportional to current multiplied by the **temperature gradient**. It was named after the British physicist William Thomson (1824–1907), later Lord Kelvin. Alternative name: Kelvin effect.

■ **thorax** *1.* In adult insects, region of the body that bears legs (and wings, if any), situated between head and **abdomen**. *2.* In land-living vertebrates, region of the body that contains the **heart** and **lungs**; the chest. In mammals it is separated from the abdomen by the **diaphragm**. [2/4/a]

thoria Alternative name for **thorium dioxide**.

thorium Th Silvery-white radioactive element in Group IIIA of the Periodic Table (one of the **actinides**). It has several isotopes, with half-lives of up to 1.39×10^{10} years. Thorium-232 captures slow, or thermal, neutrons and is used to 'breed' the fissile uranium-233. Its refractory oxide (thoria, ThO_2) is used in gas mantles. At.no. 90; r.a.m. 232.0381.

thorium dioxide ThO_2 White insoluble powder, used as a refractory and in non-silica optical glass. Alternative name: thoria.

threonine Amino acid that is essential in the diet of animals. Alternative name: 2-amino-3-hydroxybutanoic acid.

thrip Member of the insect order **Thysanoptera**.

■ **thrombin Protein** that takes part in the clotting of **blood** following injury. It converts **fibrinogen** to **fibrin** to help form the clot. [2/4/c]

■ **thrombocyte** Alternative name for a type of blood **platelet**. [2/4/c]

■ **thrombosis** Blockage of a blood vessel by a blood clot (thrombus). If the vessel is an artery serving the heart muscle, thrombosis may result in a heart attack (*i.e.* coronary thrombosis causing a cardiac infarction); if the artery serves the brain, the result may be a **stroke**. [2/4/c]

■ **thrombus** Blood clot. *See* **thrombosis**. [2/4/c]

throttle Alternative term for **accelerator** (sense *2.*).

thulium Tm Silvery-white metallic element in Group IIA of the Periodic Table (one of the **lanthanides**). Its radioactive isotopes emit gamma-rays and X-rays and are used in portable radiography equipment. At. no. 69; r.a.m. 168.934.

thymine $C_5H_6N_2O_2$ Colourless crystalline **heterocyclic** compound, one of the **pyrimidines**. Alternative names: 5-methyluracil, 5-methyl-2,4-dioxopyrimidine.

thymol $C_{10}H_{14}O$ Colourless crystallineorganiccompound found in the **essential oils** of thyme and mint. It is used in antiseptic mouth washes. Alternative name: 2-hydroxy-*p*-cymene, 2-hydroxy-1-isopropyl-4-methylbenzene.

thymus Twin-lobed **endocrine gland**, situated in the chest near the heart, that plays an important role in the **immune response**. After birth it produces many **lymphocytes** and induces them to develop into **antibody**-producing cells. It declines after puberty and atrophies in older adults.

thyratron Gas-filled thermionic **valve** that can act as a **relay**.

thyristor Semiconductor diode, often used to control a.c. electric motors.

thyroid gland Twin-lobed **endocrine gland** situated in the neck of vertebrates. Under the influence of thyroid-stimulating hormone (TSH) from the **pituitary**, it secretes the **hormone thyroxin**, which is important in growth and **metabolism**. In human beings, undersecretion (hypothyroidism) causes cretinism in children and myxoedema in adults; overproduction (hyperthyroidism) may cause thyrotoxicosis, resulting in goitre. **Iodine** is needed in the diet for efficient functioning of the thyroid; deficiency of iodine may cause the gland to swell and form a goitre.

thyroid-stimulating hormone (TSH) **Hormone** produced by the **pituitary gland** which stimulates the **thyroid gland** into activity.

thyroxin White crystalline organic compound, an **iodine**-containing **amino acid** derived from **tyrosine**. It is a **hormone**, secreted by the **thyroid gland**.

Thysanoptera Order of small (less than 2cm) slender insects characterized by short legs and sucking mouthparts, commonly known as thrips. Many are specific pests on human crops (*e.g.* pea thrips, grain thrips).

■ **tibia** *1.* One of the two bones below the knee in a tetrapod vertebrate (the other is the **fibula**); the shinbone. *2.* One of the segments in the leg of an insect. [2/3/a]

time base In a **cathode-ray tube**, a horizontal line on the screen produced by applying a **sawtooth wave** to the tube's deflection coils.

time sharing In computing, method of employing all or part of a system for more than one task or user, by allowing each of them in turn to have access for a fraction of a second at a time.

tin Sn Soft silvery-white metallic element in Group IVB of the Periodic Table, which forms three **allotropes**. It occurs mainly as tin (IV) oxide, SnO_2, in ores such as cassiterite (tinstone). It is used mainly as a protective coating for steel (tin plate) and in making alloys with lead (solder, type metal, pewter). Its compounds are used as catalysts, fungicides and mordants. At. no. 50; r.a.m. 118.69.

tin (II) Alternative name for stannous.

tin (IV) Alternative name for stannic.

tincal Naturally occurring crude **borax**.

tin (II) chloride $SnCl_2$ White soluble solid. A **reducing agent**, it is used as a **catalyst** in organic reactions and as an anti-sludge agent for oils. Alternative names: stannous chloride, tin salt.

tin (IV) chloride $SnCl_4$ Colourless fuming liquid, used to coat **glass** with **tin (IV) oxide** (to make it conductive), as a mordant and in the preparation of inorganictin and organotin compounds. Alternative name: stannic chloride.

tin dioxide Alternative name for **tin (IV) oxide**.

tin disulphide Alternative name for **tin (IV) sulphide**.

tinea Ringworm, a skin disorder characterized by raised, roughly circular and discoloured patches, that results from an infection by a **fungus**.

tin (IV) hydride SnH_4 Unstable gas, used as a **reducing agent** in organic chemistry. Alternative name: stannane.

tin hydroxide oxide Alternative name for **stannic acid**.

tin (IV) oxide SnO_2 White crystalline solid, used as a **pigment** and as a refractory material. Alternative name: tin dioxide.

tin salt Alternative name for **tin (II) chloride**.

tin (IV) sulphide SnS_2 Yellow insoluble solid, used as a **pigment**. Alternative name: tin disulphide.

tissue Amalgamation of **cells** (of usually one type) that perform a specific function. In higher organisms tissues may combine to form a highly specialized **organ**. Tissues are found in animals (*e.g.* muscle, connective tissue) and in plants (*e.g.* parenchyma).

tissue culture Process by which **cells** or **tissues** are maintained outside the body (in vitro) in a suitable medium. The material is kept at a suitable temperature, **pH** and **osmotic pressure**. The composition of the medium depends on the type of tissue cultured. **Cancer** cells do not show normal cell properties when maintained in this way.

tissue fluid Alternative name for **lymph**.

titania Alternative name for **titanium (IV) oxide**.

titanic chloride Alternative name for **titanium (IV) chloride**.

titanium Ti Silvery-white metallic element in Group IVA of the Periodic Table (a **transition element**). Its corrosion-resistant lightweight alloys are employed in the aerospace industry. Naturally occurring crystalline forms of titanium (IV) oxide (titania, TiO_2) constitute the semi-precious gemstone rutile. The powdered oxide is used as a white pigment and a dielectric in capacitors. At. no. 22; r.a.m. 47.9.

titanium (IV) chloride $TiCl_4$ Colourless fuming liquid, a source of pure **titanium (IV) oxide**. Alternative names: titanium tetrachloride, titanic chloride.

titanium dioxide Alternative name for **titanium (IV) oxide**.

titanium (IV) oxide TiO_2 White inert solid that occurs in three crystalline forms: rutile, brookite and anatase. It is used as a white **pigment** and as a component in various dielectrics for electrical capacitors. Alternative names: titanium dioxide, titania.

titanium tetrachloride Alternative name for **titanium (IV) chloride**.

titrant Chemical solution of known concentration, *i.e.* a **standard solution**, which is added during the course of a **titration**.

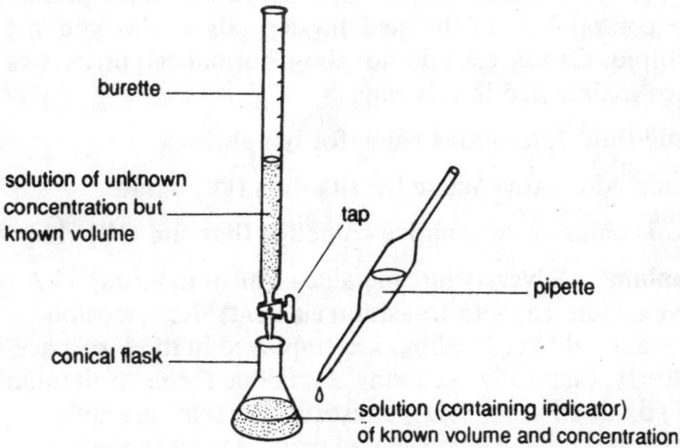

Apparatus for titration

titration Technique in **volumetric analysis** in which one chemical solution of known concentration is added (using a

burette) to a known volume of another chemical solution of unknown concentration (measured by a **pipette**), and the **chemical reaction** followed by observing changes in colour, **pH**, etc. An **indicator** may be added to indicate the end-point of the reaction, which allows the unknown concentration to be determined.

titre Concentration of a solution as determined by **titration**.

TNT Abbreviation of **trinitrotoluene**.

tobacco Plant grown for its leaves, which are dried and smoked (in a pipe or after being made into cigars or cigarettes); for many people, tobacco smoking is a form of **addiction**. Burning tobacco releases various chemicals which the smoker inhales, and some of these have been linked to lung cancer.

tocopherol Vitamin isolated from plants that increases fertility in rats. Deficiency of it causes wasting of muscles in animals. It has been found to have **antioxidant** activity, and it is important in maintaining membranes. Alternative name: vitamin E.

Tollens' reagent Ammoniacal solution of **silver oxide** used as a test for **aldehydes,** which reduce it to deposit a mirror of silver. It was named after the German chemist Bernhard Tollens (1841–1918).

toluene $C_6H_5CH_3$ Colourless aromatic organic liquid that occurs in coal-tar, used as an industrial solvent and starting point for making explosives. Alternative name: methylbenzene.

toluidine $CH_3C_6H_4NH_2$ One of three isomericaromatic **amines**, used in the manufacture of dyes and drugs. The ortho- and meta- forms are colourless liquids; the para-isomer is a colourless crystalline solid. Alternative names: aminotoluene, methylaniline.

tongue Muscular organ located in the **buccal cavity** (mouth) of some animals, used for manipulating food (and in human beings involved in speech).

tonne Unit of mass equal to 1,000 kilograms. 1 tonne = 2,204.62 lb, slightly less than the imperial ton (2,240 lb). Alternative name: metric ton.

tonsil One of a pair of **lymphoid tissue** regions located at the back of the mouth which helps to prevent infection by producing **lymphocytes**.

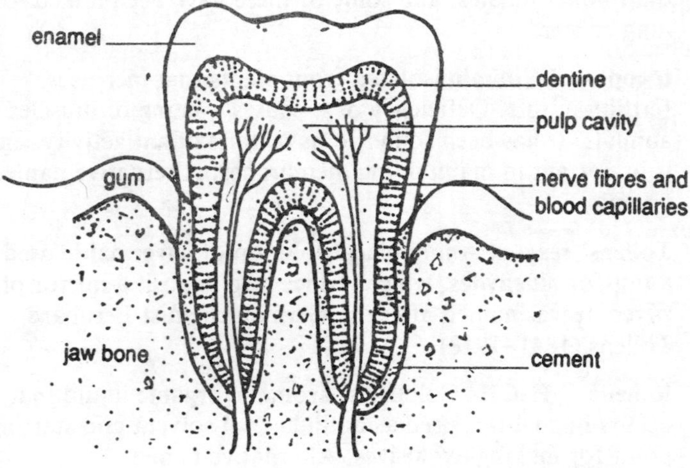

Structure of a molar tooth

tooth Hard structure embedded in the jawbone and adapted for cutting and grinding food. Teeth are composed of a cavity containing **capillaries** and **nerve** endings, a layer of **dentine** and, over the crown of the tooth, a tough outer layer of **enamel**. *See also* **canine tooth**; **carnassial tooth**; **incisor**; **molar tooth**; **premolar**.

topaz $Al_2SiO_4(OH,F)_2$. Usually pale yellow semi-precious gemstone consisting of the mineral aluminium silicate with fluoride and hydroxyl ions.

■ **torque** Turning moment produced about an axis by force acting at right angles to a radius from the axis (*see* **moment of force**). [5/7/b]

torr Unit of pressure equivalent to that produced by a 1 mm column of mercury. It is equal to 133.3 newtons m^{-2}.

torsion Twisting force exerted on a material.

torus Solid figure that resembles a tyre or doughnut. Alternative name: anchor ring.

total internal reflection Complete reflection of a light wave at a boundary between two media (*e.g.* glass and air). It occurs when the wave is incident at the **critical angle**.

toxin *1.* Poison produced by bacteria or other biological sources. *2.* Any harmful substance.

toxoid Bacterial **toxin** that has been chemically treated to make it non-poisonous, used as a vaccine.

trace element Element essential to metabolism, but necessary only in very small quantities (*e.g.* copper and cobalt in animals, molybdenum in plants). Such elements are usually toxic if large quantities are ingested.

■ **trachea** *1.* In mammals, tube through which air is drawn into the **lungs**; the windpipe. *2.* In certain arthropods (*e.g.* insects), one in a system of tubes through which air is drawn into the tissues. [2/4/a] [2/5/a] [2/7/a]

tracheid Water-carrying elongated plant cell with lignin-containing walls that occurs in **xylem** tissue.

Tracheophyta Major group in the plant kingdom that contains all vascular plants, including the **Pteridophyta** (ferns and their allies) and **Spermatophyta** (seed plants).

trachoma Disorder that affects the eyes, caused by the bacteria-like micro-organism *Chlamydia trachomatis*. Commonest in tropical Africa and India, it is worldwide the largest single cause of human sight loss.

trajectory Path followed by a moving **projectile** acted upon by gravity or other forces.

tranquillizer Drug that acts on the central nervous system (CNS), used for calming people and animals without affecting consciousness. Some types of tranquillizers, seldom prescribed now, can lead to **addiction**. *See also* **sedative**.

transamination Removal and transference of an **amino group** from one compound (usually an **amino acid**) to another.

transconductance Ratio of the change in anode current to the change in grid voltage in a **thermionic valve** or field-effect **transistor**.

transcription Process in cells by which the **genetic code** is copied into single-stranded **messenger RNA** from a **DNA** template. It occurs in the **nucleus** of **eukaryotes** and is separate from **translation** in the **cytoplasm**. In **prokaryotes** these processes are not separate.

transducer *1.* Device for transforming a physical effect into a voltage, thus allowing the effect to be measured. *2.* Device for converting **energy** from one form into another (*e.g.* electrical energy to mechanical energy).

■ **transfer RNA** Small molecule of **RNA** that acts as a carrier of specific **amino acids** in the synthesis of **proteins**. Amino acids are placed in a specific order by the transfer RNA molecules according to instructions in the messenger RNA, to form a **polypeptide** chain. [2/10/b]

transference number Alternative name for **transport number**.

transformation constant Alternative name for **disintegration constant**.

transformation, nuclear Transmutation of one atomic **nucleus** into another by means of a nuclear reaction.

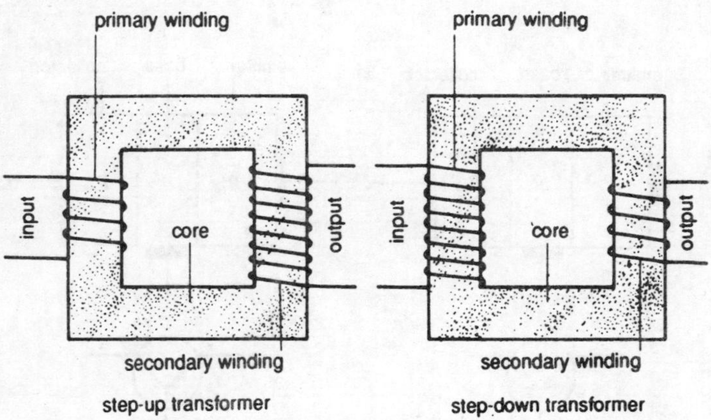

Principle of the transformer

■ **transformer** Electrical device for changing the **voltage** of an **alternating current** (a.c.), without altering its **frequency**. The change in voltage (up or down) is proportional to the ratio of the turns of wire on the primary and secondary coils of the transformer; *i.e.* if there are N_p turns on the primary coil and N_s turns on the secondary, the voltage across the secondary coil equals the voltage across the primary multiplied by N_s/N_p. [5/10/b]

transient Momentary peak in **voltage, current** or load in a electrical circuit.

transistor Semiconductor device commonly consisting of three layers, as either *p-n-p* or *n-p-n*, where *p* and *n* refer to *n-* or *p*-type semiconductors. It allows amplification of an electric current, and in this application has replaced **thermionic valves.**

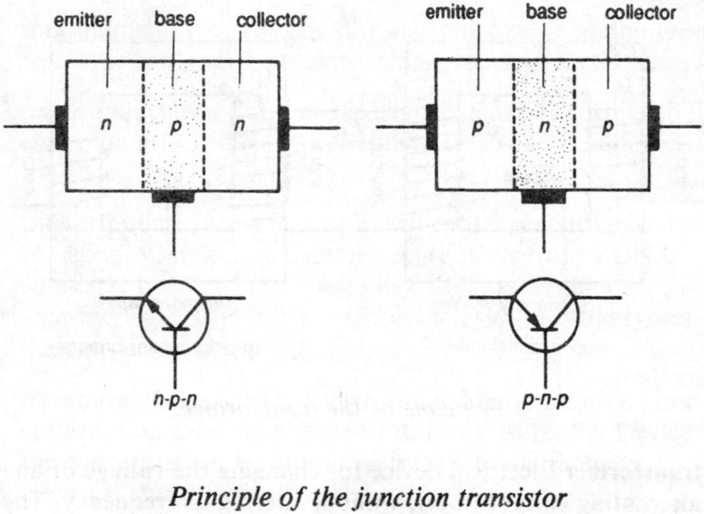

Principle of the junction transistor

transition element Member of a large group of elements that have partly filled inner electron shells, which gives them their distinctive physical and chemical properties (particularly variable valence and the tendency to form coloured compounds). They occupy Groups IIIA, IVA, VA, VIA, VIIA, VIII, IB and IIB of the Periodic Table. Many of these elements and their compounds are used as **catalysts.**

transition point *1.* **Temperature** at which the transformation of one form of a substance into another form can occur (usually one **crystalline** modification into another). *2.* Temperature at which two solid **phases** exist at **equilibrium**. *3.* Temperature at which a change to **superconductivity** happens in a substance.

transition temperature Temperature above and below which different allotropes of an element are stable.

translation Process by which **protein** is synthesized in cells. It occurs by the action of **messenger RNA,** which attaches to a **ribosome** in the **cytoplasm. Transfer RNA** molecules which are attached to a specific **amino acid** then line up according to the sequence of amino acids encoded in the messenger RNA to form a **polypeptide** chain. Alternative name: protein synthesis.

■ **translocation** Transport in plants of the soluble products of **photosynthesis**. It occurs in **phloem**. [2/3/d]

translucent Describing a substance that transmits and diffuses light, but which does not allow a well-defined image to be seen through it. *See also* **transparent**.

transmittance Ratio of transmitted **energy** to incident energy.

transmitter *1.* Apparatus that converts electrical impulses into modulated **radio waves**. *2.* Mouthpiece of telephone.

■ **transparent** Describing a substance that allows light (or other **radiation**) to pass through it with little or no diffusion. *See also* **translucent**. [5/4/d]

■ **transpiration** Process by which water is lost from plants by **evaporation**. It occurs mainly through **stomata** in the leaves, and allows movement of water and salts through the plant. It is affected by humidity and temperature, and the drying action of wind. [2/3/d]

transport number In **electrolysis**, fraction of the total current carried by a particular **ion** in the **electrolyte**. Alternative name: transference number.

transuranic element Element with a higher atomic number than **uranium** (92). Transuranic elements are artificially made and **radioactive**. Alternative names: transuranium element, uranide. *See also* **post-actinide**.

tri- Prefix meaning three.

triaminotriazine Alternative name for **melamine**.

triangle of forces Triangle that denotes three **vectors** which in turn represent three forces in **equilibrium**.

triangle of vectors Triangle whose sides represent the magnitude and direction of three **vectors** in **equilibrium** acting at a point. *See also* **parallelogram of vectors**.

triangle of velocities **Triangle of vectors** in which each vector denotes a **velocity**.

triatomic molecule Molecule of an element that consists of three **atoms**, *e.g.* **ozone**, O_3.

triazine $C_3H_3N_3$ One of a group of isomeric **heterocyclic** organic compounds with three nitrogen atoms and three carbon atoms in the ring. Triazine derivatives are used as dyes and herbicides.

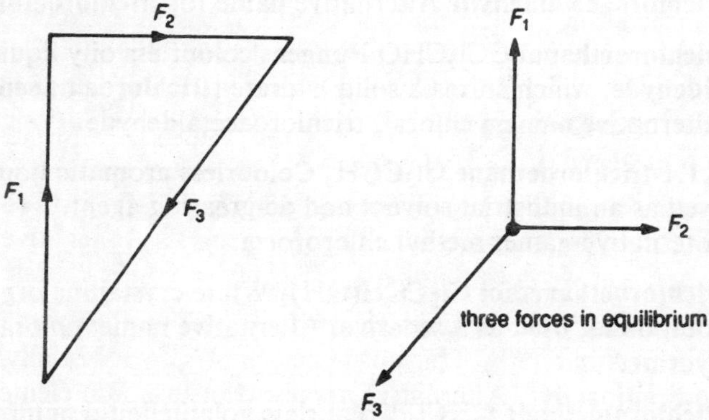

Triangle of three forces in equilibrium

triazole $C_2H_3N_3$ One of a group of isomeric **heterocyclic** organic compounds with three nitrogen atoms and two carbon atoms in the ring.

tribasic acid Acid that has three replaceable hydrogen atoms in its molecules, *e.g.* orthophosphoric acid, H_3PO_4.

tribe *1.* In botany, sub-division of a family of plants. *2.* In zoology, local and functional grouping of animals of the same species. *See also* **race; variety**.

tribromomethane $CHBr_3$ Colourless liquid **haloform,** used for the separation of minerals and in organic synthesis. Alternative name: bromoform.

tricarboxylic acid cycle Alternative name for **Krebs cycle**.

triceps Extensor muscle in the upper arm with three points of origin. It is one of an antagonistic pair of muscles, the other being the biceps.

trichloroacetaldehyde Alternative name for **trichloroethanal**.

trichloroethanal CCl_3CHO Pungent colourless oily liquid **aldehyde**, which forms a solid hydrate (**trichloroethanediol**). Alternative names: chloral, trichloroacetaldehyde.

1,1,1-trichloroethane Cl_3CCH_3 Colourless aromatic liquid, used as an industrial solvent and de-greasing agent. Alternative name: methyl chloroform.

trichloroethanediol $Cl_3CCH(OH)_2$ White crystalline organic compound, used as a sedative. Alternative name: chloral hydrate.

trichloromethane $CHCl_3$ Colourless volatileliquid **haloform**, used as an anaesthetic and as a solvent. Alternative name: chloroform.

triglyceride Ester of **glycerol**, in which all three **hydroxyl groups** have been substituted by ester groupings from **fatty acids**. Many **fats** are triglycerides.

triiodomethane CHI_3 Yellow crystalline solid **haloform**, used as an antiseptic. Alternative name: iodoform.

trilobite Extinct marine animal, a segmented **arthropod** *known only from fossils, that externally resembled a large* (3–9cm long) woodlouse, and which flourished in great numbers between about 550 million and 350 million years ago.

trimer Chemical formed by the combination of three similar (monomer) molecules; *e.g.* three molecules of **acetaldehyde** (ethanal), CH_3CHO, combine to form a single cyclic molecule of **paraldehyde** (ethanal trimer), $C_6H_{12}O_3$.

trimethylaluminium $Al(CH_3)_3$ Colourless reactive liquid, used in organic synthesis. Alternative name: aluminium trimethyl.

trinitroglycerine Alternative name for **nitroglycerine**, or glyceryl trinitrate.

trinitrophenol Alternative name for **picric acid**.

trinitrotoluene $CH_3C_6H_2(NO_2)_3$ Yellow crystalline aromatic organiccompound, used as a high explosive. Alternative names: methyl-2,4,6-trinitrobenzene, TNT.

triple bond **Covalent bond** formed by the sharing of three pairs of **electrons** between two **atoms**.

triple point Temperature and pressure at which the three phases (solid, liquid and vapour) of a substance are in equilibrium. The triple point of water occurs at 273.16 K and 611.2 Pa. *See also* **Kelvin temperature**.

triploblastic Describing an animal that has three layers of cells: **ectoderm**, **endoderm** and **mesoderm**. Most members of the **Metazoa** are triploblastic (exceptions include coelenterates, which are diploblastic).

trisaccharide **Carbohydrate** consisting of three joined **monosaccharides**.

■ **trisomy 21** A state in which a **diploid** cell nucleus has three **chromosomes** of one type. **Down's syndrome** is an example in human beings. [2/9/c]

tritium Weakly radioactive isotope of hydrogen; it has two **neutrons** and one **proton** in its nucleus. It has a relative atomic mass of 3.016.

triton Atomic nucleus of **tritium**, consisting of two neutrons and one proton.

trivalent Having a **valence** of three. Alternative name: tervalent.

■ **trophic level Energy** level found in the **food chain** of an **ecosystem**. Green plants constitute the first trophic level because they are the producers; higher levels take energy from the preceding ones and contain the consumers. The number of organisms at each trophic level becomes smaller as the number of levels increases, **carnivores** being in the highest trophic levels. *See* **pyramid of numbers**. [2/7/d]

■ **tropism** Growth movement exhibited by plants that occurs in response to a specific **stimulus,** *e.g.* light. *See also* **geotropism**; **phototropism**; **taxis**. [2/10/a]

true north Geographical north, the direction towards the North Pole and the centre of the Earth's axis.

trypsin Digestive **enzyme** secreted into the small intestine, where it catalyses the **hydrolysis** of **polypeptide** chains (of proteins) at specific sites.

tryptophan Essential amino acid that contains an aromatic group (*see* **aromatic compound**), needed in animals for proper growth and development.

TSH Abbreviation of **thyroid-stimulating hormone**.

tube-feet Arrangement of hollow tubes powered by the water vascular system that open onto the undersides of certain marine invertebrates such as starfish and sea-urchins, and which provide traction during movement. Alternative name: podia.

tuber Food-storage organ (*e.g.* in dahlias, potatoes) which develops from a plant stem or root (in which case it is called a root tuber), and which can form buds from which new plants develop (a type of **vegetative propagation**).

tumour Abnormal growth, which may be **benign** (*e.g.* a polypus) or **malignant** (*i.e.* cancerous).

tungsten W Steel-grey metallic element in Group VIA of the Periodic Table (a **transition element**) of high melting point and great hardness, used for making electric filaments and special steels for turbine blades and cutting tools. **Tungsten carbides** are also extremely hard. At. no. 74; r.a.m. 183.85. Alternative name: wolfram.

tungsten carbide WC and WC_2. Hard refractory substances, used for the tips of cutting tools and as abrasives.

Tunicata Animals that belong to a subphylum of the **Chordata**; tunicates or sea-squirts. Alternative name: Urochordata.

tuning fork Piece of metal shaped like an elongated letter U designed to produce a particular pure musical note when struck.

turbine Type of engine or motor in which jets of gas or liquid are used to drive a turbine wheel. Gas turbines (jet engines) are used extensively in aircraft, marine vessels and locomotives. Steam turbines are used to drive alternators for generating electricity.

turbulence Chaotic motion of particles in a **gas** or **liquid** when flowing through a pipe or across an aircraft wing, etc. It does not occur in laminar, or streamlined, flow.

turgid Describing something in a state of **turgor**.

turgor Inflation of plant cells by cell sap (brought about by **osmosis**), which provides the plant with rigidity and internal support. An inflated cell is said to be turgid.

turpentine Yellow sticky natural **resin** obtained from various coniferous trees. Alternative name: pine-cone oil.

turpentine oil Colourless volatile liquid; an **essential oil** obtained from the distillation of **turpentine**. It is used as a solvent for varnishes and polishes, and in medicine. Alternative name: oil of turpentine.

turquoise Blue mineral; naturally occurring copper aluminium phosphate. It is used as a gemstone. Alternative names: callaite, callanite.

twins Two offspring born to an animal that normally has only one. In maternal (or identical) twins, two embryos develop when a single fertilized ovum (egg) splits in two. In fraternal (non-identical) twins, two embryos arise when two separate ova are produced and both are fertilized. Maternal twins have the same sex and same genetic make-up; fraternal twins are no more alike than any two offspring of the same parents.

tympanum Membrane between the outer **ear** and inner ear. Alternative name: ear drum.

Tyndall effect **Scattering of light** when a light beam passes through a colloidal solution or suspended particles of matter. It was named after the British physicist John Tyndall (1820–93). Alternative name: Tyndall phenomenon.

tyrosine White crystalline organic compound, a naturally occurring **essential amino acid** found in most **proteins**, and a precursor in the body of various **hormones**.

U

■ **ulna** Rearmost (and usually larger) of the two bones in the lower forelimb of a tetrapod vertebrate (the other bone is the radius). [2/5/a]

■ **ultra-high frequency** (UHF) Band of **frequencies** in the radio spectrum between 3×10^8 and 3×10^9 **hertz**. [5/9/a,c]

ultramicroscope Instrument for viewing sub-microscopic objects, *e.g.* bacteria. It is more powerful than an ordinary optical microscope, but not so powerful as an electron microscope.

■ **ultrasonic** Describing a band of sound frequencies of about 2×10^9 hertz, which are just above the upper limit of normal human hearing. Ultrasonic energy is used in **sonar**, for degreasing (in conjunction with a suitable solvent) and for scanning soft tissues in medical diagnosis. Alternative name: supersonic. [5/9/a,c]

■ **ultraviolet radiation** (UV) **Electromagnetic radiation** with wavelengths in the range 4×10^{-7} to 4×10^{-9}m, the region between visible light and X-rays. Alternative name: ultraviolet light. [5/9/a,c]

umbilical cord *1.* In embryology, vascular structure that contains the **umbilical arteries** and **veins**, connecting the **foetus** to the **placenta**. *2.* In space engineering, flexible and easily disconnectable mechanical or electrical cable which might carry oxygen, information or power to a missile or spacecraft before launching. In space it connects a space-walking astronaut to the spacecraft.

uncertainty principle The limit of accuracy of the

measurement of two conjugate variables of a moving object, such as its position and momentum, may be given by the relation $p_x x > h/2\pi$, where p_x is the uncertainty in the momentum, x the uncertainty in position and h is Planck's constant. Alternative name: Heisenberg uncertainty principle.

ungular Describing a hoof, claw or nail.

ungulate Hoofed grazing mammal (a **herbivore**), characterized by having hoofs, grinding teeth and often horns or antlers and a comparatively long neck.

uniaxial crystal Doubly refracting crystal in which there is only one direction of single refraction (*see* **double refraction**).

unicellular Describing an organism that consists of only one **cell** (*e.g.* **protozoans, bacteria**).

unified field theory Theory that attempts to describe both the electromagnetic and the gravitational theories within a single context.

unimolecular reaction Chemical reaction that involves only one type of **molecule** as the **reactant**; *e.g.* the decomposition of mercury (II) oxide, HgO, on strong heating to give mercury and oxygen.

unisexual Describing an organism with either male or female sex organs, but not both.

unit Alternative name for kilowatt-hour, the unit that measures consumption of electricity.

unit cell Smallest group of **atoms, ions** or **molecules** whose three-dimensional repetition at regular intervals produces a **crystal lattice**.

univalent Alternative name for **monovalent**.

universal gravitation, law of *See* **Newton's law of universal gravitation**.

universal indicator Mixture of chemical **indicators** that gives a definite colour change for various values of **pH**. [4/5/a]

unsaturated *1.* Describing an organic compound with doubly or triply bonded carbon atoms; *e.g.* ethene (ethylene), $C_2{=}H_4$, and ethyne (acetylene), $C_2{\equiv}H_2$. *See also* **saturated compound**. *2.* Describing a solution that can dissolve more **solute** before reaching **saturation**.

■ **upthrust** Upward force on an object immersed in a fluid, tending to make it float. *See* **Archimedes' principle**. [5/3/a]

upwelling Rising to the surface of large lakes and seas of nutrient-rich cold water from lower depths, as a result of currents or wind action. Where this occurs, *e.g.* off the coast of Peru, it sustains a profusion of marine life.

uracil $C_4H_4N_2O_2$ **Pyrimidine base** that forms an essential part of **ribonucleic acid** (RNA). Alternative name: 2,6-dioxypyrimidine.

urania Alternative name for **uranium (IV) oxide**.

uraninite Dark-coloured, dense and slightly greasy mineral, one of the major ores of **uranium**. Alternative name: pitchblende.

■ **uranium** U Radioactive grey metallic element in Group IIIA of the Periodic Table (one of the **actinides**), obtained mainly from its ore uraninite (which contains uranium (IV) oxide, UO_2). It has three natural and several artificial **isotopes** with half-lives of up to 4.5×10^9 years. Uranium-235 undergoes **nuclear fission** and is used in nuclear weapons and reactors; uranium-238 can be converted into the fissile plutonium-239 in a **breeder reactor**. At. no. 92; r.a.m. 238.03. [4/7/f]

uranium dioxide Alternative name for **uranium (IV) oxide**.

uranium-lead dating Dating method used in calculating the geologic age of minerals based on the **radioactive decay** of **uranium**-235 to lead-207 and **uranium**-238 to lead-206.

uranium (IV) oxide UO_2 Highly toxic **radioactive** spontaneously inflammable black crystalline solid, used in photographic chemicals, ceramics, pigments and packing of nuclear fuel rods. Alternative names: urania, uranium dioxide.

uranium (VI) oxide UO_3 Highly toxic **radioactive** orange powder, used in uranium refining, as a pigment and in ceramics. Alternative names: uranium trioxide, orange oxide.

uranium trioxide Alternative name for **uranium (VI) oxide**.

uranyl Radical UO_2^{2+}, *e.g.* as in uranyl nitrate $UO_2(NO_3)_2$.

urea H_2NCONH_2. White crystalline organic compound, found in the urine of mammals as the natural end-product of the metabolism of **proteins**. It is also manufactured commercially from **carbon dioxide** and **ammonia** under high pressure. It is used in plastics, adhesives, fertilizers and animal-feed additives. Alternative name: carbamide.

urea-formaldehyde resin Colourless, non-inflammable, weather-resistant **thermosetting** plastic, manufactured by the reaction of **urea** with **formaldehyde** (methanal) or its **polymers**. Alternative name: urea resin.

urease **Enzyme** that occurs in plants (*e.g.* soya beans) and acts as a **catalyst** for the **hydrolysis** of **urea** to **ammonia** and **carbon dioxide**.

■ **ureter** One of a pair of ducts that carry **urine** from the **kidneys** to the **bladder** of mammals or the **cloaca** of reptiles and birds. It is functionally replaced by the Wolffian duct in fish and amphibians. [2/5/a] [2/7/a]

urethane $CO(NH_2)OC_2H_5$. Highly toxic inflammable organic compound, used in veterinary medicine, biochemical research and as a solvent and chemical intermediate. Alternative names: ethyl carbamate, ethyl urethane.

urethane resin *See* **polyurethane**.

■ **urethra** Tube through which urine is discharged to the exterior from the urinary bladder of most mammals. [2/5/a] [2/7/a]

uric acid $C_5H_4N_4O_3$ White crystalline organic acid of the **purine** group, the end-product of the metabolism of **amino acids** in reptiles and birds. In human beings uric acid deposition in the joints is the principal cause of gout.

■ **urine** Liquid, produced in the **kidneys** and stored in the urinary **bladder**, that contains **urea** and other excretory products. It is discharged to the outside via the **urethra** or **cloaca**. [2/7/a]

Urochordata Subphylum of animals in the phylum **Chordata**; sea squirts. Alternative name: Tunicata.

uterus Muscular organ located in the lower abdomen of female mammals, in which a fertilized **ovum** develops into a **foetus** prior to birth. Alternative name: womb.

uvula Finger-like projection that hangs from the soft palate at the back of the mouth.

V

vaccination Process of giving somebody a **vaccine**.

vaccine Suspension of killed or weakened **antigens** (such as **viruses** or **bacteria**), which is used for immunization. It is either injected (by inoculation) or ingested into the body where it stimulates the production of **antibodies** and so confers **immunity** against infection; both methods are examples of vaccination. Less commonly, vaccines are used in treating a disease.

vacuity Any gap between the bones of a skull.

vacuole Alternative name for a **vesicle**. *See also* **contractile vacuole**.

vacuum Space containing no matter. A good laboratory vacuum still contains about 10^{14} molecules of air per cubic metre (compared with about 10^{24} molecules in a litre of air at ordinary pressure); intergalactic space may have an almost perfect vacuum (although it does contain some subatomic particles).

vacuum distillation **Distillation** under reduced pressure, which helps to lower the boiling point and hence reduce the risk of thermal dissociation. Alternative name: reduced-pressure distillation.

vacuum flask Container designed to keep hot liquid hot or cold liquids cold by minimizing heat transfer with the exterior. It consists of a stoppered double-walled glass bottle, silvered on the inside and with a vacuum between the walls, supported by blocks of insulating material in an outer protective case. Alternative name: Dewar flask, after the British chemist and physicist James Dewar (1842–1923).

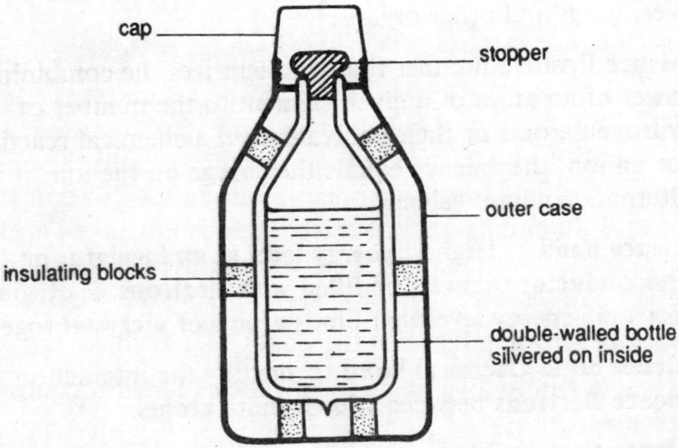

Construction of a vacuum flask

vacuum pump Alternative name for **diffusion pump**.

vacuum tube Alternative name for **thermionic valve**.

vadose zone Levels of soil above the water table that are aereated, and within which most soil chemistry takes place.

vagility Ability of an organism to adapt to changed environmental circumstances, rated on a scale between 'high' and 'low'.

vagina *1*. In most female mammals, muscular duct that extends from the **uterus** to the vulvar opening. It receives the male **penis** during mating. *2*. In plants, expanded sheath-like structure at a leaf base.

vagus nerve In vertebrates, tenth **cranial nerve**, which forms the major nerve of the **parasympathetic nervous system**, supplying motor nerve fibres to the stomach, kidneys, heart, liver, lungs and other organs.

valence Positive number that characterizes the combining power of an **atom** of a given **element** to the number of hydrogen atoms or their equivalent (in a chemical reaction). For an ion, the valence equals the charge on the ion. Alternative name: valency.

valence band *1.* Highest **energy level** in an **insulator** or **semiconductor** that can be filled with **electrons**. *2.* Region of electronic energy level that binds **atoms** of a **crystal** together.

valence bond Chemical bond formed by the interaction of **valence electrons** between two or more **atoms**.

valence electron Electron in an outer **shell** of an **atom** which participates in bonding to other atoms to form **molecules**.

valine $C_5H_{11}NO_2$ One of the **essential amino acids** required for normal growth in animals. Alternative names: 2-aminoisovaleric acid, 2-amino-3-methylbutyric acid.

valve *1.* In botany, part of a dehiscing fruit wall or **capsule**. *2.* In anatomy, a flap of tissue in the body that controls movement of fluid through a tube, duct or aperture in one direction, *e.g.* as between the chambers of the **heart** or in the **veins**. *3.* *See* **thermionic valve**.

vanadium V Silvery-grey metallic element in Group VA of the Periodic Table (a **transition element**), used to make special steels. Vanadium (V) oxide, V_2O_5, is used as an industrial **catalyst** and in ceramics. At. no. 23; r.a.m. 50.94.

Van de Graaff generator Apparatus for producing very high voltages, consisting of a large metal dome to which electrostatic charge is carried by a vertical conveyor belt. It was named after the American physicist Robert Van de Graaff (1901–67).

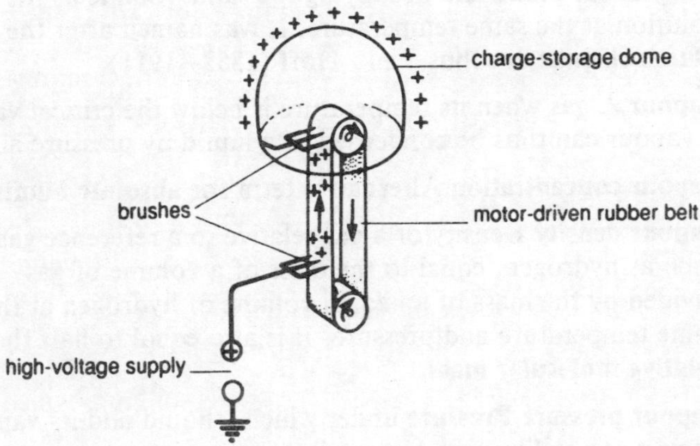

Principle of a Van de Graaff generator

van der Waals' equation (of state) Equation of state that takes into account both the volume of the gas molecules and the attractive forces between them. It may be represented as $(P + a/V^2)(V - b) = RT$, where V is the volume per mole, P the pressure, T the absolute temperature, R the **gas constant**, and a and b are constants for a given gas, evaluated by fitting the equation to experimental PVT measurements at moderate **densities**. It was named after the Dutch physicist Johannes van der Waals (1837–1923).

van der Waals' force Weak attractive force induced by interaction of **dipole moments** between atoms or non-polar molecules. It is represented by the coefficient a in **van der Waals' equation**.

Van't Hoff's law Osmotic pressure of a solution is equal to the pressure that would be exerted by the **solute** if it were in the gaseous phase and occupying the same volume as the solution at the same temperature. It was named after the Dutch chemist Jacobus Van't Hoff (1852–1911).

vapour A **gas** when its temperature is below the critical value; a vapour can thus be condensed to a liquid by pressure alone.

vapour concentration Alternative term for **absolute humidity**.

vapour density Density of a gas relative to a reference gas, such as hydrogen, equal to the mass of a volume of gas divided by the mass of an equal volume of hydrogen at the same temperature and pressure. It is also equal to half the **relative molecular mass**.

vapour pressure Pressure under which a liquid and its vapour coexist at equilibrium. Alternative name: saturation vapour pressure.

variable *1*. In experimental science, property or quantity that is altered to study the resulting change in another property or quantity. Usually all others of the system are kept constant. *2*. In computing, a block of data that is stored at different locations during the operation of a program.

■ **variation** In biology, differences between members of the same species, which may be either continuous (having a normal distribution about a species mean, *e.g.* height and weight) or discontinuous (having different specific characteristics with no intermediate forms, *e.g.* blood types). *See also* **genetic variation**. [2/6/b]

variety Any sub-division of a **species**, *e.g.* breed, race, strain, etc.

variola Any of the diseases caused by the poxviruses. Variola major is an alternative name for smallpox.

■ **vas deferens** One of a pair of ducts that carry sperm from the **testes**. In mammals it joins the **urethra** and passes along the **penis**. Plural: vasa deferentia. Alternative name: sperm duct. [2/4/a] [2/5/a]

■ **vascular bundle** Strand of fluid-conducting plant tissue that consists of cells which comprise the **xylem** and **phloem**. Separate vascular bundles are linked by interfascular cambium. [2/4/a] [2/5/a] [2/7/a]

vascular plant *See* **Tracheophyta**.

vascular system *1*. In animals, system of interlinked fluid-filled vessels, *e.g.* the **blood vascular system**. *2*. In seed plants and **pteridophytes**, system of conducting tissue consisting of **vascular bundles** that is responsible for the transport of mineral salts and water from the roots to the aerial parts of plants, and of food from the leaves to the growing points or to storage organs. It also provides mechanical support in plants with **secondary growth**.

vascular tissue Composite plant tissue, found in **angiosperms**, **gymnosperms** and **pteridophytes**, that consists of **xylem** and **phloem**. It also contains **parenchyma** and **sclerenchyma fibres**, which together form the support tissues. All vascular plants have primary vascular tissue which is formed from the **procambium**. Only vascular plants with **secondary growth** have secondary vascular tissue, formed from the **cambium**.

vasectomy Method of sterilizing a male by surgically cutting (and tying the cut ends) of each **vas deferens**, thus making it impossible for sperm from the testes to reach the urethra and penis. *See also* **salpingectomy**.

vasoconstriction Reduction in diameter of a **blood vessel** due to contraction of the **smooth muscles** in its walls. It may be induced by the secretion of **adrenaline** in response to pain, fear, decreased blood pressure, low external temperature, etc. or result from stimulation by vasoconstrictor nerve fibres.

vasodilation Increase in diameter of small **blood vessels** due to relaxation of the **smooth muscles** in their walls. It is induced in response to exercise, high blood pressure, high external temperature, etc. or results from stimulation by vasodilator nerve fibres. Alternative name: vasodilatation.

vasomotor nerve Nerve of the **autonomic nervous system** that controls the variation in the diameter of **blood vessels,** *e.g.* causing them to become constricted or dilated.

vasopressin Peptide hormone, secreted by the **pituitary gland** and **hypothalamus,** that stimulates water resorption in the **kidney** tubules, contraction of the **smooth muscles** in the walls of **blood vessels** and permeability of skin and bladder cells in **amphibians**. Alternative name: antidiuretic hormone (ADH).

vauqueline Alternative name for **strychnine**.

VDU Abbreviation of visual display unit.

■ **vector** *1.* In medicine, agent capable of mechanical or biological transference of **pathogens** from one organism to another; *e.g.* the *Anopheles* mosquito is the vector of the malaria parasite. Alternative name: carrier. [2/4/d] [2/6/a] *2.* In physics, quantity with both direction and magnitude (a quantity with magnitude only is a **scalar**).

vegetative propagation *1.* Method of **asexual reproduction** in plants in which a new plant grows from a part of the 'parent' plant – *e.g.* from a bulb, corm, rhizome, stolon (runner) or tuber. Alternative name: vegetative reproduction. *2.* Asexual reproduction in animals – *e.g.* **budding** in coelenterates such as hydra. Alternative name: vegetative reproduction.

■ **vein** *1.* In plants, any of several **vascular bundles** in a leaf. *2.* In insects, any of several chitinous tubes that provide membranous wings with support and shape. *3.* In animals, blood vessel that, with the exception of the **pulmonary vein**, carries deoxygenated blood away from cells and tissues. [2/4/a] [2/4/c] [2/5/a] [2/7/a]

velamen Layer of translucent water-retaining spongy cells that surround the **aerial roots** of **epiphytes**.

■ **velocity** (v) Rate of movement in a particular direction; distance travelled per unit of time, measured in units such as m s^{-1}, km/h or mph. If something travels a distance d in a particular direction in a time t, its velocity is given by $v = d/t$. Velocity is a **vector** quantity, unlike **speed** (which is **scalar**, for which direction is not specified). [5/7/a]

velocity constant Alternative name for **rate constant**.

velocity of light *See* **speed of light**.

■ **velocity of sound** *See* **speed of sound**. [5/4/c] [5/5/a]

■ **velocity ratio** For a simple machine (*e.g.* lever, pulley), the ratio of the distance moved by the load to the distance moved by the effort. Alternative name: distance ratio. [3/5/d]

■ **vena cava** Collective term for the precaval (anterior vena cava) and postcaval (posterior vena cava) vein. The precaval vein is paired, and carries deoxygenated blood away from the head and forelimbs (or arms); the postcaval vein is single and carries deoxygenated blood away from most of the body and hind limbs (or legs) to the heart. [2/4/a] [2/4/c] [2/5/a] [2/7/a]

venation Arrangement of veins in a plant's leaves or an insect's wings.

ventral Describing something that is on or near the undersurface of an organism and directed downward (on a human being it is directed forward). *See also* **dorsal**.

■ **ventricle** *1.* In mammals, thick-walled muscular lower chamber of the **heart**. Contraction of the right ventricle pumps deoxygenated blood into the **pulmonary artery**, and contraction of the left ventricle forces oxygenated blood into the **aorta**. *2.* In vertebrates, one of the fluid-filled interconnected cavities within the **brain**. [2/4/a] [2/4/c] [2/5/a] [2/7/a]

Venturi tube Cylindrical pipe with a constriction at its centre. When a fluid flows through the tube, its rate of flow increases and fluid pressure drops in the constriction. The rate can be calculated from the difference in pressure between the ends of the tube and at the constriction. It was named after the Italian physicist G. Venturi (1746–1822). See also **Pitot tube**.

■ **venule** *1.* In animals, small **vein** located close to **capillary blood vessels**, where it collects and conveys deoxygenated blood from the capillary network to a **vein**. *2.* In plants, small vein of a leaf. [2/4/c]

verdigris Green copper (II) carbonate, $CuCO_3.Cu(OH)_2$, formed by corrosion of metallic copper or its alloys. The term is also used for the similar basic copper (II) acetate, used as a pigment, fungicide and mordant in dyeing.

vermiform appendix Alternative term for the **appendix**.

vernation Arrangement of the outer leaves that enclose a developing bud.

vernier Arrangement for measuring that permits more precise readings than a simple calibrated scale. It consists of a small movable scale graduated in intervals that are 9/10 of those on the main scale, allowing the latter to be read to a tenth of a division. A circular vernier is used on a micrometer gauge. It was named after the French mathematician Pierre Vernier (1580–1637).

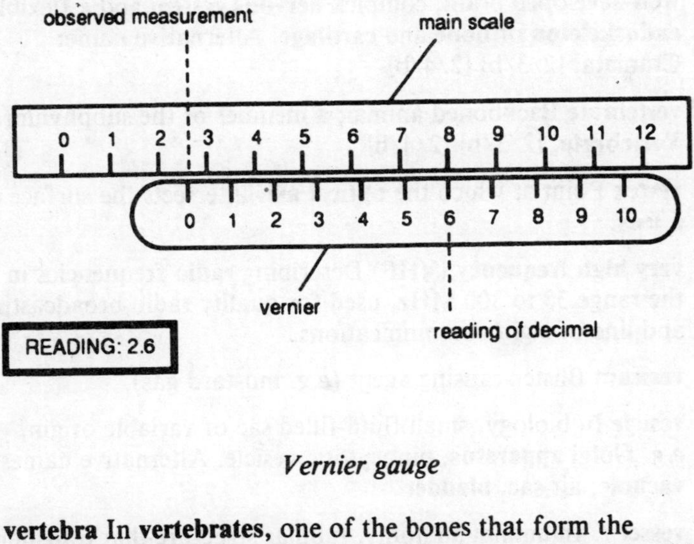

Vernier gauge

vertebra In **vertebrates,** one of the bones that form the **vertebral column**. Each vertebra typically consists of a solid block of bone (centrum) and a neural arch, which protects the **spinal cord**. In humans there are dorsal and lateral processes (projections) for the attachment of muscles. The spine is divided (from top to bottom) into cervical, thoracic and lumbar vertebrae.

■ **vertebral column** Flexible column of closely arranged **vertebrae** that form an axial **skeleton** running from the **skull** to the tail. It provides a protective channel for the **spinal cord**. The vertebral column becomes larger and stronger towards the posterior (in humans the lower end), which is the major weight-bearing region. Alternative names: spinal column, backbone. [2/3/a] [2/5/a]

■ **Vertebrata** Major subphylum of **Chordata** that contains all animals with a **vertebral column**, *i.e.* mammals, birds, fish, reptiles and amphibians. Vertebrates are characterized by a well-developed **brain**, complex **nervous system** and a flexible **endoskeleton** of **bone** and **cartilage**. Alternative name: Craniata. [2/3/b] [2/4/b]

■ **vertebrate** Backboned animal; a member of the subphylum **Vertebrata**. [2/3/b] [2/4/b]

vertex Point at which the **optical axis** intersects the surface of a **lens**.

very high frequency (VHF) Describing **radio** frequencies in the range 30 to 300 MHz, used for quality radio broadcasting and line-of-sight communications.

vesicant Blister-causing agent (*e.g.* mustard gas).

vesicle In biology, small fluid-filled sac of variable origin, *e.g.* **Golgi apparatus**, pinocytotic vesicle. Alternative names: **vacuole**, air sac, **bladder**.

vessel *1.* In animal anatomy, tubular structure that transports fluid (*e.g.* blood, lymph). *2.* In seed plants, advanced form of **xylem** made up of vessel elements.

■ **vestigial organ** Reduced structure that has lost its original function during the course of **evolution**, but resembling the corresponding fully functional organs found in a related organism, and hence manifesting an evolutionary relationship, *e.g.* wings of flightless birds, limb girdles of snakes. [2/9/c]

VHF Abbreviation of **very high frequency**.

vibration Alternative name for a regular oscillation.

Vibrio Type of Gram-positive comma-shaped **bacterium** which includes the organism that causes cholera.

vicariance Evolution of two closely related species from populations of the same species that became separated geographically. There are many examples between the Old World (Europe and Africa) and the New World (the Americas).

villi Plural of **villus**.

villus One of many finger-like structures that line the inside of the **small intestine**. Villi increase the surface area for absorption. Each villus contains a central **lacteal** and a network of blood **capillaries**, which absorb the soluble products of **digestion** into the body.

vinegar Dilute solution (about 4 per cent by volume) of **acetic acid** (ethanoic acid). Natural vinegar is made by bacterial fermentation of cider or wine.

vinyl acetate $CH_2=CHOOCCH_3$ **Monomer** from which the plastic and adhesive polyvinyl acetate (PVA) is made. Alternative name: ethenyl ethanoate.

vinyl benzene Alternative name for **styrene**.

vinyl chloride $CH_2=CHCl$ **Monomer** from which the plastic (polymer) polyvinyl chloride (PVC) is made. Alternative name: chloroethene.

vinyl cyanide Alternative name for **acrylonitrile**.

vinyl group Double-bonded organic group $CH_2=CH-$.

virgin neutron Any **neutron** from any source before collision.

viroid Smallest disease-causing agent, a tight loop of **RNA** lacking any form of capsid (outer coat).

virtual image Image brought to a focus by a lens, mirror or other optical system that can be seen by the eye but which cannot be focused on a screen (as opposed to a real image, which can).

■ **virus** Pathogenic micro-organism with a diameter between about 20 and 400 nm, visible only under an electron microscope. It consists of an outer coat (capsid) of **protein** and an inner core of deoxyribonucleic acid **(DNA)** or ribonucleic acid **(RNA)**. Viruses infect plants, animals and bacteria. Outside the host cells viruses are metabolically inactive (and may be crystallizable), and only when attached to a cell or wholly inside it does the viral DNA interfere with the metabolic activities of the cell, suppressing its normal control processes and causing it to manufacture new protein coats and nucleic acid threads identical with those of the invading virus. Viruses which actively attack and proliferate in cells are described as virulent.

■ **viscera** Internal organs, particularly the gut (intestines). [2/4/a]

visceral Relating to the **viscera**.

viscometer Instrument for measuring **viscosity**.

viscose rayon *See* **rayon**.

viscosity Property of a fluid (liquid or gases) that makes it resist flow, resulting in different velocities of flow at different points in the fluid. Alternative name: internal friction.

visible spectrum Range of **wavelengths** of visible **electromagnetic radiation** (light), between about 780 and 380 nm. [5/9/a,c]

visual display unit (VDU) Television-type screen (based on a **cathode-ray tube)** for displaying **alphanumeric** characters or graphics that represent data from a computer or word-processor. Data is entered using a **keyboard, light pen** or mouse.

visual purple Alternative name for **rhodopsin**.

vitamin A *See* **retinol**.

■ **vitamin** Any of several organic compounds that in small quantities are essential to the proper growth and regulation of metabolic processes, *e.g.* energy transformation in animal organisms. There are two major groups, water-soluble (*e.g.* vitamins C, B) and fat-soluble (*e.g.* A, D, E, K), which are present in foodstuffs and must be taken as part of a balanced diet.

vitamin B_1 *See* **thiamine**.

vitamin B_2 *See* **riboflavin**.

vitamin C *See* **ascorbic acid**.

vitamin D *See* **calciferol**.

vitamin E *See* **tocopherol**.

vitamin H *See* **biotin**.

vitreous humour In the vertebrate **eye**, firm, transparent gel-like substance that fills the the space behind the **lens**, thus maintaining the shape of the eyeball. *See also* **aqueous humour**.

vitriol Alternative name for **sulphuric acid**.

■ **viviparous** *1.* In animals, giving birth to live young rather than laying eggs which hatch later. This ability is almost entirely confined to mammals, although the term is generally used of animals other than mammals in which viviparity is unusual, *e.g.* certain snakes. *2.* In plants, having seeds that germinate and start growing within the fruit, *e.g.* mangrove. [2/7/a]

vocal cord One of a pair of membranous flaps in the **larynx** that are vibrated by air from the lungs to produce sounds.

volatile *1.* In chemistry, describing any substance that is readily changed to a **vapour** and hence lost through **evaporation**. Volatile liquids have low boiling points. *2.* In computing, describing stored information that is lost through a power cut.

■ **volt** (V) SI unit of **potential difference** (p.d) or **electromotive force** (e.m.f.), which equals the p.d. between two points when one **coulomb** of electricity produces one **joule** of work in going from one point to the other.

■ **voltage** (*V*) Value of a **potential difference**, or the potential difference itself. [5/6/b] [5/9/b]

voltage divider Resistor that can be tapped at a point along its length to give a particular fraction of the voltage across it. Alternative name: potentiometer.

voltaic cell Any device that produces an electromotive force (e.m.f.) by the conversion of chemical energy to electrical energy, *e.g.* a battery or accumulator. Alternative name: galvanic cell.

voltmeter Instrument for measuring **voltage**, or potential difference.

volumetric analysis Method of chemical analysis that relies on

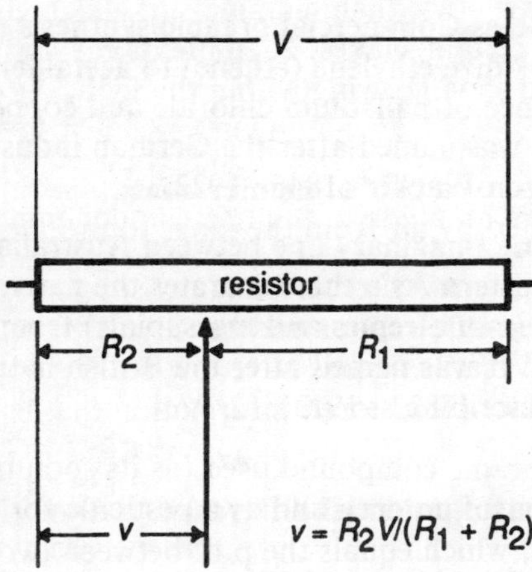

Principle of a voltage divider

the accurate measurement of the reacting volumes of substances in solution (*e.g.* by carrying out a **titration**).

■ **voluntary muscle** Type of muscle, connected to the bones, that is under conscious control. It is responsible for most body movements. Alternative name: striated muscle. [2/5/a]

volvox One of a group of ciliated freshwater **algae** (genus *Volvox*) that form well-organized spherical colonies consisting of thousands of cells.

vulcanization Method of hardening natural or artificial **rubber** by heating it with sulphur or sulphur compounds.

■ **vulva** In female mammals, the external opening of the **vagina**. [2/4/a] [2/5/a] [2/7/a]

W

Wacker process Commercial organic synthesis that uses oxygen to oxidize ethylene (ethene) to acetaldehyde (ethanal) in the presence of palladium chloride and copper (II) chloride **catalysts**. It was named after the German industrial chemist Alexander von Wacker (1846–1922).

■ **Wallace's line** Imaginary line between Australia/New Guinea and south-eastern Asia that separates the native mammals of Australasia (monotremes and marsupials) from those of Asia (placentals). It was named after the British naturalist Alfred Russel Wallace (1823–1913). [2/9/c]

warfarin Organic compound used (as its sodium derivative) as an anticoagulant drug and as a pesticide for killing rats and mice.

warm-blooded Alternative name for **homoiothermic**.

■ **warning coloration** Bright patterns that occur in some noxious animals (especially insects) which results in a predator learning to avoid them. Alternative name: aposematic coloration. [2/4/d]

washing soda Alternative name for hydrated **sodium carbonate**.

■ **water** H_2O Colourless liquid, one of the oxides of hydrogen (the other is **hydrogen peroxide**, H_2O_2) and the commonest substance on Earth. It can be made by burning hydrogen or fuels containing it in air or oxygen, or by the action of an **acid** on an **alkali** or **alcohol**. It is a good (polar) solvent, particularly for ionic compounds, with which it may form solid **hydrates**. It can be decomposed by the action of certain reactive metals (*e.g.* the **alkali metals**) or by **electrolysis**. Water is essential for life and forms the major part of most body fluids (*e.g.* blood, lymph). It freezes at 0 °C and boils at 100 °C (at normal atmospheric pressure), and has its maximum density at 3.98 °C. *See also* **hardness of water**.

■ **water cycle** Continuous movement of water between the atmosphere and the land and oceans. Water that falls as precipitation (*e.g.* rain, snow, hail) passes into the ground or runs off to form springs and rivers, which ultimately flow into the oceans. Water evaporates from the oceans as vapour, forms clouds in the atmosphere, and falls again as precipitation. Some water is also returned to the atmosphere through transpiration and respiration after having been taken in by plants and animals. Alternative name: hydrological cycle.

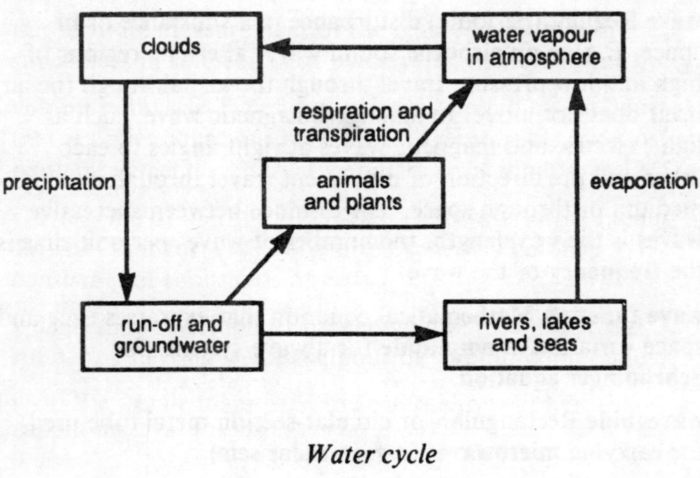

Water cycle

water glass Alternative name for **sodium silicate solution**.

water of crystallization Definite amount of water retained by a compound (usually a salt) when crystallized from solution. The chemical formula of the resulting **hydrate** shows the number of molecules of water of crystallization associated with each molecule of hydrate; *e.g.* $Na_2SO_4.7H_2O$. The water can usually be removed by heating, and the resulting compound is termed **anhydrous**. Alternative name: water of hydration.

■ **watt** (W) SI unit of **power**, equal to 1 joule per second (Js^{-1}). It was named after the British engineer James Watt (1736–1819).

watt-hour Measure of electric **power** consumption. Alternative name: unit.

wattmeter Instrument for measuring electric **power** in a circuit. Power consumption is usually expressed in watt-hours, or units.

■ **wave** Regular (periodic) disturbance in a substance or in space. *E.g.* in an airborne sound wave, alternate regions of high and low pressure travel through the air, although the air itself does not move. In an electromagnetic wave, such as light, electric and magnetic waves at right angles to each other and the direction of movement travel through a medium or through space. The distance between successive waves is the **wavelength**; the number of waves per unit time is the **frequency** of the wave.

wave function Mathematical equation that expresses time and space variation in **amplitude** for a wave system. *See* **Schrödinger equation**.

waveguide Rectangular- or circular-section metal tube used for carrying **microwaves** (*e.g.* in **radar** sets).

wavelength λ Distance between successive crests, or successive troughs, of a wave. It is equal to v/f, where v is the **velocity** of the wave and f its **frequency**. The relationship is true for all waves, whatever their nature.

wave number Reciprocal of the **wavelength** of an **electromagnetic wave**. Alternative name: reciprocal wavelength.

wave theory of light Interference and **diffraction** phenomena of **electromagnetic waves** as explained by James Clerk Maxwell, and verified by Heinrich Hertz for **radio waves**.

■ **wax** Solid or semi-solid organic substance that is *1.* an **ester** of a **fatty acid**, produced by a plant or animal (*e.g.* beeswax, tallow), or *2.* a high molecular weight hydrocarbon (*e.g.* paraffin wax, from petroleum), also called mineral wax.

weak acid Acid that shows little **ionization** or dissociation in solution; *e.g.* **carbonic acid, acetic** (ethanoic) **acid**.

weber (Wb) SI unit of magnetic flux, named after the German physicist Wilhelm Weber (1804–91).

■ **weight** *1.* On Earth, the force of gravity (9.8 m s^{-2}) acting on an object at the Earth's surface; *i.e.* weight = mass × acceleration of free fall (acceleration due to gravity). It is measured in newtons, pounds-force or dynes. *See also* **mass.** *2.* Elsewhere in the Universe, the force of gravity with which a star, planet or moon attracts a nearby object. Thus an object's weight is less on the Moon than on Earth (although its mass remains the same).

weightlessness Property of an object that is not in a **gravitational field** or is in free fall (moving with an acceleration of g, the **acceleration of free fall**).

Wheatstone bridge Electric circuit for measuring the

resistance of a **resistor** (by comparing it with three other resistors of known values). It was named after the British physicist Charles Wheatstone (1802–75).

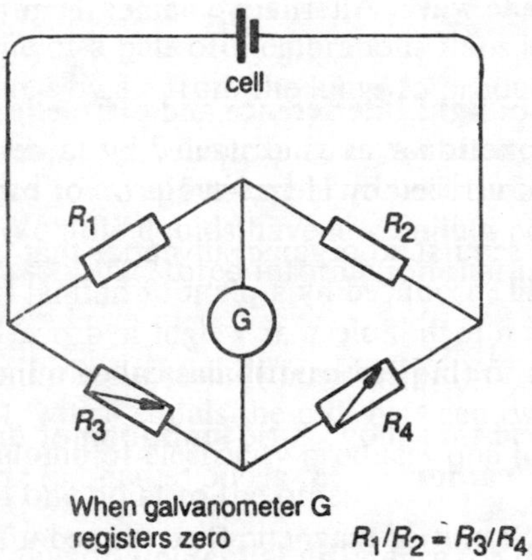

Wheatstone bridge circuit

white arsenic Alternative name for arsenic (III) oxide (arsenious oxide). *See* **arsenic**.

white spirit Petroleum distillate used as a solvent and in the manufacture of paint and varnish.

white vitriol Alternative name for **zinc sulphate**.

Wiedemann-Franz law Ratio of electrical and thermal **conductivities** of a metal is equal to a constant times the **absolute temperature**. Alternative name: Lorentz relation.

wildlife Any non-domesticated organisms, although the term is often used to refer only to wild animals.

Williamson's synthesis Method of synthesizing **ethers** from alkyl iodides and sodium alcoholates. It was named after the British chemist Alexander Williamson (1824–1904).

will-o'-the-wisp *See* **ignis fatuus**.

Wilson's cloud chamber Alternative name for **cloud chamber** (after the British physicist C.T.R. Wilson, 1869–1961).

■ **wilting** Limpness of plant tissue that occurs when there is insufficient cell sap to keep the cells rigid, and which may be caused by lack of water or disease. *See also* **flaccid**. [2/4/a] [2/5/a] [2/7/a]

Wimshurst machine Generator of static electricity consisting of two close oppositely rotating insulating discs with pieces of metal foil on their edges. It was named after the British engineer James Wimshurst (1836–1903).

winchester *1.* **Hard disk** and its drive that is small enough to use in a **microcomputer**. It was named after the city of Winchester, USA. *2.* Bottle, commonly used for liquid chemicals, with a capacity of about 2.25 litres.

Wöhler's synthesis Method of preparing **urea** by heating ammonium isocyanate. It was the first synthesis of an organic compound (at that time, 1828, thought to be produced only by living organisms) from an inorganic one. It was named after the German chemist Friedrich Wöhler (1800–82).

wood spirit Alternative name for **methanol**.

wood sugar Alternative name for **xylose**.

wool fat Alternative name for **lanolin**.

word processor Microcomputer that is programmed to help in the preparation of text, for data transmission or printing.

work (W) Energy transfer that occurs when a force f causes an object to move a distance d in the direction of the force; $W = fd$. It is measured in **joules** (newton-metres).

wound-tumour virus One of a group of **viruses** that cause the formation of galls (*e.g.* oak apples) on many plants, and which are often transmitted by insects.

write head Electromagnetic device that records signals – audio, video or computer data – onto a magnetic storage medium (*e.g.* magnetic tape or disk).

Wurtz reaction Method of synthesizing **hydrocarbons** using alkyl iodides and sodium metal. It was named after the French chemist Charles-Adolphe Wurtz (1817–84).

X

xanthate Salt or **ester** of **xanthic acid**. Sodium and potassium xanthates are used as ore flotation collectors; cellulose xanthate is used in the manufacture of **rayon**.

xanthene $CH_2(C_6H_4)_2O$ Yellow crystalline organic compound, used as a fungicide and in making dyes. Alternative name: tricyclicdibenzopyran.

xanthic acid $CS(OC_2H_5)SH$ Organic acid whose salts and esters (**xanthates**) have various industrial applications.

xanthine $C_5H_4N_2O_2$ Toxic organic compound that occurs in small amounts in potatoes, coffee beans, blood and urine, used industrially as a chemical intermediate. Alternative name: 3,7-dihydro-1H-purine-2,6-dione, 2,6-dihydroxypurine.

xanthone $CO(C_6H_4)_2O$ Plant **pigment** that occurs in gentians and other flowers, used commercially as an insecticide and dye intermediate. Alternative name: 9H-xanthen-9-one.

■ **xanthophyll** $C_{40}H_{56}O_2$ Yellow to ora: ge **pigment** present in the normal **chlorophyll** mixture of green plants. [2/3/d]

■ **X-chromosome Sex chromosome** that occurs paired in human females (*i.e.* as XX), and coupled with a **Y-chromosome** (XY) in males. It is the larger of the two sex chromosomes, and carries many **sex-linked genes**. [2/8/b]

■ **xenon** Xe Unreactive gaseous element in Group 0 of the Periodic Table (the **rare gases**) which occurs as traces in the atmosphere, from which it is extracted. It is used in electronic flash tubes and high-intensity arc lamps. The **isotope** xenon-135 is a uranium fission product and a troublesome 'poison' in **nuclear reactors** (because it captures slow neutrons). At. no. 54; r.a.m. 131.30. [4/8/a] [4/9/a]

xeric Describing environments that are predominantly dry.

xerography Printing process usually employed in a photocopier. In the most common method, a projected image of the page to be copied causes loss of electric charge from a drum where the light falls on it. Resinous carbon powder (called toner) adheres to the charged areas of the drum (corresponding to the black areas of the image) and is transferred to paper and 'fixed' to it with heat.

■ **xerophyte** Plant that grows in dry habitats. Adaptations that enable it to do this include swollen stems that store water and leaf modifications (copious hairs, a thick cuticle, reduction of leaves to spines as in cacti) that reduce loss of water. [2/6/a]

xerosere Pattern of plant **succession** that is characteristic of arid environments.

X-ray crystallography Study of **crystal** structure by examination of the **diffraction** pattern obtained when a beam of X-rays is passed through the crystal **lattice**.

X-ray diffraction Pattern of variable intensities produced by **diffraction** of **X-rays** when passed through a **diffraction grating** consisting of spacings of about 10^{-8} cm, in particular that formed by the **lattice** of a crystal.

X-ray fluorescence Less penetrating, secondary X-rays emitted by a substance when subjected to primary X-rays or high-energy electrons. The secondary X-rays are characteristic of the bombarded substance.

■ **X-rays Electromagnetic (and ionizing) radiation** produced in partial vacuum by the sudden arrest of high-energy bombarding electrons as they collide with the heavy atom nuclei of a target metal. The X-rays produced are thus characteristic of the target's atoms. X-rays have very short wavelengths (10^{-3} to 1 nm) and can penetrate solids to varying degrees; this characteristic has made them useful in medicine, dentistry and X-ray crystallography. Overexposure to X-rays can be harmful, possibly causing cancer. Alternative name: röntgen (roentgen) rays; X-radiation.

X-ray spectrum Line spectrum of the intensity of **X-rays** emitted when a solid target is bombarded with electrons. It consists of sharp superimposed lines, which are characteristic of the target atoms, on a continuous background.

X-ray tube Vacuum tube designed to produce **X-rays** by using an **electrostatic field** which accelerates and directs **electrons** on to a target.

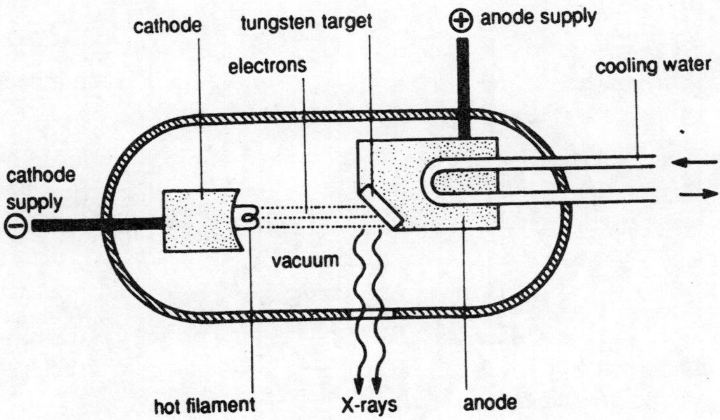

Principle of an X-ray tube

■ **xylem** In higher plants, water-conducting tissue that consists of lignified vessel elements, **tracheids**, fibre and **parenchyma** tissue, which together provide mechanical support. It is the woody part of a plant. [2/4/a] [2/5/a] [2/7/a]

xylene $C_6H_4(CH_3)_2$ Aromatic liquid organic compound that exists in three isomeric forms (ortho-, meta- and para-xylene), obtained from coal-tar and petroleum. They are used as solvents in **polyester** synthesis, in microscopy for preparation of specimens and as cleaning agents. Alternative name: dimethylbenzene.

xylose $C_5H_{10}O_5$ Naturally occurring **pentose sugar**, found in the form of xylan or as **glycosides** in many plants (*e.g.* cherry and maple wood, straw, pecan shell, corn cobs and cotton-seed hulls). Alternative name: wood sugar.

Y

yaws Tropical ulcerative disease caused by the bacterium *Treponema pertenue*, characterized by raspberry-like skin eruptions. Alternative name: framboesia.

■ **Y-chromosome** Smaller of the two **sex chromosomes**, found only in the **heterogametic** sex; *e.g.* human males have one Y-chromosome and one X-chromosome (XY). [2/8/b]

■ **yeast** Collective name for **fungi** whose vegetative bodies consist of single individual cells. Yeasts contain **enzymes** (*e.g.* **zymase**) which bring about **fermentation**. They are used in making bread (when they ferment sugar to produce carbon dioxide gas, which causes the bread to rise) and in making wine and beer (when they ferment sugar to produce alcohol).

yellow spot (macula lutea) Alternative name for **fovea**.

yoke In electromagnetism, piece of **ferromagnetic** material that permanently connects two or more magnet cores.

yolk Part of an ovum that stores the nutritive materials, or the yellow central portion of the egg of birds and reptiles.

Young's modulus Elastic modulus equal to longitudinal (tensile) stress divided by longitudinal strain. It was named after the British physicist Thomas Young (1773–1829).

ytterbium Yb Silvery-white metallic element in Group IIIA of the Periodic Table (one of the **lanthanides**), with no commercial uses. At. no. 70; r.a.m. 173.04.

yttrium Y Grey metallic element in Group IIIA of the Periodic Table (one of the **alkaline earths**, but often classed with the **lanthanides**). It is used in alloys for superconductors and magnets, and yttrium (VI) oxide, Y_2O_6, is employed in lasers and **phosphors**. At. no.39; r.a.m. 88.905.

Z

Zeeman effect Splitting of **spectral lines** of a substance into components of different **frequency** when placed in a **magnetic field**. It was named after the Dutch physicist Pieter Zeeman (1865–1943).

Zener current Current produced in an **insulator** in a strong electric field when its **valence band electrons** are raised to the **conduction band**.

Zener diode Semiconductor diode which at a certain negative voltage produces a sharp breakdown of current and hence may be used as a voltage control device. Alternative names: avalanche diode, breakdown diode.

zeolite Hydrated aluminosilicate mineral, from which the water is easily removed, used for making **molecular sieves** and **ion exchange** columns (*e.g.* for water softeners).

zero energy thermonuclear apparatus (ZETA) Apparatus at the British atomic research establishment at Harwell used for studies of **fusion reactions** and **plasma** physics.

zero method Alternative name for **null method**.

zeroth law of thermodynamics *See* **thermodynamics, laws of**.

ZETA Abbreviation for **zero energy thermonuclear apparatus**.

zeta potential Potential difference that exists across the interface between a solid particle and a liquid in which it is immersed. Alternative name: electrokinetic potential.

Ziegler catalyst Catalyst made by mixing certain **transition element** salts (*e.g.* titanium (IV) chloride, $TiCl_4$) with an organometallic compound (*e.g.* triethylaluminium, $Al(C_2H_5)_3$). It is used in the **Ziegler process**. It was named after the German chemist Karl Ziegler (1898–1973).

Ziegler process Commercial process for making high-density **polymers** (*e.g.* polyethylene, stereospecific rubbers) using **Ziegler catalyst**.

zinc Zn Bluish-white metallic element in Group IIB of the Periodic Table (a **transition element**), used to give a corrosion-resistant coating to steel (**galvanizing**), to make dry batteries and in various alloys (*e.g.* brass, bronze). Its oxide, ZnO, is used as a white pigment (Chinese white). At. no. 30; r.a.m. 65.38.

zincate Compound formed by the reaction of metallic **zinc** or zinc oxide with an alkali; *e.g.* Na_2ZnO_2.

zinc chloride $ZnCl_2$ White **hygroscopic** salt produced commercially by heating metallic **zinc** in dry **chlorine** gas. It is used to fireproof timber, in battery making, vulcanizing and galvanizing, in oil refining, and as a fungicide and catalyst.

zincite Deep red-orange mineral, mainly zinc oxide, which is an important **zinc** ore; it often also contains some manganese. Alternative names: red oxide of zinc, spartalite.

zinc oxide ZnO White crystalline solid (yellow when hot) which can be produced directly by heating zinc in air. It is used as a white pigment (Chinese white), in ceramics, cosmetics, pharmaceuticals and floor coverings, and in the manufacture of tyres. It dissolves in alkalis to form **zincates**.

zinc sulphate $ZnSO_4.7H_2O$ Colourless crystalline salt prepared by dissolving metallic **zinc** in dilute **sulphuric acid**. It is used in manufacture of rayon, glue, fertilizers, fungicides, wood preservatives, rubber, paint and varnishes. Alternative names: white copperas, white vitriol, zinc vitriol.

zinc sulphide ZnS Occurs naturally as blende (an important **zinc** ore) and can be prepared as a white precipitate by adding ammonium sulphide or hydrogen sulphide to a solution of a zinc salt. It is used as the pigmentary base for white zinc sulphide (lithopone), which contains up to 60% zinc sulphide and a balance of **barium sulphate**. It is also used in fungicides and **phosphors**.

zircon $ZrSiO_4$ Mineral form of zirconium silicate, a pale blue, golden-yellow, red or greyish substance which is the chief source of **zirconium**. Colourless crystals are employed as semi-precious gemstones.

zirconium Zr Silvery-grey metallic element in Group IVA of the Periodic Table (a **transition element**). It is used to clad uranium fuel rods in **nuclear reactors**. Naturally occurring crystalline zirconium (IV) oxide, ZrO_2, is the semi-precious gemstone zircon; the oxide is also used as an **electrolyte** in **fuel cells**. At. no. 40; r.a.m. 91.22.

zone *1*. In analytical chemistry, orientation of solute molecules in a series of tubes in a liquid-liquid extraction procedure. *2*. In crystallography, crystal faces that intersect along parallel edges.

zooid Individual invertebrate animal that lives as part of a **colony**.

zoology Systematic study of animals with relation to other animals, plants and their non-living environment.

zooneuston Narrow layer above and below the surface of open water containing organisms that exploit the effects of **surface tension**, *e.g.* pond skaters.

zoonosis Disease that can be transmitted from animals to human beings (*e.g.* rabies).

zooplankton In an aquatic **ecosystem**, group of passively floating and drifting microscopic animals. *See also* **plankton**.

■ **zygote Diploid cell** that results from the fertilization of a female **gamete** by a male gamete. It sometimes undergoes immediate cleavage, but may also develop a thick resistant outer coat to form a zygospore. [2/10/b]

zymase Enzyme that catalyses the **fermentation** of **carbohydrates** to **ethanol** (ethyl alcohol).

zymogen Inactive precursor of an **enzyme** formed by plants and animals. It is activated by the action of a kinase. Alternative name: proenzyme.

zymotic Describing an agent that causes an infectious disease.

APPENDIX I

SI Units

Basic unit	*Symbol*	*Quantity*	*Standard*
metre	m	length	Distance light travels in vacuum in 1/299792458 of a second
kilogram	kg	mass	Mass of the international prototype kilogram, a cylinder of platinum-iridium alloy (kept at Sèvres, France)
second	s	time	Time taken for 9,192,631,770 resonance vibrations of an atom of caesium-133
kelvin	K	temperature	1/273.16 of the temperature of the triple point of water
ampere	A	electric current	Current that produces a force of 2×10^{-7} newtons per metre between two parallel conductors of infinite length and negligible cross-section placed a metre apart in vacuum
mole	mol	amount of substance	Amount of substance that contains as many atoms (or molecules, ions, or subatomic

			particles) as 12 grams of carbon-12 has atoms
candela	cd	luminous intensity	Luminous intensity of a source that emits monochromatic light of frequency 540 × 10^{12} hertz of radiant intensity 1/683 watt per steradian in a given direction
Supplementary units			
radian	rad	plane angle	Angle subtended at the centre of a circle by an arc whose length is the radius of the circle
steradian	sr	solid angle	Solid angle subtended at the centre of a sphere by a part of the surface whose area is equal to the square of the radius of the sphere
Derived units			
becquerel	Bq	radioactivity	Activity of a quantity of a radioisotope in which 1 nucleus decays every second (on average)
coulomb	C	electric charge	Charge that is carried by a current of 1 ampere flowing for 1 second
farad	F	electric capacitance	Capacitance that holds a charge of 1 coulomb when it is charged by a potential difference of 1 volt

gray	Gy	absorbed dose	Dosage of ionizing radiation corresponding to 1 joule of energy per kilogram
henry	H	inductance	Mutual inductance in a closed circuit in which an electromotive force of 1 volt is produced by a current that varies at 1 ampere per second
hertz	Hz	frequency	Frequency of 1 cycle per second
joule	J	energy	Work done when a force of 1 newton moves its point of application 1 metre in its direction of application
lumen	lm	luminous flux	Amount of light emitted per unit solid angle by a source of 1 candela intensity
lux	lx	illuminance	Amount of light that illuminates 1 square metre with a flux of 1 lumen
newton	N	force	Force that gives a mass of 1 kilogram an acceleration of 1 metre per second squared
ohm	Ω	electric resistance	Resistance of a conductor across which a potential of 1 volt produces a current of 1 ampere

pascal	Pa	pressure	Pressure exerted when a force of 1 newton acts on an area of 1 square metre
siemens	S	electric conductance	Conductance of a material or circuit component that has a resistance of 1 ohm
sievert	Sv	dose equivalent	Radiation dosage equal to 1 joule of radiant energy per kilogram
tesla	T	magnetic flux density	Flux density (or magnetic induction) of 1 weber of magnetic flux per square metre
volt	V	electric potential difference	Potential difference across a conductor in which a constant current of 1 ampere dissipates 1 watt of power
watt	W	power	Amount of power equal to a rate of energy transfer of (or of doing work at) 1 joule per second
weber	Wb	magnetic flux	Amount of magnetic flux that, decaying to zero in 1 second, induces an electromotive force of 1 volt in a circuit of one turn

APPENDIX II

Periodic Table

IA	IIA	IIIA	IVA	VA	VIA	VIIA	VIII	
1 H								
3 Li	4 Be							
11 Na	12 Mg							
19 K	20 Ca	21 Sc	22 Ti	23 V	24 Cr	25 Mn	26 Fe	27 Co
37 Rb	38 Sr	39 Y	40 Zr	41 Nb	42 Mo	43 Tc	44 Ru	45 Rh
55 Cs	56 Ba	57 La	72 Hf	73 Ta	74 W	75 Re	76 Os	77 Ir
87 Fr	88 Ra	89 Ac						
				58 Ce	59 Pr	60 Nd	61 Pm	62 Sm
				90 Th	91 Pa	92 U	93 Np	94 Pu
		IB	IIB	IIIB	IVB	VB	VIB	VIIB	0

								2 He
			5 B	6 C	7 N	8 O	9 F	10 Ne
			13 Al	14 Si	15 P	16 S	17 Cl	18 Ar
28 Ni	29 Cu	30 Zn	31 Ga	32 Ge	33 As	34 Se	35 Br	36 Kr
46 Pd	47 Ag	48 Cd	49 In	50 Sn	51 Sb	52 Te	53 I	54 Xe
78 Pt	79 Au	80 Hg	81 Ti	82 Pb	83 Bi	84 Po	85 At	86 Rn

63 Eu	64 Gd	65 Tb	66 Dy	67 Ho	68 Er	69 Tm	70 Yb	71 Lu
95 Am	96 Cm	97 Bk	98 Cf	99 Es	100 Fm	101 Md	102 No	103 Lr